IMPOSSIBLE HEIGHTS

IMPOSSIBLE HEIGHTS

Skyscrapers, Flight, and the Master Builder

ADNAN MORSHED

MINNESOTA

University of Minnesota Press
Minneapolis • London

The University of Minnesota Press gratefully acknowledges financial assistance provided for the publication of this book from the Wyeth Foundation for American Art.

Published by the University of Minnesota Press
111 Third Avenue South, Suite 290
Minneapolis, MN 55401–2520
http://www.upress.umn.edu

Library of Congress Cataloging-in-Publication Data
Morshed, Adnan.
Impossible heights : skyscrapers, flight, and the master builder / Adnan Morshed.
Includes bibliographical references and index.
ISBN 978-0-8166-7318-6 (hc)
ISBN 978-0-8166-7319-3 (pb)
1. Architecture and society—United States—History—20th century. 2. Technological innovations—Social aspects—United States—History—20th century. 3. Architecture, Modern—20th century—Philosophy. 4. Aeronautics and civilization. I. Title.
NA2543.S6M67 2015
720.1′08—dc23
2014024833

Printed in the United States of America on acid-free paper

20 19 18 17 16 15 10 9 8 7 6 5 4 3 2 1

For my parents,
ZOHARA *and* MATIN
my sister, TAHMINA
my wife, SADIA

Contents

Acknowledgments

I was inspired to write this book after exploring Le Corbusier's 1929 experience of flight in South America. As I was reading Le Corbusier's *Precisions* and, a little later, his lesser-known book *Aircraft,* I became fascinated with the Franco-Swiss architect's insight that the view from an airplane would propel the modern planner's understanding of how to reorder the world. Further reading on the topic revealed that the phenomenon of human flight was related to broader discussions of modernist visuality and, in particular, a modern politics of seeing. At its core was the ideation of a modern protagonist as a new type of observer. Later, as I expanded my research, I noticed that a parallel debate on the philosophical meaning of seeing things from above—in particular, from the vantage point of a tall urban building—was also crucial to the skyscraper literature of early twentieth-century America. I began to group these research observations loosely under the theme of the "aesthetics of ascension." When I started exploring this aesthetic consciousness in the cultural context of the 1920s and 1930s, I became interested in the ways the phenomenon of seeing things from above was intimately related to the radical practice of envisioning what was then popularly called the world of tomorrow.

To sustain a love affair with a book project for several years, one must have—alongside unflinching devotion to it—a network of mentors, readers, and colleagues. I was fortunate to receive the support of many people. When I began this project, I was a graduate student of architectural history, theory, and criticism at the Massachusetts Institute of Technology. James Ackerman, Sibel Bozdogan, Henry Millon, and Akos Moravansky offered many helpful suggestions, one of which was to study the subject from various theoretical and historical angles. Jorge Liernur at Harvard University was an intellectual stimulant.

As I delved deeper into the topic, Stanford Anderson, Mark Jarzombek, and Leo Marx provided me with invaluable intellectual guidance. I have learned so much from these astute scholars, and it has been a privilege to be able to research and write under their mentorship. Gail Fenske reviewed multiple drafts of the manuscript, and I am most grateful for her insights, suggestions, and friendship.

The Nancy Royal Grant from MIT and a travel grant from the New England Chapter of the Society of Architectural Historians helped defray the cost of initial research. The Wyeth Fellowship for American Art, from the Center for Advanced Study in the Visual Arts at the National Gallery of Art in Washington, D.C., provided an excellent academic setting in which to write a partial draft of the book. There, I benefited from the helpful encouragement of Elizabeth Cropper, Peter Lukehart, Therese O'Malley, Caroline Elam, Sheryl Reiss, Nancy Yeide, and Al Acres. A fellowship from the Wolfsonian–Florida International University (Wolfsonian–FIU) in Miami Beach enabled me to continue researching and writing. Cathy Leff, Jon Mogul, and Micky Wolfson were excellent hosts at the Wolfsonian–FIU. The Verville Fellowship from the National Air and Space Museum (NASM) helped move this project forward by allowing me to have an uninterrupted year in which to write the first draft of the book. Dominick Pisano and Peter Jakab, at the NASM, have been strong supporters of my work.

A Graham Foundation Publication Grant afforded me the opportunity to pursue crucial research that buttressed the first full draft of the book. I also gratefully acknowledge the support of the National Endowment for the Humanities Summer Stipends program, which enabled me to complete the final stage of research for the book. A Wyeth Foundation for American Art Publication Grant covered the reproduction costs of the book's illustrations.

Over the years, I have greatly benefited from conversations with various scholars and colleagues. My gratitude goes to Randall Ott, Shamsul Wares, Jeffrey Meikle, Eric Jenkins, Nancy Stieber, Francesco Passanti, Paul Emmons, Kathy Plummer, Cynthia Field, Walter Driscoll, and Roger Lipsey.

I greatly appreciate the privileges and courtesies extended to me by several archives. Among them are the Hugh Ferriss Collection at the Avery Library, Columbia University; the Richard Buckminster Fuller Papers at the Green Library, Stanford University; the Norman Bel Geddes Collection at the Harry Ransom Humanities Research Center, University of Texas, Austin; the Beinecke Rare Book and Manuscript Library, Yale University; the Library of Congress; the American Philosophical Society in Philadelphia; the National Air and Space Museum in Washington, D.C.; the Library of Congress; the National Museum of American History in Washington, D.C.; and the New York Public Library.

I fondly recall the help of many archivists who graciously responded to my incessant requests for access to historical documents. Helen Baer, at the Harry Ransom Center, was tireless in her efforts to find the materials that I needed. Although unaffiliated with the Buckminster Fuller Papers at Stanford, Bonnie DeVarco knew everything about Fuller's mammoth archive. She was most accommodating and shared her vast knowledge of Fuller's life. Mattie Taormina, at Stanford, helped me identify many images that I had requested. Janet Parks, at Columbia's Avery Library, was exceedingly gracious to grant me access to the Ferriss Papers. Kristi Finefield, at the Prints and Photographs Division, Library of Congress, went out of her way to locate a number of images and, particularly, their copyright information.

I greatly benefited from the close reading of my manuscript by Kate Solomonson at the University of Minnesota Press. I am grateful to Pieter Martin, also at the University of Minnesota Press, for maintaining his faith in my project and bringing it to fruition. Kristian Tvedten's management of the project's nitty-gritty was noteworthy. Judy Selhorst was an excellent copy editor.

Jayson Hait read many versions of the manuscript and sharpened my writing. She embraced the project with much enthusiasm from the beginning, and I cannot be more appreciative. Over the years, my research assistants played an integral role in the book's development. Emily Pierson and Lillian Heryak were outstanding in their acquisition of permissions to reproduce copyrighted images. Kathleen Lane, Joanna Beres, Deanna Keil, Yulia Beltikova, Jobi Jones, Nina Africa, Eva Frenz, Sydney Katz, and Timothy Dowdy conducted extensive research on my behalf. I couldn't have asked for more from them.

I owe my deepest gratitude to my family, for supporting me through emotional ups and downs during the writing of this book. My parents, Matin and Zohara, have been the pillars of my life. Even though my father passed away when the project was in its infancy, I can imagine how much interest he would have taken in the book as it evolved. During our frequent overseas telephone conversations, my mother's first question would invariably relate to how the book was progressing. It would not be an exaggeration to say that her persistence kept the ascension project aloft. My sister, Tahmina, despite her immensely busy life as a pediatric surgeon, never ceased to demand the completion of the book on time. My siblings, Hasan, Hossain, Mohsin, and Adina, have always inspired me to do more. My wife, Sadia, endured much as I was writing this book, including the tragic loss of her mother to cancer. Yet her interest in and support of my project never wavered. To boost my morale, she kept buying me Superman memorabilia!

INTRODUCTION

The Aesthetics of Ascension

> Perhaps the reason that New York was a great city a few minutes the other day when Wilbur Wright was there was that Wilbur Wright had a new vision in the presence of all those men of something that they could do. He touched the imagination of men about themselves. They were profoundly moved because they saw him in their presence inventing a new kind and new size of human being. He raised the standard of impossibility, and built an annex on to the planet while they looked; took a great strip off of space three miles wide and folded it softly on to the planet all the way round before their eyes. For three miles more—three miles farther up above the ground—there was a space where human beings would have to stop saying, "I can't," and "You can't," and "We can't." The modern imagination takes to impossibilities naturally with Wilbur Wright against the horizon.
>
> —Gerald Stanley Lee, *Crowds: A Moving-Picture of Democracy,* 1913

A distinct cultural consciousness came into focus in America during the interwar years with the excitement over airplanes and skyscrapers and their anticipated roles in the creation of an ideal world of tomorrow. For many observers, aerial movement and height opened up new perspectives from which to rediscover the world. This, in turn, emboldened many visionary designers and thinkers to imagine the future in their own terms. The mobile "eye" of the airplane seemed to distinguish the twentieth century from earlier times by virtue of its promise to transform fragmentary earthbound experiences into the "mingling lines of Picasso," as noted art collector Gertrude Stein recalled after her first flight over America in the 1930s.[1] Aerial photographs of cities and landscapes—accompanying magazines articles with such alluring titles as "Aeronautics Will Develop a Broader Vision," "A Bit of Philosophy on Flying," and "Seeing Things from Above"—often fused the airplane view with a Promethean seer to whom the Earth promised full disclosure (Figure I.1).[2]

In this romantic narrative of the skyward spirit, the skyscraper was popularly seen in the role of the airplane's urban alter ego. The Chicago architect and architectural historian Thomas Tallmadge, encapsulating the heightened pride that radiated from the peak of the skyscraper, wrote in 1927: "[The skyscraper's] eyes gaze down from immeasurable heights on a welter of humanity and machinery. Its shining flanks are dappled with shadows of aeroplanes that 'laugh as they pass in thunder,' while 'sublime on their towers' the mysterious antennae

Figure I.1. The aviator as an aerial observer, drawn by the American illustrator Herbert Paus, on the cover of Popular Science *112, no. 4 (New York: Popular Science Publishing Co., Inc., April 1928).*

'join cape to cape over a torrent sea.'"[3] Public discussions on skyscrapers popularized words like *ascent, pinnacle,* and *skyward* as fitting expressions of modern city life. As New York City became the premier theater of metropolitan modernity in the early twentieth century, an artistic and literary generation brought to the fore the idea of a metropolitan spectator who, from the skyscraper loggia, inspected the city below with a sense of psychological and visual domination over it. The photographs of Alvin Coburn, Margaret Bourke-White, Lewis Hine, and Alfred Stieglitz, the drawings and paintings of Hugh Ferriss and Georgia O'Keeffe, and the writings of F. Scott Fitzgerald and John Dos Passos captured such sentiments.

The captivating image of an airplane flying over the rising metropolis led many Americans to believe that a new civilization had dawned. Witness a typical caption to such an image in a 1930 book: "Almost a symbol of civilization is this picture—the fantastic towers of a great city rearing from the earth, and above them a machine that flies—new ways of living and traveling" (Figure I.2).[4] Such a moment of nationalistic cheerleading was already foreshadowed in

© *Wide World Photos.*

PROGRESS.

Almost a symbol of civilization is this picture—the fantastic towers of a great city rearing from the earth, and above them a machine that flies—new ways of living and traveling.

Figure I.2. An airplane flies over the towers of Manhattan. Harry F. Guggenheim, The Seven Skies *(New York: G. P. Putnam's Sons, The Knickerbocker Press, 1930), 36. Photograph by Wide World Photos.*

New York City after Wilbur Wright's aerial stunts, as Gerald Stanley Lee tells us in the above epigraph. The arch provocateur and aviation enthusiast Le Corbusier seemed to have chosen America as the representation of a new kind of vertical space with a telling photomontage on the back cover of his 1935 book *Aircraft* (Figure I.3).[5] With the dramatic synthesis of two signifiers of modern life—a Douglas DC-3 gliding over the upward urban form of Manhattan—Le Corbusier perhaps could not find a more suitable example than New York City as the focus of a quintessentially modernist gaze that simultaneously contemplated new forms of mobility and new forms of living.

Two situations arose from this type of heightened visual practice. First, modernist designers exploited the gaze from hitherto impossible heights as a source of inspiration for envisioning the so-called World of Tomorrow, a theme that recurred during the interwar period in a range of popular venues. From the pages of the first American science fiction pulp magazine *Amazing Stories* to the official slogan—"Building the World of Tomorrow"—of the 1939 New York World's Fair, the imagination of Tomorrow assumed the status of a cultural fetish. Second, the World of Tomorrow was all too often seen as facilitated by the supposedly benevolent role of airplanes and skyscrapers.

Although the observer on the airplane and the one atop the skyscraper represented different modalities of viewing, one kinetic and the other static, an overarching theme of aerial vision signaled the advent of a modern spectator.

Figure I.3. A DC-3 airplane flies over Manhattan. Back cover of Le Corbusier's Aircraft *(New York: The Studio Publications Inc., 1935). © 2012 Artists Rights Society (ARS), New York / ADAGP, Paris / F.L.C.*

In significant ways, this modern spectator's perches, the airplane cockpit and the skyscraper observatory, became suitable pulpits for new kinds of urban analysis and reform (Figure I.4). This panoptic figure, observing the world from an idealized platform—be it a skyscraper loggia or the airplane or the human body itself depicted as an airplane—offered reform-minded American designers the illusion of a messianic character, a *master builder* of sorts, able to create a brighter tomorrow (see Plate 1). Expressing a utopian zeal, many designers even went so far as to consider this "ascending" character as the emblem of highest human evolution. In the 1930s, the American pop hero Superman became an apt example of this cultural attitude. No other protagonist of interwar America exemplified the populist dream of a righteous creator of a just civilization more perfectly than the flying superhero.

An "aesthetic of ascension" emerged from a wide range of reactions to powered human flight and skyscrapers. These were variously expressed in the interwar period's vast terrain of cultural products, including urbanistic and architectural projections of the future, science fiction, aerial photography, painting, and film. The aesthetics of ascension implied a peculiar blend of godlike spectatorship, technological utopianism, and evolutionary idealism—all converging

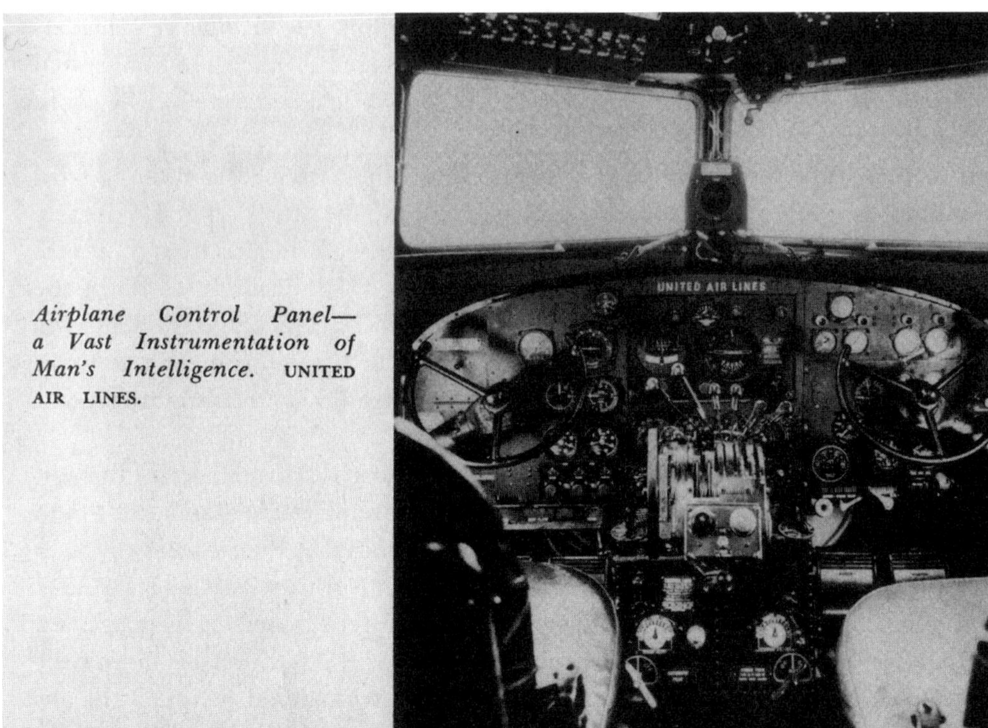

Figure I.4. The cockpit of a United Air Lines airplane. Reproduced in Gerald Wendt, Science for the World of Tomorrow *(New York: W. W. Norton & Company, 1939), 143.*

to create the seductive myth of a master builder, able to redeem a chaotic world from his high perch of authority (this was a masculinist discourse). While the word *ascension* readily invokes the theological narratives of the ascent of prophets to heaven or conjures up images of divine providence, I use the word primarily in the sense defined by the *Oxford English Dictionary* as a "rise from the inferior to the superior," or a "rise or advancement in thought or feeling."[6] Ascension in a secular sense, however, does not necessarily discount the characteristic mystical dimensions embedded in the scriptural notion of ascension, as an American theologian wrote in 1917: "We see that man is not confined to the materialistic level: he is more than matter, and is able to soar to heights from which he may read the meaning of the universe."[7] The spiritual content of ascension was not antithetical to the aviator's or the skyscraper dweller's meditative experience of altitude, revealing an often overlooked cross-pollination between technology and spirituality. Thus, I note an important moralizing and spiritualizing undercurrent in technology-enabled ascension, which, as the historian Stephen Kern suggests—while observing the cultural impact of the advent of the airplane—is marked by "deeply rooted values associated with the up-down axis. Low suggests immorality, vulgarity, poverty, and deceit. High is the direction of growth and hope, the source of light, the heavenly abode of angels and gods."[8] As we shall see in the following chapters, the meditative consideration of altitude provided a fertile symbolic space in which the conceptual development of an ideal human type as a precursor to a better tomorrow could be variously contemplated. One of my key motivations for writing this book is to probe the discursive ways the concept of the master builder, along with all of its semantic trappings and philosophical ramifications, offered a suitable vantage point for a wide range of utopian imaginings, architectural fantasies, and spiritualizing introspections. By connecting the cultural histories of the airplane and the skyscraper with the prophecies of a new civilization and the protagonist who would initiate it, this book both alters and expands the standard set of assumptions identified with the architectural and urban histories of the period.

This book focuses on three major figures whose work exemplifies the aesthetics of ascension in American culture: the architectural illustrator Hugh Macomber Ferriss (1889–1962), the visionary innovator Richard Buckminster Fuller (1895–1983), and the industrial designer Norman Bel Geddes (1893–1958).[9] They were born at the dawn of the cultural excitement over heavier-than-air flying machines and skyscrapers. Maturing professionally during the 1920s, Ferriss, Fuller, and Bel Geddes represent what could be called the airplane and skyscraper generation. Their youths spanned the years between the Wright brothers' aerial experiments and Charles Lindbergh's transatlantic crossing in 1927, or the years between the rise of the Woolworth Building (1913) and that of the Empire State Building (1931).

Like millions of American adolescents in the early twentieth century, Ferriss, Fuller, and Bel Geddes gawked at the first-generation airplanes and skyscrapers with both curiosity and admiration. "How to Make a Model Aeroplane," in *Scientific American* in 1911, exemplified early twentieth-century technology-oriented educational programs chalked out to train the "Winged Superchildren of Tomorrow."[10] The motto of the "Superchildren"—ascension—made an indelible impression on Ferriss, Fuller, and Bel Geddes, which, in turn, resulted in philosophic, aesthetic, and even mystical reincarnations later in their adulthoods. Fuller and Bel Geddes designed airplanes (although not in professional capacities) and tall structures, while airplanes glide by, within, and above the urban canyons in Ferriss's representation of the metropolis of tomorrow (Figure I.5). For his entire professional life Ferriss lived in a rooftop studio of the seventeen-story Architects Building in Manhattan, an elevated habitat that demonstrably affected his perceptions of urban futurity, while Fuller and Bel Geddes employed their actual flight experience to both utopian and practical ends. Yet airplanes and skyscrapers recur in their imaginations not just as technological marvels that they merely extolled and used as efficient models to vitalize their design thinking but also as epistemic instruments they used to look into the future. In keeping with the common excitement affiliated with the era's technological utopianism, they viewed the ascending subject as the architect-seer of a brave new world, as well as the most convincing model on which to fashion their own imposing images.

This book positions the work of Ferriss, Fuller, and Bel Geddes in the context of visionary ideas prevalent among many architects, urbanists, artists, novelists, photographers, science fiction writers, and social scientists during the 1920s and 1930s. Conflicting views of progress and modernity notwithstanding, this period was marked by a romantic optimism that viewing the world from above would usher in new spatial dynamics, introducing the city of the future (see Plate 2). Popular American magazines such as *Aviation, Aviator, Aerial Age, Popular Mechanics, The American City,* and *Scientific American* frequently carried laudatory essays on the contributions of aerial perspectives to the planning of ideal cities.[11] Despite their scientific pretensions, these essays were often tinged with moralistic and utopian beliefs that the view from above would broaden human vision for enlightened living while imposing spatial discipline on all physical and social disorder. Le Corbusier—whose seminal writings of the 1920s Ferriss, Fuller, and Bel Geddes read—paralleled this populist sentiment in the section titled "Airplane" in his *Towards a New Architecture* when he wrote: "To search for a means of suspension in the air and a means of propulsion was to put the problem properly" and "The man who is intelligent, cold, and calm has grown wings to himself."[12] The modernist fascination with a winged avatar as the embodiment of godlike authority over a vast visual field appealed to Ferriss, Fuller, and Bel Geddes.

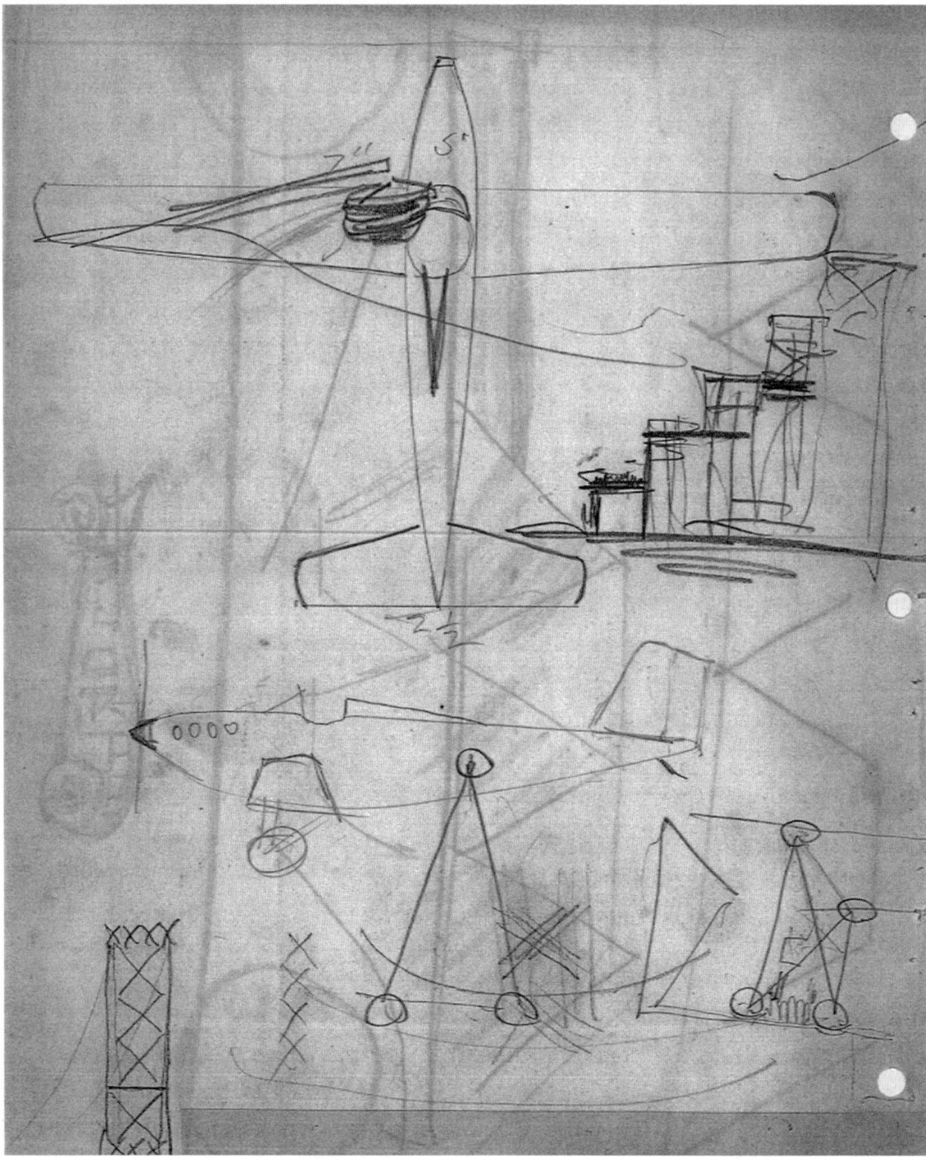

Figure I.5. An airplane designed by Buckminster Fuller, ca. 1928. The Special Collections of Green Library, Stanford University. Courtesy, The Estate of R. Buckminster Fuller.

They were fired up by a common indignation after World War I that, as authors like Madison Grant, Edward East, Raymond Fosdick, Walter Lippmann, and H. G. Wells noted, Western civilization had reached a precarious crossroads due to population growth, bureaucratization of social relationships at the expense of individuals' spiritual development, and unhygienic urban growth allegedly corrupting the moral fabric of city life.[13] In particular, their idealistic mission resonated within a metropolitan culture of 1920s America, one that was partly shaped by, as the American historian William Leuchtenburg suggested, a new fraternity of urban writers rebelling against what they perceived as traditional American values and moral standards.[14] These writers revolted against what Ezra Pound called "a botched civilization" or what Waldo Frank loathingly termed the "cold lethal simplicities of American business culture." In 1922, T. S. Eliot devastatingly exposed what was dubbed the sterility and emptiness of modern industrial civilization with the modernist poem *The Waste Land,* which became the standard literary account of the despair felt by intellectuals disillusioned with modern life. Ferriss, Fuller, and Bel Geddes embraced the "crisis of civilization" pathos as the justification for their own grand plan for the future. An appropriate response to prevailing pessimism, they reasoned, was not to be trapped in, say, a *Waste Land*–type despair, but to envision a master class or an intellectual oligarchy that would lead civilization out of the morass. One of the authors that Fuller read, the American journalist William Harlan Hale, articulated this worldview: "Only by going to the heart of the modern chaos is there any hope. Many are stuck there. Some will go on. It is worth the sadness and the labor. Civilization is created and given by men, not by masses."[15] In their nuanced ways of depicting the savior of civilization, Ferriss mused about the "harmonious development of man" while Fuller developed a utopian character called "Phantom Captain" and Bel Geddes prophesied the advent of the "man of tomorrow."[16] One way or the other, these pursuits of a heroic leader could be seen as different iterations of a master builder.

My study of Ferriss, Fuller, and Bel Geddes elaborates on the concept of the master builder as the creator of epoch-making ideas or structures, or as a powerful agent of transformation in the physical fabric of an urban area or a region. The term *master builder* has an archetypal appeal that cuts across historical eras, cultures, and disciplines. Phidias, the chief of Athenian statesman Pericles's building program, was called a master overseer. Roman emperor Trajan's chief architect Apollodorus invoked the notion of a master builder, a term also frequently used to describe the creators of Gothic cathedrals during the Middle Ages. In the humanist context of fifteenth-century Italy, Leon Battista Alberti adopted as his personal symbol a winged eye, referring to God, the all-seeing master builder of the universe. In modern times, the term has been used variously to denote a builder or a designer whose personality combines erudition, authority, hubris, and a superior vision, all packaged within an aura of grandeur and tragedy, as in master architect Halvard Solness, the title character of Henrik Ibsen's 1892 play *The Master Builder.* New York City's park commissioner Robert Moses is another example of the modern era's attempt to portray its hero (or antihero in this case) as a master builder. I am interested in the various ways the concept was represented within design discourses during the interwar period. As the professions of architecture, design, and physical planning gained recognition and credibility, a revitalized notion of the master builder appeared in multiple forms and expressions, sometimes referred to as master builder and sometimes as new man, master planner, master architect, or new engineer.[17] I employ *master builder* as a general term to characterize this notion of a lofty figure who was seen as capable of transforming civilization by means of the diagrammatic imposition of an all-encompassing plan from the summits of power. The master builder revealed a natural alliance between heights and authority.

The union of heights and authority struck a popular chord with modern architects and planners. Committed to visual cleanliness, formalist aesthetics, and related premises of social functionalism, architect-planners like Daniel Burnham, Clarence Stein, Henry Wright, and Frank Lloyd Wright and bureaucrats like Robert Moses promoted planning as a "logical diagram," the visualization of which called for a heightened perspective. "Make big plans," Burnham said, "[and] aim high in hope and work, remembering that a noble, logical diagram once recorded will never die."[18] In many ways, Burnham's lack of patience for "little plans," because "they have no magic to stir men's blood," foreshadowed the grandiosity of the master builder's panoptic desire to see the world as a manageable picture. Frank Lloyd Wright's promotion of decentralization in his Broadacre City (1932) as a way to create a new America that meshed the mythos of individualism with the collectivities of city life exemplified the towering legacy of the master builder. "After the First World War there was a strong surge of enthusiasm for a better world," recalled Stein.[19] What he also implied was that a better world needed a better, more gallant builder. Whether the early twentieth-century master builders actually flew in airplanes or scaled skyscrapers during their planning processes is not the point of contention here. My purpose is rather to ponder whether and how analyses of big plans and grand visions evince the nature of their creators, ideologically perched on a high pedestal and unencumbered by the humbling minutiae of the ground, recalling the visual regime of the aviator or the observer on the skyscraper. The detached viewpoint of the master builder and his modus operandi of "planning from above" have drawn much postmodernist derision for favoring the utopian simplicity of big plans at the expense of the messy realities of the ground and participation of people. My study is about a period when the image of the master builder as a figure of authority with a probing downward gaze on the world found a consummate match in the aviator or the spectator atop the skyscraper.

The concept of the master builder owes its intellectual fertility to two key modernist imaginations that are crucial to understanding the work of Ferriss, Fuller, and Bel Geddes. First, ascension, real or metaphorical, promised an enlightened awareness and a superior economy of vision, revealing in the process a modernist logic of looking at the world. Namely, if the viewer saw the *total* picture, with all the linkages, separations, and relations among its various components, then he would know how to order or even discipline that picture. What seemed cluttered and claustrophobic from the pedestrian's point of view might appear transparent and liberating to the eyes of the observer on high. Take, for example, the labyrinth: if the labyrinth's terror lies in its inescapable interiority, then ascending *above* it is to collapse its very meaning and to see its mazy internal routes resolve into a mere optical adventure. This understanding—rooted in the Enlightenment philosophy of oculocentric clarity and transparency articulated, for instance, by the French philosopher Jean le Rond D'Alembert—granted an enhanced moment of visual authority to the ascending seer.[20] In his 1751 book *Preliminary Discourse to the Encyclopedia of Diderot*, d'Alembert proposed the view from above as the philosopher's operating tool for the "encyclopedic arrangement of knowledge." As an intellectual descendant of the Enlightenment project, the ascending observer of the early twentieth century was seen as capable of discerning the world as a simplified diagram of lines and shapes. This godlike privilege harked back to medieval roots in the depiction of God as *deus architectus mundi*, revealing a long history of the sustained relationship between heights and divine providence.[21] The visual domination afforded to an elevated seer appealed to the grand reformist ambition of the master builder, particularly in the context of early twentieth-century modernism's formalist promotion of a rationalized order and geometric grid of the city as a panacea for social pathologies.

The second imagination was that the master builder, empowered by his putative ability to see things in their entire breadth of connectivity, invoked a highly plausible image of an evolved human type. An advanced evolutionary symbol distilled from popular utopianism, social Darwinism, and Nietzsche's idea of the *Übermensch*, the aggrandizing concept of a heroic builder was already embedded in the discourse of architecture by the 1920s.[22] As we shall see through the lenses of Ferriss, Fuller, and Bel Geddes, the aggrandizement of the master builder was also couched in the broader discussions of eugenics (a derivative of the Greek word *eugenes*, meaning "good in birth"), the pseudoscience of engineering a superior human breed. Charles Darwin's cousin Francis Galton had coined the term *eugenics* in 1883, and by the 1920s, a wide cross section of American intelligentsia, designers, scientists, social reformers, industrialists, and artists had been influenced in various degrees by the doctrines of the eugenics movement.[23] They exhibited a sustained zeal for the idea of the sociobiological betterment of the people as an antidote to the prevailing specter of social decline due to immigration, unhygienic urbanization, and consumerism. While Ferriss, Fuller, and Bel Geddes certainly did not participate in the exclusionary politics of eugenics, their idea of a master builder exhibited their interest in the project of social betterment. Ferriss and Fuller were exposed to esoteric teachings in self-propelled evolution as promoted by the Russian mystic George Ivanovich Gurdjieff and his acolytes in New York City and read, among others, Alexis Carrel's best-selling eugenics propaganda book, *Man, the*

Unknown (1935). Bel Geddes's personal library contained a sizable number of books related to evolutionary principles, including those by H. G. Wells. The fascination of these ideas for the master builder—as if it were possible to produce a human analogue of the airplane and the skyscraper, which were deemed the most advanced stages in technological evolution—provided the consummate prototype of an ideal leadership class.

My motivation to focus on Ferriss, Fuller, and Bel Geddes was sparked, to some degree, by the fact that these three dynamic American thinker-designers have traditionally received marginal coverage in standard design histories, despite their ubiquitous contributions to the interwar visual culture. Scholarship on Ferriss, surprisingly, has not gone much beyond his daughter Jean Ferriss Leich's 1980 essay *Architectural Visions* and Carol Willis's useful introduction to the 1986 reprint of *The Metropolis of Tomorrow* (1929).[24] Symptomatic representations of Ferriss as a "prophet" of the modern metropolis are found in the histories and theories of American urbanism, such as *The American City: From the Civil War to the New Deal* (1973), by Giorgio Ciucci et al., and *Delirious New York* (1978), by Rem Koolhaas.[25] Critics and architectural historians—such as Lewis Mumford, Manfredo Tafuri, and Vincent Scully—have commented on Ferriss's drawings only as part of their broader discussion of American urban transformation.[26] Mumford was anxious about the dehumanizing scale and the total lack of human presence in Ferriss's depiction of the future metropolis, while Tafuri and Scully detected in his drawings various organic influences of American culture. Tafuri viewed Ferriss's metropolis as "savage primitivism," even though it was derived from a desire for a "formal optimum of massing from the regulations of the [1916] New York zoning ordinance" and its setback rules for tall structures. For Scully, it was an urbanistic transmogrification of the American wilderness and the Grand Canyon. The flip side of Scully's observation also merits attention—that is, Ferriss's drawings are so visually seductive that they are likely to be viewed less as social documents of an extraordinary time than as graphic fantasies, rolling America's dual fascination with nature and city into one grandiose image of the future.

Fuller has been the subject of many biographical studies since the 1960s. However, most of these studies appear to be too invested in Fuller's later heroic mythos and have been written by his enamored pupils or acquaintances, such as Jay Baldwin, Alden Hatch, Robert Marks, John McHale, Martin Pawley, and Lloyd Steven Sieden.[27] In *Theory and Design in the First Machine Age* (1960), Reyner Banham first presented a critical assessment of Fuller's view of technology: Contrary to the pioneers of modern architecture who incorporated the machine mostly as a tool of aesthetic rationalism, Fuller conceived his work *as* machine.[28] Fortunately, Fuller has received considerable scholarly attention in recent times, partly because of the inclusion of his work within the ongoing debates on sustainable design and the relatively late availability of his archive.[29]

Yunn Chii Wong's doctoral dissertation of 1999 titled "The Geodesic Works of Richard Buckminster Fuller, 1948–68" and two recent exhibitions, *Your Private Sky* (Zurich, 1999) and *Buckminster Fuller: Starting with the Universe* (New York, 2008) begin to approach Fuller's oeuvre more comprehensively and analytically.[30] Loretta Lorance's book *Becoming Bucky Fuller* (2009) offers a provocative argument pertaining to Fuller's early life: in the early 1920s, Fuller pursued the career of an industrial entrepreneur, but once he realized that his proposed standardized single-family unit, the Dymaxion House, would not come to fruition, he switched gears and sought to create a visionary self-image by manipulating his personal story.[31] Lorance's characterization of Fuller ignores the conflicted nature of her subject's personality, a puzzling combination of practical motivations and antiestablishment "outsider" attitude.

As for Bel Geddes, design historians like Donald Bush, Jeffrey Meikle, and Arthur Pulos have explored his seminal contribution to the development of stage set design, industrial design, and, in particular, 1930s streamlined aesthetics. Roland Marchand has discussed Bel Geddes's work to show the corporate manipulation of display technique as a way to drum up a consumerist ideology during the Great Depression.[32] Another group of writers have identified Bel Geddes with experimentations in express highway design during the New Deal era.[33] Christina Cogdell's work has sought to unearth a eugenicist impulse behind Bel Geddes's streamline aesthetics.[34] Despite the useful writings mentioned here, scholarship on Bel Geddes's multifaceted design work remains somewhat limited to the domain of industrial design and highway engineering.

Two main reasons can be identified for the tangential presence of Ferriss, Fuller, and Bel Geddes within the period's design histories. First, the pervasive visibility and seemingly ahistorical nature of their signature design projects—Ferriss's metropolis of tomorrow drawings published collectively in 1929; Fuller's unorthodox residential unit, Dymaxion House, finalized through the waning years of the 1920s; and Bel Geddes's Futurama, the most visited exhibit at the 1939 New York World's Fair—may have discouraged design historians from focusing on them discursively. Unfortunately, this lack precipitated the reduction of these designers' contributions to a utopian design genre that was seen either invested in subjective excesses or all too often left outside the traditional history proper.

Second, because the multifaceted professional identities of these three designers often elude normative art historical classifications, cultural theorists and historians appear to struggle with how to include them within the standard thematic structuring of historiography. For instance, Fuller called himself, rather dubiously, a "comprehensive anticipatory design scientist," while Bel Geddes was a stage set designer, an industrial designer, and a self-proclaimed architect, after failing the architectural registration examination. Of the three, Ferriss was the only one who was an architect by training, and he fashioned himself as a modern "clairvoyant" of the metropolis of tomorrow; Fuller and Bel Geddes were self-taught after they were ousted from school for misdemeanor. Their maverick personas, compounded by their common interest in the spiritual philosophies of the mystically inclined architect Claude Bragdon and Gurdjieff as a cure-all for the rampant materialism and moral decadence ailing the industrial societies, seem to have prevented architectural historians from paying extended and scholarly attention to their design work. I propose that an investigation of the work of Ferriss, Fuller, and Bel Geddes helps to build a rich historiography of the interwar period. The desire to create a personal myth led each of them to maintain a self-conscious distance from the architectural mainstream while both drawing inspiration from and challenging some of the heroes of the modernist movement in architecture.

Ferriss, Fuller, and Bel Geddes, in disparate ways, grappled with the presumption that to desire to create utopia was also to dream of its builder.[35] The concept of tomorrow's builder was, however, not exclusively an American phenomenon. Le Corbusier's metaphoric reflection on the aviator as the harbinger of a new civilization is just one example. While it is very possible to see narratives of ascension and associated visuality as phenomena of broader modernist contestations on both sides of the Atlantic, I focus here on the ways the modernist positioning of the aerial perspective as a kind of parable of the future resonated with what authors Herbert Croly and Waldo Frank, writing in 1909 and 1919, respectively, called a quixotic American impulse to search perpetually for an ideal future.[36] The factors that prompted this startling affiliation were as varied as they were ambivalent.

First, the cultural responses to the airplane and the skyscraper both calcified and challenged an enduring American myth concerning the role of the frontier in American history, especially in the context of a prevailing perception of the frontier as an unfamiliar but promising space—a space of adventure, heroism, and conquest—on the edge of the known territory. As we shall see later in the book, Bel Geddes framed the Futurama spectator's simulated aerial journey over a scale model of an American utopia with the rhetoric of frontier spirit, while at the same time alluding to the disappearance of the frontier in the age of aerial transportation.[37] Ferriss and Fuller, too, addressed the ongoing debate about the frontier in expressing the broader philosophy behind their work. Thus, the issue of whether and how dominant nineteenth-century frontier values informed the trope of a "vertical frontier" created by the airplane and skyscraper provokes an intellectual challenge. In his famous "frontier thesis," presented at the annual meeting of the American Historical Association in Chicago in 1893, the American historian Frederick Jackson Turner gave rise to a tenaciously sanitized view of the frontier: an uncharted territory explored by industrious pioneers, a space of abundant opportunities that fostered democratic values of fair competition.[38] The frontier, Turner argued, was a vast natural tabula rasa, and the experience of developing it created an American character of "rugged individualism." This Turnerian vision of the frontier, however, brushed off from it the troubling histories of conquest, violence, and lawless machismo, focusing instead on the frontier's natural bounty that supposedly inspired the ethos of diligence and the enduring "frontiering" tendencies of the nation. The political geography of the frontier continued to be fiercely debated through the interwar years. Turner's book *The Frontier in American History*, including his seminal essay of 1893, was published in 1920. Authors such as Charles Beard, Isaiah Bowman, Percy Boynton, John Dewey, Robert Dripps, Guy Emerson, Waldo Frank, Archer Hulbert, Oswald Villard, and Benjamin Wright Jr. variously deliberated the significance of "frontier values" for an America undergoing rapid changes in the early twentieth century.[39] Even American presidents Herbert Hoover and Franklin Roosevelt joined the fray.[40] The presumption of innocence and American exceptionalism that was interwoven with the Turnerian frontier continued to complicate emerging thoughts on the notion of the frontier, especially the urban.

In many ways, Ferriss, Fuller, and Bel Geddes embraced the romantic myth of the frontier, recalibrating it, however, for their own time and mission. It was not surprising that after his solo conquest of the Atlantic, Lindbergh, one of their heroes, was called a "Daniel Boone or Davy Crockett of the air," among other things, in the popular press.[41] Even when America had plunged into the Great Depression, many authors continued, although not in any reference to aviation or skyscrapers, to dabble with the allegory of America being a permanent frontier that demanded unceasing deliberations on the nature of its future.[42] Ferriss, Fuller, and Bel Geddes participated in, and molded, this contentious debate for their visionary projects. The country's rapid urbanization and the various social ills that allegedly resulted from it provoked them to deliberate on the "world of tomorrow" with the innocence of a pioneering bravado reminiscent of Turner's. Their work offers fascinating cases for an examination of the nature of American utopianism at a particular point in its history and as it unraveled within the cultural contestations over the futuristic significance of the frontier.

Second, the euphoric public response to Lindbergh's 1927 flight sheds light on a then popular American perception of the aviator as an individualist, doggedly in control of his flying machine in the air and, ultimately, of his destiny. In contrast, when the French aviator Louis Bleriot successfully flew across the English Channel in 1909, the French viewed Bleriot's heroic feat not necessarily as the gallant achievement of an individual citizen but as the collective representation of the French nation and its technical ingenuity.[43] The mythology of rugged individualism that colored the popular description of Lindbergh seemed untenable in the case of a contemporaneous European aviator. Although this observation may not be broadly generalized, it gives rise to a provocative opportunity to examine the enduring myth of individualism in interwar America. I am particularly intrigued by the way that populist American celebration of the individualist aviator or the solitary urban voyeur atop the skyscraper could also conjure up the image of the master builder. We shall see in the works of Ferriss, Fuller, and Bel Geddes that during the interwar period, conceiving utopias through the omniscient gaze of a lofty spectator became a common artistic narrative device for gazing into the so-called World of Tomorrow.

Finally, from the late nineteenth century the rapid urbanization of America—and the intensifying debates on its urban future—was a highly ambivalent project. By 1920, more than 50 percent of the U.S. population lived in cities. Alleged antiurban sentiments stemming from agrarian-pastoralist worldviews clashed with new appreciations of urban modernity.[44] On one end of the spectrum were those who continued to view the city as an escalating threat of social turmoil. The American clergyman Josiah Strong's representative book *The Challenge of the City* (1907) demonstrates that Puritans in America long promoted the provocative idea that the worldly city could corrupt the virtuous rural dweller.[45] This kind of distrust of city life further fed on widespread Spenglerian prophecies of doom. John Dewey's hope that the city would provide greater opportunities for the populace was dashed by the haunting growth of urban impersonality and the eclipse of community, something the Wisconsin farm boy Frank Lloyd Wright also took as his point of departure for his irascible critique of the American city.

On the other end of the spectrum were the civic-minded bourgeoisie, urban planners, and social reformers who sought to salvage the decaying industrial city as a pathway to a more optimistic tomorrow. The 1925 *Titan City* exhibition in New York, among others, was a high-spirited example of attempts to bring futuristic monumentality, spatial order, and technological excellence to the American city. However, the fear and optimism of the city were not by any stretch of the imagination distinct, as they sometimes fed off each other, creating an intricate tapestry of attitudes toward the American city. In such an equivocal context, to examine the challenge of creating the Metropolis of Tomorrow (as in Ferriss) or the highly urbanized World of Tomorrow (as in Bel Geddes) signified no less a window onto interwar America.

A crucial question still remains: How are aerial views, or their simulation, from the airplane and skyscraper, as articulated in the work of Ferriss, Fuller, and Bel Geddes during the interwar period, different from earlier elevated views of cities and landscapes? Whether the visual practices of the modern subject deliberated in this book are a new episode in the history of seeing and representation is less interesting as an intellectual inquiry than how American visionary designers appropriated these practices and, out of them, developed the aesthetics of ascension as a rhetorical tool to outline a bold future and its putative creator. Historians interested in the history of visuality and city representations have shown that relatively accurate renditions of cities and landscapes from elevated points of view—by means of imagination and perspectival plotting—were common beginning in the early sixteenth century. Jacopo de Barbari's woodcut *View of Venice* (1500) is a good example.[46] The nineteenth-century balloonist Nadar photographed Haussmann's Paris from the outsized balloon called *Le Géant* in 1858, in search of a viable alternative to the painstaking and inaccurate processes of mapping.[47] Beaumont Newhall reported that one of the earliest uses of balloon observation was for the purposes of battle strategy during the American Civil War.[48] A Union balloon observer named Thaddeus Lowe created an "aerial telegraph" by means of the cables tethered to his balloon *Intrepid,* with reference to the duplicate gridded maps located in the balloon and on the ground. Coburn's aerial photograph titled *The Octopus* (1912), taken from the then tallest building in New York City, the Metropolitan Life Tower, illustrated the fledgling consciousness of a metropolitan observer of the heights.

"Bird's-eye views" existed prior to World War I, but an aesthetic consciousness of heights, and related artistic activities, in response to aviation and skyscrapers came into sharper focus as part of a mainstream visual culture after the war. While Nadar's balloon photography was certainly an important landmark in the urban history of Paris, never before the 1920s had an aerial subject been so thoroughly articulated and theorized as a new type of self-conscious spectator of modern life. The dominant visual practice in nineteenth-century Paris was best represented not by Nadar's elevated eye, for it was exclusive, but by Baudelaire's flaneur, the solitary stroller within the urban maze, not above it. It was not until after World War I that the aerial spectator's viewpoint became typical, not only because of the broad availability of airplanes and access to skyscraper observatories as part of the modern experience but also because aerial photography reproduced and circulated the erstwhile exclusive aerial eye in the realm of mass culture. As the American entrepreneur Sherman Fairchild popularized aerial photography and reconnaissance—a battlefield strategy developed during World War I, for it provided enough ground coverage and detail—an enthusiastic civic culture of aerial observation developed during the 1920s.[49] Furthermore, the technologies of flight and skyscrapers came to wider public view during the 1920s—exemplified by such iconic events as Lindbergh's celebrated flight and the construction of the tallest skyscraper to date, the Empire State Building—inspiring a wide spectrum of artistic, urbanistic, and utopian reactions. All of these materials offer a fertile cultural milieu, warranting fresh interpretations of the dynamic relationship between aerial perspectives and the formation of a modern spectator.

My goal is not to belabor an argument that the aviator and the skyscraper dweller emblematized a wholly new paradigm of seeing or that they presented a radical departure from the historic tradition of elevated representational techniques. Instead, I aim to examine the crucial ways the symbolic meaning of ascension percolated within the particular cultural conditions of the interwar era. A critical history of visual perception defies the suggestions of the universal continuity of a seeing subject. As such, the aviator or the skyscraper dweller constituted a modern subject, not simply because this subject could see the world from hitherto impossible heights, but also because he could be articulated as a distinct protagonist of an interwar American narrative.

The three chapters that follow focus on Ferriss, Fuller, and Bel Geddes, respectively.[50] Not all chapters discuss skyscrapers and airplanes with equal weight. Chapter 1, on Ferriss, deals with skyscrapers, while chapters 2 and 3, on Fuller and Bel Geddes, revolve primarily around airplanes. However, flying machines glide through the urban canyons of Ferriss's nighttime metropolis, whereas Fuller gazed at the newly constructed Empire State Building with awe from the roof of a high-rise warehouse, and Bel Geddes's model cities showcase ultramodern, streamlined skyscrapers in downtown locations.

The chapter on Ferriss examines his imaginary city views, along with his insightful captions accompanying them, published collectively in his book *The Metropolis of Tomorrow* (1929). I argue that Ferriss's chiaroscuro drawings, frequently depicting upward cities, could be seen as social commentary on the mind-set of the burgeoning metropolitan population. Tracing Ferriss's biographical moments in his drawings, the chapter situates Ferriss's rendition of the metropolis, often conceived through the eyes of a lonely spectator perched atop the skyscraper, within emerging forms of urbanistic, artistic, journalistic, and photographic inquiries into the nature of metropolitan morphology. Against the backdrop of an urbanizing America, recasting the dialectic between pastoral ideals and industrial reality into a hybrid cultural space from which to conjecture the future, Ferriss's drawings are instructive of how they narrate both the scope of the future metropolis and the people who would inhabit it. The chapter elaborates on how Ferriss's ideation of the modern man was related to his brief subscription in the 1920s to mystical spiritualism and his enthusiastic participation in the lectures of the English theosophist Alfred Orage, an acolyte of Gurdjieff. Ferriss's drawings of the metropolis share the period's mystical anguish, allegedly caused by a growing discord between science and the spiritual health of society, and seem to celebrate the ways theosophical advocacy of the "harmonious development of man" sought to allay that anguish. Analogous to the teachings of theosophy, Ferriss believed that the loss of the "mystic spirit" in the emulative and pastiche architecture of the early twentieth century somehow paralleled the psychic fragmentation of the modern man. The possibility of a meaningful architecture, therefore, hinged on the reconstitution of the spiritual wholesomeness of man.

Focusing on a little-explored period of Richard Buckminster Fuller's life—from his self-published manifesto *4D Time Lock* (1928) to his first book, *Nine Chains to the Moon* (1938)—chapter 2 examines the "ascensional" roots of this visionary entrepreneur's worldviews. This was a crucial period in the quintessentially "vertical" philosopher's life, when he was arguably more intertwined with the social currents of his time than he was during the period from the 1950s onward, when he became a "think tank" in his own right with the proliferation of his geodesic structures and his radical reputation as a solver of planetary problems. Fuller's philosophical interpretation of human flight provided a theoretical framework for his industrially reproducible house, which he considered an ideal venue for the exploration of the notion of social betterment. He self-published *4D Time Lock* on the first anniversary of Charles Lindbergh's transatlantic flight, which he hailed as an American watershed heralding the frontier of transnational industrial economy and its efficient management. Fraught with references to aviation (and other means of telecommunication), Fuller's writings during this period presented the flying machine not just as an apotheosis of efficiency, mobility, and lightness, but also as a useful metaphor for a highly advanced human breed, thanks to his reading of mystics like Gurdjieff and Bragdon as well as scientists like Einstein and Alexis Carrel. Hardened by his professional failures and personal tragedies during the late 1920s and early 1930s, Fuller consciously began to craft a self-image committed to serving humanity. The myth of the aviator was his ruse in this pursuit.

The final chapter focuses on Norman Bel Geddes's Futurama—the most visited show at the 1939 New York World's Fair, presented as part of auto giant General Motors' corporatist prediction of America's future. With a one-acre animated model, Bel Geddes prognosticated the growth of an American highway system and other infrastructures necessary to transform the continent into a massive production machine that appealed to an American audience in the final throes of the Great Depression. Canonical design histories marvel at Futurama's panoramic display of the future. I take a different route to understand Futurama. Why was it that Bel Geddes made his American utopia visually legible only to aerial spectators hovering over the model by means of a gliding conveyor belt emulating an airplane flight? Is it possible to see Futurama's flying audience as part of Bel Geddes's very conjecture of America's future? Did Bel Geddes replicate the aviator's aerial perspective to construct a modernist spectator who was the practitioner of the same idealistic gaze that compelled the early twentieth-century planner's aspiration of redeeming civilization through spatial organization? I address these questions by expanding the discussion of Futurama's visualization technique into the American valorization of the aviator as a cultural protagonist. The contemporaneous advent of the comic-strip hero Superman—along with his heady moralist view of the world and ability to police the boundaries of good and evil in his metropolitan precinct—provides a foil for Futurama's politics of renewing the American promise, from above.

This book is meant to be read on multiple levels. As much as it deals with the cultural meanings of technology, it is also a history of ideas. It is an analytical account of a cultural attitude that permeated American aesthetic thought during the interwar period. This attitude could be experienced and illuminated, at one level, quite directly through the various cultural products and pronouncements of the period, such as drawings, architectural artifacts, popular magazines, and world's fair exhibits, all accessible in the archives of history. On another level, only an interpretive approach, or even an intuitive one, toward these products and pronouncements could spark new ways of looking at the period's broader visual culture. Here, I was influenced by what the British historian E. H. Carr has called "imaginative understanding" of facts gathered from the past.[51] History, Carr posits, is hardly an objective collection of facts, for the facts are altogether less autonomous and less innocent than they are often taken to be. Facts illuminate a broader story when some of them are selected, arranged, and positioned in a variety of shifting interrelationships. This debate concerning the role of objectivity in historical narratives has been sharpened by scholars such as Northrop Frye, who wrote in *The Educated Imagination* (1964) that inventive interpretations of facts and their impressions can embroider ever newer cultural and literary tapestries.[52] To develop my argument, I rely on archival research, but I also read the historical data with the working assumption that texts, images, buildings, and photographs fundamentally warrant multidisciplinary interpretations, for they are related inseparably to broader patterns, patches, and grids of culture, society, and politics.

ONE

Hugh Ferriss and the "Harmonious Development of Man"

A NEW URBAN SPECTATOR

In a series of sublime views of modern cities and their high-rise buildings, the New York architect, illustrator, and poet Hugh Macomber Ferriss (1889–1962) grappled with what was a subject of intense literary and sociocultural inquiry in the first decades of the twentieth century: the psychological conditions of modern city life (Figure 1.1).[1] Published collectively in his book *The Metropolis of Tomorrow* (1929), Ferriss's chiaroscuro drawings of the 1920s offered a two-pronged commentary on the rapid urbanization of early twentieth-century America.[2] First, his renderings explored the vertiginous environment of skyscrapers and brought to the fore the multifaceted ways "upward urbanism" influenced social and aesthetic perceptions. Second, solitary observers—or symbolic representations of them—appeared in Ferriss's tranquil cityscapes. Sometimes humbled by the city's vertical thrust and sometimes superhuman in their aerial gaze from skyscraper loggias or airborne vantage points, these figures offered a coded visual saga of modernity's broader effects on the mind-set of the burgeoning city population.

Drawn with pencil, crayon, and charcoal, often in black and gray rubbed tones, Ferriss's drawings highlighted an America that was transforming into an urban society.[3] His illustrations were in many ways the products of a cultural shift during which the American public began to see the city as a site for all kinds of social experiments, avant-garde imaginations, and rags-to-riches odysseys. Having migrated, in 1912, from America's heartland to its littoral urban center, New York City, Ferriss understood firsthand how the image of the rising city signified the dramatic reversal of nineteenth-century America's pastoral ideals as embodied by the horizontal sweep of the hinterland. In *The New World Architecture* (1930), the American author Sheldon Warren Cheney (1886–1980) discussed Ferriss's visual oeuvre to contemplate the broader meanings of urban verticality in interwar America.[4] Including one of Ferriss's drawings

Figure 1.1. A staged photo of Hugh Ferriss in his studio veranda atop the Architects Building in New York City. Reprinted in Hugh Ferriss, The Metropolis of Tomorrow *(New York: Princeton Architectural Press, 1986), 182.*

employing the period's characteristic bird's-eye view to delineate the vision of the city of tomorrow, Cheney was hardly remiss in noticing how the impulse of the Ferrissian city's "lift toward the skies" was already encoded in an invisible airborne spectator's magisterial gaze at the city (Figure 1.2).[5]

Many of the illustrations in *The Metropolis of Tomorrow* are from a pedestrian point of view, especially those in the book's opening section, "Cities of Today," in which a stand-alone skyscraper is depicted (Figure 1.3). Examples include the Radiator Building, the Chicago Tribune Building, the Shelton Hotel, and the Waldorf-Astoria Office Building. In these "look-up" drawings, we get the impression of an awestruck pedestrian visually scaling the height of a tall landmark structure, all set in a chimerical display of light and shadow. These drawings portray the quiet beauty of a soaring building, devoid of people and removed from the frenzied city life. Despite their visual manipulation, Ferriss's look-up drawings appear to be primarily documentary in their representation, each embodying the vertical spirit of a single high-rise building (constructed recently or to be constructed shortly) rather than the wide swath of an urban ensemble. Thus, they are somewhat limited in their ability to narrate the multilayered

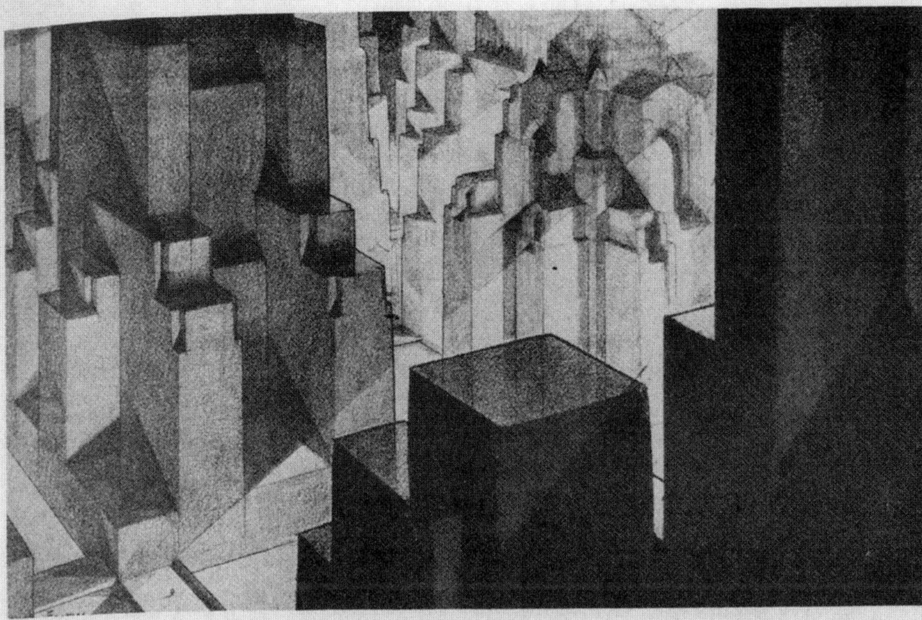

From a drawing by Hugh Ferriss

CHAPTER I

THE NEW WORLD ARCHITECTURE

AFTER civilization, what? The question is not an idle one. Granted that civilization is what we have, humanity is about ready to move on to something else. Signs are not lacking that the present generations are living at one of the three or four major turning-points in human history. The latest centuries-long slope has brought us to a summit. Almost suddenly a different prospect widens before us. An old age, a human epoch, and its methods of organization and its beliefs, have proved their utmost. As we look back at the achievements that men prized most during the last five hundred years of that age — the military victories, the regal pomp, the arts, the religions — we begin to discern that these were less important than the incidental building of foundations for a different unforeseen world. A new faith, a new power, and a new way of living have been born. The new conception of Space-time has found wide acceptance.

1

Figure 1.2. A bird's-eye drawing by Hugh Ferriss on the opening page of Sheldon Cheney's The New World Architecture *(New York: AMS Press, 1930), 1. The drawing had also appeared in* Pencil Points *(1925).*

Figure 1.3. Hugh Ferriss's sketch of a single skyscraper in The Metropolis of Tomorrow *(New York: Ives Washburn, 1929), 137.*

spatial story of the modern metropolis and are not the ones that best represent the futuristic ambitions of Ferriss, or the era.[6]

Even if look-up drawings abounded in *The Metropolis of Tomorrow,* Ferriss preferred to represent the spatial complexities of the metropolis from the vantage point of an elevated observer, navigating the aerial distances above or between skyscrapers (Figure 1.4).[7] From the heights, this observer inspected the city with a broad visual sweep, both zooming out to the broad urban mass and zooming in to the details of individual edifices. In Ferriss's bird's-eye view, the observer experienced a paradoxical effect of omnipotence and removal from the city, unlike, for instance, the turn-of-the-century Ashcan School artists, such as George Bellows and John Sloan, who painted New York's urban life from within the frenzied environment of sidewalks (see Plate 3).[8] Ferriss's drawings brushed away all people and the surrounding urban cacophony to retain only the essential building morphology, dramatized by artificial light. "Bird's-Eye View, the City at Dawn," the introductory drawing of the book's first section, encapsulated this tendency set in a mysterious tapestry of light and darkness. This drawing and others of its kind demonstrated the Ferrissian observer's visual authority more persuasively than the ones in which the observer gazes upward at a stand-alone, object-like building. In Ferriss's bird's-eye renditions, we encounter not only his meditative inquiries into cities but also a paradigmatic visual practice of negotiating the spatial intrigue of the modern metropolis. "The majority of the public," David Nye tells us, "take less interest in the [look-up] views than in the panoramas available from the observation deck at the top of the building. For most people, the skyscraper achieves a good deal of its meaning as a vantage point, rather than as a view. In popular culture the skyscraper is not a thing in itself, but a platform from which other things can be seen and evaluated."[9] In many ways, Ferriss's skyscraper viewing of the metropolis appeared to be a commentary on how, in the early twentieth century, modernist artists and architects sought to come to terms with conflicted deliberations on metropolitan life, both its optimism and its alienation.

Popular media depictions of Ferriss during the 1920s as "architecture's most grandiose seer" frequently highlighted his strategic use of the bird's-eye view to discern the meaning of the modern city. Newspaper articles featuring Ferriss's drawings abounded, with headlines such as "New York from a Studio Rooftop," "Architects Dream of a Pinnacle City," and "City Yearns to Sky in Architect's Vision."[10] Using his aerial perspectives, these articles implied that the full scope of the early twentieth-century metropolis—in terms of its network of rising buildings, their interrelationships, and the in-between spaces of circulation and aerial transit—could be legible perhaps only to an omniscient observer on the heights. The same year that Ferriss published *The Metropolis of Tomorrow,* the American critic Stuart Chase brought out his book *Men and Machines.* In this book, Chase wondered about the peculiar visual legibility of skyscrapers—that

Figure 1.4. Ferriss's birds-eye view sketch of the metropolis showing a group of skyscrapers at night in The Metropolis of Tomorrow *(New York: Ives Washburn, 1929), 63.*

is, he asked whether the modern American city's "choicest aesthetic treasures [were] primarily for the inspection of seagulls."[11] The March 1930 issue of *The American City* magazine used Ferriss's bird's-eye drawing of the metropolis of tomorrow as its frontispiece. Not only did such aerial representation demonstrate Ferriss's public image as an iconic American seer of the future, but it could also be understood, more broadly, as a linchpin of the 1920s visual techniques with which the city of tomorrow was envisioned most convincingly.

Ferriss's chiaroscuro rendition of the metropolis of tomorrow was as much about articulating the nature of a metropolitan spectator as it was about illustrating the future city. By self-consciously positioning a human figure gazing down on the city, he showed that aesthetic modernity was simultaneously a saga of transformation in the urban experience and the image of the "man" who envisions it (Figure 1.5). Contrary to the common interpretation that Ferriss's oeuvre represented what Joshua Taylor called 1920s "urban optimism," Ferriss's drawings were entangled in a litany of existential conflicts experienced by the metropolitan man, the alleged victim of a chaotically modernizing world.[12] His nighttime iteration of the metropolis as a sanitized and meditative aesthetic product could, then, be understood as an attempt to come to terms with the modern man's inner struggles that could be in some ways alleviated by his idealization as the überseer of the city.

Ferriss's ideation of the modern man as a conflicted protagonist elevated above the tempestuous urbanity of metropolitan streets was related to his tentative embrace of the mystical spiritualism of the controversial Russian mystic George Ivanovich Gurdjieff in the mid-1920s. Gurdjieff and his British acolyte Alfred Richard Orage (the acclaimed editor of the political weekly *New Age*) descended on the New York scene in 1924 to disseminate the spiritual doctrine of "harmonious" self-development. During this period a broad cross section of New York's literary and artistic avant-garde embraced, contemplated, or at least dabbled in the subject of "cosmic consciousness" as an alternative to modern life allegedly corrupted by unbridled consumerism, overrationalized worldviews, and a false division between "Western science" and "Eastern spirituality." Echoing the teachings of Gurdjieff and Orage, Ferriss believed that the loss of the "mystic spirit" in the pastiche architecture of the early twentieth century paralleled the psychic fragmentation of the modern man. The possibility of a meaningful architecture, therefore, hinged on the reconstitution of the spiritual wholesomeness of man. If mystical spiritualism found a peculiar resonance in the cultural excitement created by the rise of skyscrapers, then Gurdjieff's project of reconstituting the human race paralleled Ferriss's metropolis as a stage set for the creation of an ideal urban citizenry.

Ferriss's drawing "The Lure of the City" in *The Metropolis of Tomorrow* reflected the intellectual crisis stemming from the American experience of what William R.

Figure 1.5. The aerial observer of the nighttime metropolis in The Metropolis of Tomorrow *(New York: Ives Washburn, 1929), 101.*

Taylor has called the two colliding frontiers: the tranquil horizontality of a pastoral idyll and the skyward thrust of the modern metropolis (Figure 1.6).[13] The radiant machine and the serene garden, to paraphrase Leo Marx, form a precarious juxtaposition.[14] The drawing did not reminisce about nineteenth-century technology's or industrial modernity's encroachment into the American wilderness. Rather, it traced the agrarian hinterland's arrival at the fringe of vertiginous urban modernity and its promised, gleaming future. The American historian Frederick Jackson Turner's nostalgic lament of the frontier closure in the last decade of the nineteenth century now seemed mitigated by the mystical aura of a rising urban frontier.[15]

A country bumpkin—standing on the prairie and gawking at the escalating metropolis, his body arched backward and his hands stretched outward—seemed awestruck. His back was turned obliquely toward the frontier he left behind. In the language of the era's visual clichés, this dramatic portrayal signified that he was looking irreversibly into the future. The cluster of skyscrapers contrasted with the pioneer's hut, perhaps presented here as a metaphor for the bygone simplicity of frontier life. The skyscrapers suggested a dreamscape that glowed with the "urban optimism" of the 1920s. The glaring light from below obscured the details of the skyscrapers, emphasizing only the formal outline of an urban "eruption." Ferriss justifiably called this phenomenon the "lure of

Figure 1.6. Hugh Ferriss, "The Lure of the City" in The Metropolis of Tomorrow *(New York: Ives Washburn, 1929), 58.*

the city." His drawing appeared to be a succinct visual narrative of the country yokel's first encounter with the metropolis, the momentous event Ferriss described as "the rural youth . . . arising to his dream of 'the big city'—the unformulated yet gleaming metropolis."[16] The drama of the chiaroscuro representational technique hardly concealed what Ferriss presented here as a familiar trope of early twentieth-century American life: the country youth's paradigmatic pursuit of the American Dream in the metropolis.[17] The Irish American painter Thomas Hovenden's 1890 painting *Breaking Home Ties* (Plate 4)—depicting a melancholic familial tableau in which a young man is about to leave his rural home and his downcast mother to make his fame and fortune in the big city—was voted the most popular work of art at the Chicago World's Columbian Exposition in 1893.[18]

Ferriss's visual articulation of the technology-driven, futurist city, in contrast to a landscape of agrarian traditionalism, abounded in corporate advertisements of the 1920s.[19] An advertisement in the March 6, 1926, issue of the *Saturday Evening Post* showed a frontier farmer standing on a rocky prairie with his livestock. He appeared conflicted between the comfort of his agrarian livelihood and the promise of a distant future, symbolized by the fabled city of glistening skyscrapers (Figure 1.7). Philadelphia-based advertising agency N. W. Ayer & Son's message in "The Widened Vision" denied this conflict, however, in favor of a deterministic view of progress: the city is a logical outgrowth of the pioneering spirit of the hardworking frontiersmen "who dared conceive of a mighty nation rising from tangled woods and fertile plains." In a similar example from the February 27, 1926, issue of the same magazine, a hallowed, rugged yet impatient frontiersman—a twentieth-century Daniel Boone—rose above the seeming banality of the western trail and swung open the door to the city, the future's epicenter that radiated light, symbolizing endless prosperity (Figure 1.8). The visual tableaux of these advertisements—frontier in the foreground, as a thing of yore, with the gleaming city of towers in the background, representing an unequivocal future—compelled the viewer to participate in the frontiersmen's ambitious gaze toward the city. With the fading glory of the nineteenth-century frontier myth, the city of tomorrow, with its towers and bright lights, was presented as the theater of modernity that required a captivated rural youth with a particular type of gaze that both mirrored and aggrandized the promise of the future itself.

In many ways, "The Lure of the City" was autobiographical.[20] A son of Missouri Supreme Court judge Franklin Ferriss, who served the court from 1910 to 1912, Ferriss spent his childhood in St. Louis. In 1912, one year after receiving his bachelor of science degree in architecture from Washington University in St. Louis, the ambitious Ferriss, at the age of twenty-three, forsook the comfort of his hometown. He quit his job as a junior draftsman at the St. Louis architectural office of Mariner and LaBeaume to move to New York City in search of a

Figure 1.7. "The widened vision" of the frontiersman—from the advertising agency N. W. Ayer & Son in the Saturday Evening Post, *March 6, 1926.*

Figure 1.8. "The tremendous national development . . . brought about the change from a few frontier trading stations to the many mercantile centers of modern times," and the frontiersman opens the door to the metropolis of tomorrow! A Certain-teed (formerly General Roofing Manufacturing Company) advertisement from the Saturday Evening Post, *February 27, 1926. The ad's art was the work of the famed American illustrator Herbert Paus, who created most of Certain-teed's ads from 1925 to 1929.*

more fulfilling architectural career. His new career, Ferriss resolved, would be guided not by Beaux-Arts nostalgia, but by "an indigenous American architecture." He wrote: "One wanted to get to the Metropolis. In New York . . . an indigenous American architecture would be in the making, with engineers and artists working enthusiastically together—and maybe even with the populace warmly appreciating and applauding their alliance."[21] Ferriss's journey would reverberate a decade later in the sentimental dramatization of "The Lure of the City." New York was not unkind to him. From 1912 until 1915 he worked as a draftsman in the New York office of famed architect Cass Gilbert. He then left to set up his own practice in "specialized . . . drawings visualizing future buildings and cities."[22]

After setting up his own architectural drafting practice, Ferriss gained a reputation for being an excellent architectural illustrator. Architects, builders, and engineers frequently commissioned him to visualize their projects, as did the U.S. War and Navy Departments and architectural firms and advertising agencies across America. Ferriss was firmly on his way to architectural stardom, making important contributions to the nation's first comprehensive zoning law, the 1916 New York City statute, and having worked on visionary city ideas with such luminaries as Harvey Wiley Corbett and Raymond Hood.[23] In 1925, throngs of New Yorkers were captivated by Ferriss's renderings of the high-density metropolis of crystalline skyscrapers and vertically stratified circulatory systems at *The Titan City: A Pictorial Prophesy of New York, 1926–2026,* an exhibition held at the John Wanamaker department store in New York City.[24] By this time, Ferriss was a seasoned urbanite and a sought-after architect. On one occasion, Corbett solicited Ferriss to lecture at Princeton University.[25] And when Ferriss applied for the architect's registration in 1929, his former employer Cass Gilbert wrote a recommendation letter on his behalf.[26] In the midst of this surging reputation, Ferriss's rendition of "The Lure of the City" was somewhat anachronistic. He imagined the metropolis of tomorrow not through his current cosmopolitan New Yorker eyes, but through the bewildered eyes of an American youth who had left the country to pursue his ambition in the big city, as if to dramatize his own middle-class story with a Horatio Alger-esque fable of boyish dreams and perseverance.

A different Ferrissian gaze was evident in another drawing in *The Metropolis of Tomorrow,* "The City at Night: Descent into the Street" (Figure 1.9).[27] The Ferrissian youth was now perched atop a skyscraper loggia, a panoptic vantage point from which he inspected the city at will. The lonely seer here was a totem for the reader's omniscient gaze, whose visual power was mirrored by the sprawling reaches of the urban roofscape. Leaning on the parapet high above the city, the solitary figure gazed down on the eerily quiet and oneiric metropolis of the night, lit from below by glaring streetlights. The seer alone negotiated the luminous drama unfolding at the source of the light on the street. The barren roofscape offered a tranquilized opposite of the rancorous street life. Even if it was inspired by the ascending morphology of 1920s New York, the drawing proposed a sanitized and fictive New York, while the "real" city underwent sweeping social, cultural, and urban transformations during the 1920s. In this sense, "The City at Night" is the conceptual opposite of, say, Ashcan School artist George Bellows's *New York* (1911) (Plate 3), which revealed common urban experiences of most New Yorkers in the early twentieth century: the challenge of navigating city streets teeming with people, carriages, and automobiles. The former transformed the urban spectacle into a clean slate on which to pen another narrative, while the latter plunged the viewer into the dizzying congestion of the metropolis.

Ferriss's representation of the city offers a range of interpretive possibilities. His perspective from the skyscraper could be viewed as a twentieth-century urban descendant of what Albert Boime calls the "magisterial gaze" of the nineteenth century. This gaze experienced the sublimity of the American wilderness from the uplands, a gaze covertly articulating a "larger national will to power in the form of Manifest Destiny."[28] The Ferrissian gaze exercised a different kind of visual power, not by colonizing the wilderness but by fictionalizing the urban spectacle into a stage set for new interventions. Ferriss's aerial perspective is best understood, within the context of early twentieth-century

Figure 1.9. Hugh Ferriss, "The City at Night: Descent into the Street" in The Metropolis of Tomorrow *(New York: Ives Washburn, 1929), 19.*

modernism, as a type of urban spectatorship in which the gaze itself was an instrument, in both the mitigation of a cultural angst resulting from the onslaught of modernity on traditional values and the rhetorical production of *tomorrow*. Ferriss positioned the metropolitan observer at great heights as a trope for how the complex spatial network of the modern metropolis called for a new type of urban negotiator, one who confronted modernity in both its coherence and its contradictions.

The argument resonated with Ferriss's own life. Until his death in 1962, Ferriss lived in a studio on the roof of the seventeen-story Architects Building located in Manhattan at the intersection of Park Avenue and 40th Street.[29] Positioned more than two hundred feet above the ground, he simultaneously surveyed and romanticized the city at night, a vertiginous fascination captured succinctly by his own photograph published in the 1930 alumni bulletin of his alma mater (Figure 1.10). In a 1923 *Christian Science Monitor* article titled "New York from a Studio Rooftop," Ferriss wrote: "Leaning, at dusk, against the dark rail, one will regard roofs, spread out below for miles and miles—blue and silent . . . at midnight, no longer disturbed by the little people, [the distant office buildings] will resume—so it will seem to the watcher on the roof—their night-long communion—low, calm, and sculptural."[30] Ferriss's alter ego, the

Figure 1.10. Hugh Ferriss looking at the city from a skyscraper balcony, in "Hugh Ferriss: Nation's No. 1 Architectural Artist," Washington University Alumni Bulletin, *1939, 26.*

naive "rural youth" seemingly overwhelmed upon his arrival in the city, here appeared to have come of age. Not only did he master the art of navigating the tricky streets of the rising metropolis, but he also positioned himself, both literally and figuratively, in the upper echelons of society to decode the clotted spatial language of the city, as well as to recast it according to his own will.[31] Thus, underneath Ferriss's dramatic portrayals of the metropolis lay the much mythologized story of the country youth who outgrew his initial bewilderment at the city's spectacle and became "successful" after trekking the metropolitan path of trial and tribulation. The sentimental message here was that the pursuit of the American Dream, with all of its values of the self-made man, no longer occurred in the rural heartland; rather, it took place in the theater of the modern metropolis.[32] But such optimistic views were hardly immune to societal ambivalence. A 1928 cartoon with the title "So This Is Progress!" exemplified the ironic social effects of the collision between two value systems: a filial attachment to pastoral America and the postfrontier American youth's pursuit of opportunities in the city, even if it meant living anonymously in a skyscraper pigeonhole (Figure 1.11).

Literary progenitors of the Ferrissian protagonist abounded in classic American novels of the nineteenth century. Horatio Alger described the labyrinthine city life through the befuddled eyes of a yokel named Frank, and Herman Melville employed the character of a neurotic country prince, Pierre, as a prism through which to view the hysteric or neurasthenic effects of civilization.[33] Continuing this literary tradition into the twentieth century, Sinclair Lewis's hero (or rather antihero) Martin Arrowsmith in the 1920s embarked on an idealistic journey from midwestern parochialism to the tempestuous urbanity of New York City.[34] It would not be a reach to draw a parallel between Ferriss's "rural youth"—gawking at, and ultimately coming to terms with, the city's spectacular growth—and these literary figures negotiating the city's mazy environment and its urban pathologies. Just as American novelists deployed literary characters to comment on social flourishes of city life, Ferriss, too, proposed his protagonist as a prop, the quintessential observer-narrator of metropolitan modernity. Ferriss's seer on the heights, as in "The City at Night," was infused with a rags-to-riches subtext and the ethos of the self-made man who reached his apotheosis during the 1920s, as Roderick Nash has demonstrated with such examples as Henry Ford and Herbert Hoover.[35] Ferriss contended with the American myth of success—characteristically described as the self-made man's upward social mobility through diligence, perseverance, and mental fortitude—with a modern allegory of verticality.[36] As Thomas A. P. van Leeuwen notes, entrepreneur Frank W. Woolworth, "the product of a modest upbringing," sought to re-create himself as Napoleon with the "imperial" decor of his executive office on the fortieth floor of his namesake building: "This is perhaps one of the most important aspects of the American skyscraper phenomenon: if

Figure 1.11. Clarence D. Batchelor's cartoon depicting the "progress" in housing from one generation to the next appeared in the New York Evening Post *in 1928.*

power could be transferred to matter, this matter could then be transformed into buildings."[37]

For Ferriss, however, the vertical city was much more than the self-made man's celebratory parable. His was also a conscious choreography of a modernist logic of looking at and comprehending the world. The ability to survey a vast swath of the city from the skyscraper observatory granted the seer an illusion of power that, in turn, facilitated his romantic desire to act like a master builder who could transform the disorderly city of the early twentieth century into a planned utopia.

THE SKYSCRAPER HEIGHT AS URBAN MYSTERY

After World War I, the escalating urban form of New York City made the debate surrounding the meaning of verticality an essential element in the discussion of America's self-image.[38] In *The Meaning of Architecture* (1918), Irving K. Pond associated horizontality with "intellectual poise" and linked verticality with "spiritual ecstasy," providing an architectural analogy for the rise of the skyscraper city in relation to the American prairie.[39] For Ferriss and his generation, the skyscraper was no less the quasi-sacrosanct muse of a new urban frontier. The mystical philosopher and Rochester-based architect Claude Fayette Bragdon (1866–1946), a friend of Ferriss, viewed the skyscraper as an authentic symbol of "the American Spirit—restless, centrifugal, perilously poised" and an organic American development "in the field of architecture to which we can lay unchallenged claim."[40] Featuring an aerial view of a city of tomorrow drawn by Ferriss, a characteristic article in the *New York Evening Post* in 1925—"City Yearns to Sky in Architect's Vision"—encapsulated the cultural climate in which the skyscraper achieved a widespread popular appeal.[41] The article's subtitle, "New York to Scorn Gravity," was corroborated by Ferriss himself with an autobiographical tinge: "In the future, when the evolution of the city is accomplished the people of New York will practically live in the sky." The skyscraper became affiliated with a range of cultural deliberations and, as Meir Wigoder's study of photographer Alvin Langdon Coburn shows, with new types of urban spectatorship.[42]

A number of modernist photographers, novelists, artists, and architects—including Alfred Stieglitz, Alvin Coburn, Edward Steichen, Margaret Bourke-White, Theodore Dreiser, F. Scott Fitzgerald, Max Weber, Georgia O'Keeffe, Bragdon, and Ferriss—lived in these tall buildings or went up to the observatories to see the city from a new angle. Their artistic production bore the influence of living high above the city, a novel form of urban residency. From the heights, they romanticized the city below.[43] In *The Story of Architecture in America* (1927), Thomas Tallmadge captured the prevalent romanticism of the skyscraper view: "Its eyes gaze down from immeasurable heights on a welter of humanity and machinery."[44] One of Ferriss's aerial views, "Overhead Traffic-Ways," arrested, with uncanny accuracy, Tallmadge's maudlin depiction of the city seen from the skyscraper, the soaring aspiration of which was further reinforced by the airplanes navigating the canyons of the metropolis (Figure 1.12).

Ferriss's intimate relationship with New York's theosophical circle and artistic communities sheds light on his mystical interpretation of living high up in the city.[45] Ferriss was introduced to theosophy by Bragdon, who was a leader of the esoteric movement. He also knew fellow theosophists O'Keeffe and Stieglitz, who lived in an apartment, first on the twenty-seventh floor and then on the thirtieth, in the thirty-four-story Shelton Hotel.[46] Bragdon was the couple's neighbor at the Shelton Hotel, while mystically inclined authors Theodore Dreiser, Jean Toomer, Waldo Frank, and Jane Heap were common acquaintances. As the art historian Anna Chave has demonstrated, the magisterial vista from O'Keeffe's apartment at the Shelton Hotel enabled her to seize a new "American Spirit" in her urbanscapes painted between 1925 and 1930.[47] Chave posits: "O'Keeffe's acute sense of the skyscraper's height was undoubtedly enhanced by her living and working in one, as she was the first artist—and among the first people ever—to reside in a skyscraper. She was able to find an apartment atop a tall building in 1925, whereas Duchamp had failed in his efforts to do so a decade earlier."[48] Both O'Keeffe and Stieglitz viewed their Shelton apartment as a refuge aloft, a quiet sanctuary—or even a retreat—claiming it allowed them to stay sane in the midst of dizzying modernity unfurling below on the ground. Living high up at the Shelton was "the most successful escape from the dirt, ugliness, noise, promiscuity of the city."[49] As Benita Eisler argues, for Stieglitz and O'Keeffe, the Shelton's elevation provided an opportunity "of the city but not in it."[50] Stieglitz wrote to the American novelist Sherwood Anderson in 1925 with a sense of existential anguish and transcendence: "New York is madder than ever. . . . But Georgia and I somehow don't seem to be of New York—nor of anywhere. We live high up in the Shelton Hotel. . . . We feel as if we were out at midocean—All is so quiet except the wind—the trembling shaking hulk of steel in which we live—It's a wonderful place."[51] The same year Bragdon, too, wrote of his experience of discovering what he would term a spiritual communion between the Shelton's top floors and pure nature: "The high-perched denizen of one of its thousand cubicles . . . receives the sun's first rays long before they penetrate into the city canyons, and all day long he gets the bright radiance of an unobstructed and unafflicted sky."[52] As for O'Keeffe, her skyscraper paintings of the 1920s bore the marks of her monastic isolation high up in the Shelton and evinced a similar grafting of serene "nature" onto the industrial modernity of urban America. Her New York, as in *Radiator Building—Night, New York* (1927) (Plate 5), was in many ways a "pastoralized" cityscape that mirrored the detachment and quiet of her own apartment.

Ferriss, too, self-consciously expressed the "transcendence" of his own studio on the seventeenth floor of the Architects Building. In a number of catalogs for

Figure 1.12. Hugh Ferriss, "Overhead Traffic-Ways" in The Metropolis of Tomorrow *(New York: Ives Washburn, 1929), 64.*

his *Metropolis* drawings, he asserted his comfort in being separated vertically from the streets, which he, like many other Americans at the time, viewed as plagued by all kinds of social pathologies of modern life. He wrote, indignantly: "Going down into the streets of a modern city must seem—to the newcomer, at least—a little like Dante's descent into Hades. Certainly so unacclimated a visitor would find, in the dense atmosphere, in the kaleidoscopic sights, the confused noise and the complex physical contacts, something very reminiscent of the lower realms."[53] "The habitual city dweller" was stuck in the quagmire of the "lower realms," the congested and unhygienic atmosphere of the ground, traversed by a host of shady urban figures, the exploitative businessman, the crooked, and the vagabond. If the streets were the city's arteries and veins, Ferriss reasoned, then the healthy American body was in a perilous condition because of poor blood circulation in the body. His anxiety over the treacherous and noisy streets, however, slowly faded away as he ascended to the upper floors of the skyscraper, a psychological erasure that, as Carol Willis posits, "strongly affected his perception of architecture and his attitude toward the metropolis" and "inspired both romance and analysis."[54] Willis suggests: "From [a] distance the aural and visual cacophony of the city softened, just as buildings appeared only in broad outline and large masses. From here, [Ferriss] observed the city at night, its disorder disguised by darkness and its mystery illuminated by electric lights."[55] Ferriss's vantage point provided a forceful abstraction of a seemingly chaotic spectacle, a sort of visual cleansing that brushed away all of the purported anomalies within the picture frame. From on high, Ferriss's gaze morphed the city into a sublime unity, a kind of authoritarian visual discipline that, decades later, Jane Jacobs would accuse modern urban planners of trying to impose from above on existing cities at the expense of the pedestrian's "real" experience of urban life.[56]

Even if Ferriss's aesthetic abstraction was tinged with unqualified utopianism and hubris, its cultural significance must not be pigeonholed with the "extravagant fantasies" of the "megalo-planners."[57] In applying a contemporary art historical framework, we might overlook the kind of zeitgeist to which Ferriss and his contemporaries reacted. American architects and planners of the time typically expressed ambivalence at best and were aghast at worst at the laissez-faire growth of the industrial city: between 1880 and 1920, the United States witnessed intensely accelerated urbanization. The meager 7 percent of the population living in urban areas in 1820 rose to more than 50 percent by 1920. The massive influx of immigrants dramatically transformed the demographic profiles of American cities. The cultural transformation from small-town values to those of an emerging bourgeois middle class shaped a new American outlook of conformity, regularity, functionality, and management.[58] Industrial capitalism and the rapid expansion of a market economy spawned an unprecedented construction boom. Teeming with people of immense ethnic variety and displaying a skyline punctuated by the belching spires of the capitalist machine and iconoclastic skyscrapers, New York City presented an unmistakable image of progress while challenging the centrality of the western frontier in the American experience. This image of national progress was, however, infused with an uneasy nostalgia, rekindling the long-standing American fear of, and hostility toward, the city.[59] Many Americans responded to emerging urban modernity with lament and a yearning for a simpler life on one hand and optimism for the future on the other. The American experience of urbanity could be described as what Marshall Berman calls a double-edged condition, one in which modernity fueled life's intellectual liberation and empowerment with the benefits of science and technology while disrupting the seamless comforts supposedly provided by a traditional lifestyle and values.[60] The debate on the metropolitan man emerged out of this situation fraught with existential rifts.[61]

As such, the urban gaze of Ferriss's modern protagonist was problematic. Initially, it seemed that the strength of his urban seer, aloft in the skyscraper, lay in his ability to see the city in totality and to fictionalize it to suit his grand futurist scheme. The gaze on the city was, however, hardly unequivocal, for the significance of the city in American life was a topic of much contention in the early twentieth century. It is tempting to think that the slow emergence of buildings in Ferriss's drawings from behind a curtain of mist, fog, or a kind of uncertainty of knowledge alluded to the ambivalent responses to the rapid growth of cities in America (Figure 1.13). As Ferriss wrote:

> There are occasional mornings when, with an early fog not yet dispersed, one finds oneself, on stepping onto the parapet, the spectator of an even more nebulous panorama. Literally, there is nothing to be seen but mist; not a tower has yet been revealed below, and except for the immediate parapet rail . . . there is no suggestion of either locality or solidity for the coming scene. To an imaginative spectator, it might seem that he is perched in some elevated stage box to witness some gigantic spectacle, some cyclopean drama of forms; and that the curtain has not yet risen . . . there could not fail to be at least a moment of wonder. What apocalypse is about to be revealed? What is its setting? And what will be the purport of this modern metropolitan drama?[62]

Ferriss's graphic technique was symptomatic of the general cultural uncertainty as to what the city promised to a nation historically steeped in bucolic sentiments. Was his "spectator of a nebulous panorama" a semantic hint of the fragility of the concept of modernity itself?

This dilemma is understandable because a crucial shift occurred in the image of the city around the end of the nineteenth century: from the Protestant

Figure 1.13. Hugh Ferriss, "A First Impression" in The Metropolis of Tomorrow *(New York: Ives Washburn, 1929), 14.*

imagination of the worldly city as a realm of sin to the city as an ambivalent, somewhat secularized object of inquiry, a great mystery.[63] The Puritan view of the city as a domain of moral destitution was originally articulated in the seventeenth century in the English Christian preacher John Bunyan's biblical allegory *The Pilgrim's Progress* (1678).[64] Unfurling the whole panorama of the tragicomic experience of man, Bunyan's story features a hero, cast as a pilgrim, who journeys along with his companions from the earthly City of Destruction to the Celestial City, a journey of redemption in which the man-made city is presented as a sinful trap that the virtuous pilgrim must overcome with a view to reaching the heavenly destination.

This denigrating view of the city, entrenched in the Puritans' worldview, began to shift as the twentieth century set its course. The new image of the city belonged less to the domain of religious righteousness and more to mystery and curiosity, sparked by the city's bewildering physical growth, hustle and bustle, fast pace of life, industrialization, rampant commercialism, and so on. As Alan Trachtenberg explains:

> The great city had enlarged the scope and scale of mystery itself, bursting the conventional biblical and Gothic tropes to form a new figure, a fusion of social, political, and technological peril. Mystery had been raised to the level of spectacle, the daily performances of city life now seemed to more and more commentators to be parades of obscurity, of enigma, of silent sphinxes challenging the puzzled citizen.[65]

If indeed the city's "mystery had been raised to the level of spectacle," solving the mystery also called for new forms of intellectual engagement with the city. The "mysterious spectacle" entailed new terms of commitment to the concept of urbanity, new methodologies of examination to decode the so-called enigma of the modern metropolis.

One such engagement was in the emerging field of urban journalism. From the late decades of the nineteenth century, American authors of travel texts increasingly turned their attention to the city as an object of inquiry instead of traveling abroad.[66] Thanks to the demand for intriguing stories sparked by a burgeoning mass culture fed by newspapers and magazines, a new generation of dogged journalists, inspired by Joseph Pulitzer, scoured urban areas in pursuit of new trends and examples of city life. Reform-minded journalists like Stephen Crane uncovered the despair of the city's blighted areas, which often ghettoized marginal communities. Jacob A. Riis, another muckraking journalist, leveled his microscopic gaze at the unhygienic, overcrowded, and dilapidated tenement housing of New York City as a way to address the social ills of the city's "other half."[67] Riis's disturbing portrayal of unhygienic, congested life in the tenements and sweatshops of immigrant New York, in lantern lectures, newspaper reports, photographs, and books, jolted a complacent city hall and others concerned with urban affairs. Investigative journalism as a way to build civic awareness to ameliorate the pathologies of urban life became a common crusade before World War I. Lincoln Steffens's 1904 book *The Shame of the Cities* is a good example of the urban investigative genre that uncovered American municipal corruption.[68] The indignant gaze of the muckraking journalist was a reaction to the new urban scenario—environmental degradation, corporate corruption, and urban poverty—that, according to many culturally conservative commentators, was a threat to the moral coherence of traditional American society.[69]

By the 1920s, however, the indignation of urban reformers became tempered as anxious reformatory agendas blended with what Paul S. Boyer suggests was a "frank delight in the totality of urban life."[70] The Protestant clergyman Josiah Strong's angst over moral deterioration in an urbanizing America, as spelled out in his books *Our Country: Its Possible Future and Present Crisis* (1885) and *The Challenge of the City* (1907), was repudiated by John G. Thompson, who, in *Urbanization* (1927), drummed up the promise of the city as the harbinger of a new era.[71] The city's physical complexity was matched by its equally complex ideation. Like modernity itself, the city could no longer be represented by one dominant narrative; it meant many things to many people. The affirmative outlook on the pledge of urban life continued to be challenged by lingering antimodern sentiments, various Spenglerian "decline" theories, beleaguered racial ideologies, and bucolic nostalgias of the 1920s. In their *Principles of Rural–Urban Sociology* (1929), Pitirim Sorokin and Carle C. Zimmerman explored the dynamic tension stemming from the coexistence of urban futurity and rural sentimentality, of technological determinism and industrial peril.[72] This fusion of contradictory ideas and goals revitalized the mystery of the great city and its "parades of obscurity, of enigma, of silent sphinxes challenging the puzzled citizen."

It is safe to say that in the early twentieth century, no American building typology accentuated the sense of urban intrigue more than the skyscraper.[73] Taking stock of the mysterious power of skyscrapers, one author has noted their "quality of magnifying and obscuring . . . [the] tension of serenity in the accomplishment of the potential and the latent violence represented in the power that was capable of creating such solid masses. These are also the qualities in the photographs of Alfred Stieglitz, Edward Steichen and Berenice Abbott. The mist, the haze, clouds, escaping steam, night, the cubist imprecision of intersecting planes of light."[74] The skyscraper's height was a central element in the whole spectrum of urban mystery. Later, Winston Weisman wrote:

> Height had prestige value both for an individual and a business firm. It symbolized position, power, and prosperity. It had public

> relations value. Height also had an esthetic appeal. Looking up at these great buildings was an awe-inspiring experience. Looking down from on high was spine-tingling. Height thus provoked and provided an opportunity for new sensations. View made possible by height has become a saleable commodity to tenants and tourists alike.[75]

If the mystery of the city was an overwhelming sensation to the "puzzled citizen" on the street, then the view from the skyscraper created a powerful eye not only to unlock the mystery of the city but also to transform the mystery itself into a field of planning interventions.[76] The desire to decode the mystery of the city became intimately tied with a generous impulse for reform. Ferriss's observer on the heights could then be viewed as one of the visual symbols of how urban thinkers of the period negotiated the city's mystery—the visual cacophony set in a dynamic dialogue with the city's promise and peril—with the intent of a master builder, the putative creator of a new civilization. If the enigmatic city offered a tantalizing spectacle and an occasion for renewal, nowhere did it find a more poignant visualization than in Ferriss's *Metropolis* drawings of the night. His portrayal of the future city, both dramatized by artificial light and sanitized by a nighttime vision, sought to come to terms with the mystery of the modern city. It resonated, too, with the kind of grandiose reformatory attitudes prevalent among the visionary urban planners of the time.[77] In this metropolitan narrative of intrigue and urban reform, the night became the alter ego of the skyscraper's height.

FICTIONALIZING THE NIGHT

Even a casual reader of Ferriss's *Metropolis of Tomorrow* would notice that the Ferrissian city was essentially a nocturnal fantasy, awash in a theatrical mix of artificial light, shadow, and darkness. "The genius of Ferriss' production," Rem Koolhaas notes in *Delirious New York* (1978), "is in the medium of his renderings itself, the creation of an artificial night that leaves all architectural incidents vague and ambiguous in a mist of charcoal particles that thickens or thins whenever necessary."[78] Ferriss's urban spectator, lodged in the skyscraper, was always a silhouetted figure observing the metropolis of the night from the darkness, his gaze conditioned by both the new social symbolism and the mystery of artificial light. His vista was a pixilated landscape with selective illuminated foci, revealed only in outline, without any distracting minutiae. Ferriss's chiaroscuro drawings produced the romantic intrigue of the nighttime metropolis and highlighted the ways in which the nighttime could inspire the imaginary, the uncanny, and the unknown as the visual world, dependent on the projection of light, morphed into broad shapes with fading details. The night was conceptually situated outside the realm of rational specularity, thus promising a peculiar complicity with future-gazing. Ferriss's rendering created the illusion that the nighttime was a natural ally of the visionary who acted when lack of light dissolved the visual stability of the world into a fertile field of imagination. The nighttime accentuated the power of the aerial spectator.

Corporate America exploited the popular appeal of the well-lit nocturnal city in the early twentieth century, making implicit suggestions that conquest of the dark night would engender a brighter future for the nation. A 1931 Westinghouse advertisement for street lighting in *The American City* included a characteristic aerial view of an urban intersection cosmetically split between a brightly illuminated section and a low-visibility urban noir, or what the company ominously called the "twilight zone" (Figure 1.14).[79] A powerhouse in the history of American artificial illumination, with a prestigious commission to light the 1893 World's Columbian Exposition in Chicago, Westinghouse merely rehashed the early twentieth-century public fascination with night lighting in this ad, revealing how electrical illumination was closely tied with the rise of corporate capitalism.[80] The ad neatly aligned light and darkness with the archetypal binary composition of good and evil, of progression and regression. The message was unmistakable: an illuminated environment equals a virtuous environment where good visibility plays the semantic role of rationality, morality, and progress. The abstracted lamppost was shown not so much as an urban utility but rather as a beacon of enlightenment, a hallowed emblem of modernity that vanquished the dark forces of evil. The company was not simply selling a functional product; rather, it was ensuring progress by banishing the sinister night and all its accompanying tragic mythologies. Another Westinghouse ad, with similar aerial views of organized nighttime cities of tomorrow, proposed that "planned lighting . . . means looking ahead."[81] These Westinghouse ads were typical in the 1920s and 1930s, when American corporations promoted the illuminated metropolis as a surefire symbol of both human ingenuity and a new urban hygiene; that is, the well-lit night drove criminals away, ensured pedestrian and vehicular traffic safety, and encouraged good city planning.[82] Not surprisingly, Ferriss's nighttime rendition of the city struck a popular chord with American corporations and institutions that promoted their products and services by including alluring views of the metropolis of tomorrow (Figure 1.15).

Corporate America's promotion of urban illumination was part of a broader modernist experience, a socioaesthetic phenomenon that has been studied by David E. Nye, Wolfgang Schivelbusch, Dietrich Neumann, John A. Jackle, Mark Caldwell, and others.[83] Many early twentieth-century artists, photographers, architects, and novelists—among them Ferriss, Raymond Hood, Stieglitz, Lewis Hine, Weegee, Berenice Abbott, Joseph Stella, O'Keeffe, and John Dos Passos—viewed the nocturnal city and its incandescent mystique as a quintessential

Figure 1.14. A Westinghouse advertisement asserts that "programmed lighting is a part of city planning," in The American City, *February 1931, 225.*

28 ARCHITECTURE JANUARY, 1931

STRUCTURAL STEEL CREATED THE SKYSCRAPER

STEEL CARRIES ACRES TO THE SKY

A CROWDED CITY pleads for space . . . swiftly, floor on floor, the sure steel climbs—and thirty-five or forty city "plots" stand where there was one before.

Structural steel not only multiplies the precious bit of ground. It increases *rentable* floor area. Its great strength is not handicapped by excessive bulk, so interiors may be larger without conspicuous construction members. Steel comes to a job ready to go into place. It is unaffected by rain, freezing or intense heat. Erected quickly, wherever and whenever men can work, it saves time, labor, interest charges.

More and more homes, small apartment and mercantile houses, small industrial plants and small as well as large bridges are being built with *structural* steel. Architects and builders are realizing that the employment of steel merely to give strength and security to weaker materials is a compromise with its many advantages when used in the form of structural shapes.

Before building anything, find out what steel can do for you. The Institute serves as a clearing house for technical and economic information on steel construction, and offers full and free co-operation in the use of such data to architects, engineers and all others interested.

The co-operative non-profit service organization of the structural steel industry of North America. Through its extensive test and research program, the Institute aims to establish the full facts regarding steel in relation to every type of construction. The Institute's many publications, covering every phase of steel construction, are available on request. Please address all inquiries to 200 Madison Avenue, New York City.—In Canada, to 710 Bank of Hamilton Bldg., Toronto, Ontario. District offices in New York, Worcester, Philadelphia, Birmingham, Cleveland, Chicago, Milwaukee, St. Louis, Topeka, Dallas, San Francisco and Toronto.

"SKYSCRAPER HANGAR IN A METROPOLIS," BY HUGH FERRISS. AN ENLARGEMENT, ON SPECIAL STOCK FOR FRAMING, WILL BE MAILED WITHOUT CHARGE TO ANY ARCHITECT, ENGINEER OR BUSINESS EXECUTIVE.

AMERICAN INSTITUTE OF STEEL CONSTRUCTION

STEEL INSURES STRENGTH AND SECURITY

Figure 1.15. An advertisement for the American Institute of Steel Construction includes Hugh Ferriss's drawing "Skyscraper Hangar in a Metropolis," in The American City, *January 1931, 28.*

frontier for modernist deliberations.[84] In *Manhattan Transfer* (1925), John Dos Passos painted a picture of the night in which the urban morphology acquired a chameleonic identity, awash in artificial light: "Night crushes bright milk out of arclights, squeezes the sullen blocks until they drip red, yellow, green."[85] Dos Passos's hallucinatory pronouncement was just one instance of how the city of the illuminated night was embraced as a foil for fantastic projections. Frank Lloyd Wright's vitriol against the skyscraper, the "moral, economic, aesthetic, ethical monstrosity," dissolved into a tranquil love story when night fell: "Seen at night, heedless of stampede, the haphazard monster has myriad beauties of silhouette; light streaming—the light punctuated by reflected or refracted lights."[86] The early twentieth-century night provided an occasion for a new sensation, a new spectator of urban noir, or what Dietrich Neumann calls the "nocturnal *flaneur,*" a voyeur of the urban night whose gaze from precarious heights symbolized a primal modernist experience (Figure 1.16).[87]

The fascination with the illuminated environment was concurrent with the rise of modern architecture and urbanization, and therefore the nocturnal spectacle became a proxy battlefield where urban enthusiasts tried to figure out both the broader meanings of modernity and the nitty-gritty aspects of buildings' nighttime visuality. Designers realized that nighttime perception of the built environment ought to be integrated with the design process itself. They responded eagerly to corporate America's invitation to collectively "launch a movement for cooperation in obtaining artistic effects in the exterior lighting of buildings in New York, Chicago and other big cities."[88] In 1930, the collaboration between architects and lighting engineers, at the behest of General Electric, resulted in a small publication titled *Architecture of the Night,* after a phrase coined by Raymond Hood.[89] The cover of the booklet featured a silhouetted aerial view of a Ferriss-esque city block lit from below with powerful golden light and punctuated by reddish skyscraper finials and kiosks (see Plate 6).[90]

Here, the illuminated night and height conflated to articulate one of the most poignant narratives of metropolitan modernity. Both the visual and intellectual quality of this modernist narrative stemmed from the expressionistic, and sometimes competing, blending of the night and verticality, a merging that abounded in corporate advertisements, artistic practices, and populist representations. The urban noir found its sublime legibility and legitimacy as a projection screen for reverie, as Wright noted, in the eyes of an aerial beholder:

> The nocturnal monster yields rhythmical perspectives, glowing spotted walls of light, dotted lines, a world of fascinating reflections hung upon other reflections ranging along vistas of the street or pendent as the wisteria hangs its violet racemes on a trellis or the trees. Then the skyscraper is, in the dusk, a shimmering, prismatic verticality; gossamer veil of a festive

Figure 1.16. Anonymous, "A Night Scene from the Metropolitan Tower during the Hudson-Fulton Celebration" (1909), in Thirty Years of New York, 1882–1912: Being a History of Electrical Development in Manhattan and the Bronx *(1913).*

> scene, hanging there against the backdrop of a black night sky to dazzle, entertain, and amaze, in great masses. Lighted interiors come through the veil with a sense of life and well-being. The City then seems alive. It does live as illusion lives.[91]

The nighttime, when aided by the skyscraper height, provided a highbrow choice to purge the metropolis of the impurities of consumerism and cheap entertainment. Discussing the aerial, nocturnal exploration of New York by the Stieglitz circle, Mary Woods notes:

> Night was an ideal working time for the Stieglitz circle, artists determined to redeem photography and skyscrapers from American materialism and commercialism. Stieglitz and his circle shunned the other illuminated New York, Broadway's White Way and Coney Island, with crowds intent on the commodities and pleasures of consumer culture. Their quest was an elite and rarified calling to make art from the deserted and illuminated skyscrapers of lower Manhattan.
>
> Eventually, they viewed the skyscraper from rooftops, balconies, and rooms. Coburn ascended to the Singer Tower's observation balcony to gaze down on the Liberty Tower, calling the image *The House of a Thousand Windows: A Cubist Fantasy.* Stieglitz viewed the changing city first from within his Fifth

> Avenue gallery near Madison Square and then from the midtown skyscrapers in the forties and fifties. Now the skyscraper became a form not only to see but also to see from.[92]

Woods outlines a visual culture in which verticality became pivotal to the nighttime observer's peculiar mental mapping of the metropolis. At the heart of the "skyscraper noir" were double pleasures: first, the awe-inspiring vertical and pixilated mass of the skyscraper itself, and second, the aerial panorama dotted with illuminated building facades, bridges, and streets to be seen from the observation deck of the skyscraper.[93]

While Ferriss's French-inspired Beaux-Arts training at Washington University, St. Louis, accounts somewhat for the chiaroscuro quality of his drawings, his consistent fascination with night cities from the heights suggests that nighttime figured in his imagination more ontologically than as a mere temporal condition or a graphic convention.[94] In explaining this consistency of the nighttime rendition in Ferriss's nocturnes, the architectural critic Douglas Haskell, in a 1931 article titled "Architecture: The Bright Lights," identified an organic American quality built on a unique cultural appreciation of nocturnal modernity, unlike the Europeans who only "get the Day." Haskell wrote: "In his [Ferriss's] 'Metropolis of Tomorrow' are sixty illustrations, among which a bare ten seem to represent day light. I say they seem to, because even in these the days were often distinctly cloudy. There are a few sunsets and mists, and the rest is solid night."[95] In *The Metropolis of Tomorrow,* Ferriss himself described a nocturnal sublime: "One can easily fancy himself perched up somewhere on the hundredth floor; one looks down, at a dizzy angle, along the flanks of adjoining precipices; one is tempted to imagine the scene at night, with geometrical lights flaring in the abyss."[96] Here is an example of Ferriss's richly fractured view of the night. On one hand was the residual Victorian fear of the dark night, the "abyss," incriminated by the alleged immorality of urban life, or by what Josiah Strong called a social "menace" to Puritanist sensibilities.[97] On the other hand, with the penetration of "geometrical lights"—surely an allusion to the new culture of electric lights on city streets and buildings—the abyss slowly opened up to the visionary gaze laden with a spectrum of emotions, ranging from the fantastic and fictive to the salutary and reformatory.

Understanding this dualism in the perception of the night requires reflection on a shift in the early twentieth century. Within the rigid moral codes of nineteenth-century American Victorianism, night was a time for, as the New York City reporter George Foster wrote in 1850, "the festivities of prostitution, the orgies of pauperism, the haunts of theft and murder, the scenes of drunkenness and beastly debauch and all the sad realities that go to make up the lower stratum—the underground story—of life in New York."[98] Night diminished the scope of the metropolis by instilling law-abiding citizens with a fear of the unknown lurking behind darkness: "To penetrate beneath the thick veil of night and lay bare the fearful mysteries of darkness in the metropolis . . ."[99] A woodcut, first published in 1885 in the *Electric Review,* with the caption "The Powers of Evil Are Fleeing before the Light of Civilization" captured the Victorian paranoia of the city night, the perfect cover for the devil, dramatically larger than petty criminals (Figure 1.17). Only visibility could subjugate the devil of urban noir. Police, the savior, masqueraded as a streetlight, equating lighting with policing and, ultimately, with the moral control of society.

The fear of darkness also found an expressive outlet in the indignant writings and photographs of urban journalists who wrote about the miseries of the urban poor living in the tenements of lower Manhattan. For the journalist and police reporter Jacob Riis, an entire community, the miserable "other half" of society—those on the lowest rung of the economic ladder—existed in "an atmosphere of actual darkness, moral and physical."[100] Riis saw a direct correlation between the absence of light, dark rooms, unlit staircases, blank walls, and basements untouched by sanitizing sunlight on one hand and the deterioration of moral health among the poor working class, the proliferation of disease, and a vicious cycle of gloom and lawlessness on the other. Aghast at the squalid living conditions of lower Manhattan's poor, health inspectors and urban reformers "concluded that light was an essential ingredient in health and an agent for morality, every bit as much as soap and water."[101] From the middle of the nineteenth century, urban observers in fact began to articulate the well-being of city life and moral decay with the binary pairing of sunlight and darkness, or "gaslight" and "shadow." Matthew Hale Smith, in *Sunshine and Shadow in New York* (1868), and James D. McCabe Jr., in *Lights and Shadows of New York Life* (1872), assumed that lack of light breeds poverty and shelters what the dust jacket of McCabe's book called "the anonymous thousands of drunks, ladies of easy virtue, shoplifters, blackmailers, the whole range of professional criminals, gamblers, swindlers, and all the unfortunates who did anything they could to keep from starving in a city where there was no class between rich and poor."[102]

In the early twentieth century, however, the Victorian denigration of the city night was permeated with a new appreciation of nocturnal modernity, one in which the illuminated night became an acceptable extension of the public realm. Exemplified by the night city par excellence—New York—the new American night accommodated a middle-class life and its bourgeois excesses, or what Lewis Erenberg has called a new liberal culture of "steppin' out" from the claustrophobic cloisters of Victorian probity.[103] Nightlife now included an urban sphere broadened beyond mere entertainment, spawning a new citizenry concerned more with creative vitality and less with social conformity, a culture perhaps best exemplified in the 1920s by F. Scott and Zelda Fitzgerald, who "flew, addled with youth and fame, through Manhattan's theatres, nightclubs,

Figure 1.17. "The Powers of Evil Are Fleeing before the Light of Civilization," woodcut as reprinted in Electrical Review and Western Electrician *56 (May 21, 1910).*

speakeasies, a world to which Prohibition had now lent a seductive perfume of lawlessness."[104] A new liberal social ethos and a break from older forms of gentility resulted in the popularization of such nocturnal urban institutions as restaurants, nightclubs, dance halls, and cinemas, the kinds of places that formerly would have been considered breeding grounds for deviant social behavior and moral depravity. In the period between 1890 and 1930, this new trend, Erenberg argues, "broke from older forms of gentility in which individuals were to subordinate themselves voluntarily to a social code. . . . starting in the 1890s, values became more informal, and the restrictions placed on the individual's personal desires and impulses were lessened. Greater emphasis was placed on self-fulfillment, self-expression, and the development of 'personality.'"[105] Nothing offered more avenues for creative self-expression than the oneiric environment of the new night, often transformed into a mystical tapestry of light and darkness. Illuminated with selective foci, the metropolitan night provided a fecund conceptual site for toying with the idea of the future: "The night is a time for dreaming. Fantasies and hidden desires seek realization in an urban world whose very anonymity permits them."[106] In the early decades of the twentieth century, visionary architects and planners often portrayed the modern city as a chiaroscuro dreamscape of the night.

Artificial illumination hastened the decline of the Victorian dread of the night and brought to the fore its new middle-class respectability.[107] The rapid proliferation of night lighting opened up spaces for socially acceptable nighttime occupations while inflating the mystique of the modern metropolis. Bright streetlights directly influenced how city people navigated the nocturnal city and redefined the scope of the city's visibility.[108] The night was no longer a spooky time when danger lurked in poor visibility; rather, it was a manifestation of progress. In 1903, New York was the premier night city, with 17,000 electric streetlamps, followed by Chicago with nearly 9,000; Berlin remained a "day city," with only 735 streetlamps.[109] Such statistics prompted American critics like Douglas Haskell to argue that the nocturnal metropolis was a unique American development. Two classic American novels of 1900, L. Frank Baum's *The Wizard of Oz* and Theodore Dreiser's *Sister Carrie,* presented images of the metropolis as a blazing epicenter of artificial light, a leitmotif also prevalent in corporate America's vision of the future city during the 1920s. Urban illumination promised a new social hygiene in the form of safety, decency, and visibility. David Nye, a historian of technology, has noted: "If by day poor or unsightly sections called out for social reform, by night the city was a purified world of light, simplified into a spectacular pattern, interspersed with now-unimportant blanks."[110] Conducive to wistful voyeurism, the city of the night was seen to be a malleable blob: one just needed to sculpt it with selective illumination to suit a particular vision of the future. To a great extent, this novelty lay in the ways the city's selective visibility leveraged its very own reimagination.

Ferriss's romance of the night was consistent with shifting American attitudes toward nocturnal public places, thanks to his own experience of the mystique of the night at its epicenter: New York. When he was working in the office of Cass Gilbert from 1912 to 1915, Gilbert's masterpiece was the Woolworth Building—then the tallest building in the world—which assumed the protagonist role in the theatricality of New York's night, especially because by then, lighting engineers had learned how to use floodlighting on skyscrapers to make them glow against the dark night sky (Figure 1.18).[111] The drama of the New York night during the 1920s climaxed with the Art Deco masterpiece Chrysler Building (1929) and the Empire State Building (1931); these skyscrapers and their colorfully lit finials framed the panorama of the illuminated night.

Ferriss's rendering of the metropolis seemed to partake of the promise of the night, seeking to invert the kind of nocturnal anxiety that haunted American Victorians in the nineteenth century. The solitude of the night and its spectacular black-and-white pixilation granted him the useful illusion of releasing the American city from modernity's disruptive social effects. The interplay of light and darkness seen from above provided him with a kind of graphic polemic to dramatize the building silhouette and heighten its bulk in relation to the larger urban context. In a way, height was the night's natural ally: "The only comprehensive views of nighttime Gotham at full scale are from the air where the experience is the quiet containment of an airplane."[112]

If height offered Ferriss a magisterial gaze over the city, nighttime accentuated this sense of empowerment by giving him the ability to be selective about what to illuminate and what to hide in darkness. Like the streetlight, his gaze from the heights penetrated selective dark areas, cleansing the alleged social filth of the night and creating a blank slate of sorts for new urban interventions. Douglas Haskell identified a dormant patriotism in Ferriss's nocturnal voyeurism from the skyscraper:

> Here again is a terrace. It is at about the fortieth story. The time, of course, is night. There are solid mountains and chasms, all man-built. This reaches into illimitable distance and loneliness, as vast and solitary as the surface of the moon, all done by steel and electricity. Whatever chaos there may be in the forms disappears in the uniform grill of this star-spangled banner, and patriotism catches at the throat. Here is modernism indeed. Thousands of years went by with their changes of style, but not until this century was there electric light, which, far, far more than the familiar triad of steel, glass, and concrete, has changed the basis of all architecture. This is us.[113]

Ferriss imagined the nighttime city as a mystically meditative opposite of its chaotic daytime counterpart. But it was much more. Ferriss's appropriation of

Figure 1.18. Anonymous, The Woolworth Building at Night, *1913. Library of Congress, Prints and Photographs Division (LC-D4-73062).*

the spectacle of the night into "the uniform grill of this star-spangled banner [when] patriotism catches at the throat," as Haskell noted, resonated with a sense of "conquest" embedded in the prevalent frontier history.

The westward expansion of settlements in America between 1830 and 1880 coincided with "the colonization of the night" when the installation of gaslight brought the major cities of the East and Midwest within a new domain of visibility. Against the backdrop of the so-called closure of the western frontier and the rising metropolis purportedly turning into a "new frontier," the electrification of urban illumination emboldened a desire to appropriate the night for new activities, new imaginations, and new ocular practices.[114] Not unlike the frontiersman's colonialist gaze, surveying the vast reaches of the frontier from elevated points in the landscape, the nocturnal seer gazing at the illuminated urban frontier sought to colonize the domain of darkness for what sociologist Murray Melbin calls "a large-scale migration of wakeful activity."[115] Like space, Melbin posits, time could be conquered, populated, and filled with activities that were otherwise constrained by the limits of the daytime and a moral fear of darkness. Melbin implicitly argues that the triumph of night light was closely allied with the advent of a new regime of visibility and, by extension, the development of an empowered spectator, whose initial terror of darkness was overcome by a subconscious of conquest.[116] Melbin's argument was foretold by a 1930 advertisement for Graybar Street Lighting in *The American City* (Figure 1.19).[117] With street lighting, the advertisement stated, "our cities no longer hide under a bushel." The dark urban night, like being under a bushel with zero visibility, meant wasted time, a vacuum in space and time. The creation of artificial light dispelled the dark cover over the metropolis, transforming its nighttime inaction into a time of new opportunities and actions.

The advertisement presented an aerial view of an urbanscape reminiscent of Ferriss's, except that it waited to be conquered with artificial light. For Ferriss, the conquered night was a fictionalized venue for the renovation of both the modern metropolis and its seer. All of the troubling minutiae of the daytime city could be made to disappear behind the curtain of darkness. Ferriss's detachment from the ground-level reality of modernity was twofold: first by height and then by the redeeming filter of the night, both animating his überbuilder desire to purge the American city of all kinds of physical and social disorder.

SPIRITUALIZING VERTICALITY, VERTICALIZING SPIRITUALITY

Ferriss's überbuilder consciousness cannot be explained adequately by the cultural logic and illogic of verticality and nocturnal aesthetics alone, as these two modernist experiences were intricately intertwined with his contemporaneous interest in mystical spiritualism, a widespread fascination among the rebellious

Figure 1.19. A Graybar Street Lighting advertisement, "Our Cities No Longer Hide under a Bushel," in The American City, *February 1930, 61.*

intelligentsia of 1920s New York. The American literary critic Gorham Munson described this intelligentsia as "the public that in 1924 read the *Dial* and the *New Republic* and *Vanity Fair,* listened to Stravinsky and Schoenberg, looked at Picasso and Matisse, discussed psychoanalysis and the progressive education of John Dewey, and inclined toward socialism."[118] The notorious Russian mystic Gurdjieff and his British emissary to America, Alfred Orage, galvanized the New York intelligentsia with contentious mantras of psychic renovation from 1924 to the end of the decade. Dissatisfied with how rampant materialism and consumerist culture took precedence over spiritual evolution in the industrial West, many well-known New Yorkers gravitated toward Gurdjieff's mystical teaching of knowing the self. Among others who attended the séances offered by Gurdjieff and Orage in New York were authors John Dos Passos, Theodore Dreiser, Jane Heap (editor of the *Little Review*), Herbert Croly, Waldo Frank, Stuart Chase, and Jean Toomer; architects Harvey Wiley Corbett, Bragdon, and Ferriss; photographers Walker Evans and Stieglitz; painter O'Keeffe; and socialite Muriel Draper.[119]

Given Ferriss's poetic persona—from the beginning of his career, he wrote poems steeped in esoteric expressions and recited them in both formal and informal settings—and his sustained interest in the "subconscious" and the "immeasurable," it was not surprising that he would gravitate to Gurdjieffite mysticism and esoteric wisdom.[120] He attended Orage's lectures as an active member of the group. Years after the publication of *The Metropolis of Tomorrow,* Ferriss reiterated his earlier quest for "spiritual values" in a lecture at Columbia University: "As though a pendulum were swinging away from the 'measurable quantities' which are the concern of science and technology and toward these esthetic and spiritual values which, if not immeasurable, have at least not as yet been measured by scientists."[121] For Ferriss, realization of an ideal metropolis was tantamount to attaining spiritual contentment. When Ferriss choreographed a gradual emergence of the metropolis from a foggy environment or a nocturnal void, or what Rem Koolhaas calls a "cosmic container, the murky Ferrissian Void, a pitch black architectural womb," one wonders whether it was his conscious allegory of the tortured path toward spiritual redemption: the utopian metropolis as a metaphor for the *final* destination.[122] While this was not unlike how John Bunyan's pilgrims reached the Celestial City, Ferriss replaced Bunyan's happily-ever-after heavenly destination with a modernist utopia that engaged a much more ambiguous view of spirituality, a conflict commensurate with the mental anguish of the modern man allegedly confronting a chasm between practical life and spiritual sustenance.

Ferriss's captions to the drawings in *Metropolis* employed vocabulary tinged with mysticism, accentuated by ecstasies of height or nighttime or the chiaroscuro rendering technique itself. In a 1923 article in the *Christian Science Monitor,* Ferriss provided a vivid description of "New York from a studio rooftop":

> If one arrives at the studio at dawn, and if it be a day of fog characteristic of early spring, he finds himself . . . the lone spectator of a nebulous panorama of mist. The immediate parapet rail, dark and wet as that of a liner in mid-ocean, is the sole suggestion of locality. Beyond it, the mist flows pale and enveloping; it is more void than the sea—it is as though he had entered a hushed balcony overhanging elementary space.
>
> Looking into this space with eyes focusing as an astronomer's, he distinguishes, distant and lofty, a faint point of light. It is the first gold high light of the tip of the Metropolitan tower. It announces the sun. It is the sole indication that in the depths beneath is a city.
>
> In an instant, a subtle distinction is perceived in the monotone of gray; vertical lines, of a tone but a degree more luminous, have magically appeared beneath one: every eastern façade in the city is pale with light.
>
> Contrast increases; as mysteriously as though being created, a city becomes distinguishable, pristine, and still.[123]

The meditative impulse of the "lone spectator of a nebulous panorama of mist" is unmistakable here. As much as it presented a slow-motion visualization of the city beneath the apex of the skyscraper, it also offered the mental mapping, or operation, of a spectator, trying to make sense of a world as if hidden behind nebulae. The spectator's spiritual reckoning, allegorized in the birth of the future city "as mysteriously as though being created," was akin to the mystical parable of self-discovery central to the Gurdjieffite pedagogy. The Ferrissian spectator provided a commentary on the nature of the modern man, seeking to examine the self against a new urban frontier. The aerial gaze of the lone seer remained invested, so to speak, in some sort of introspection, a mental process in which the misty city offered a catalytic framework for reaching some kind of higher consciousness. Ferriss's mysterious metamorphosis of the city from silence and intangibility to reality—from "elementary space" to "distinguishable" entity—was a theatrical revue of self-assessment, an inquiry that revealed a wider interest in the mutually informing relationship between what Ferriss's promoter Sheldon Cheney called "the new world architecture" and the spiritual regeneration of man. Cheney posited that "a new universal art of building [is emerging] . . . because the Spirit of Man is coming forth, because a new religion of courage, faith, human power, of the divine and miraculous in the man himself, is taking form."[124]

Ferriss's colleague Claude Bragdon introduced him to ideas related to mystical spirituality. Like Ferriss, Bragdon often gazed out from his apartment at the Shelton Hotel and mentally transformed the "weedy wilderness of roofs, water tanks and chimneys" into a platform for ruminations on the modern conditions of man.[125] Since the first decades of the twentieth century, Bragdon had discussed—often in impassioned and enigmatic tones—how physical phenomena were manifestations of unseen universal laws whose contemplation could provide insights into the spiritual meanings invested in physical forms and symbols.[126] In a 1901 article titled "Mysticism and Architecture," Bragdon argued that societies across time and space had employed architecture's visual language to articulate a physical representation of the immeasurable and spiritualized systems of the cosmos and human mind:

> The mystic imagination uses nature only as an alphabet with which to spell spiritual meanings; it creates such forms of wonder, mystery and beauty as are sculptured on the walls of Egyptian temples; it assembles the stones of a cathedral according to laws as organic as those which determine the courses of the stars. When the mystic spirit departs from a people, the forms of its creating survive by reason of their beauty, but they are meaninglessly employed.[127]

The loss of the "mystic spirit" in the emulative architecture of the period, according to Bragdon, paralleled the psychic fragmentation of the modern man. A meaningful architecture and, eventually, the spiritual wholesomeness of man were tenable insofar as man remained connected to the hidden, subconscious laws—or what Bragdon called the "infinitely simple, infinitely subtle, incommunicable, evanescent . . . beautiful necessities"—that govern nature and its irrefutable harmony.

Bragdon's advocacy of spiritualism as an antidote to the rampant individualism and materialism of late nineteenth-century industrial capitalism was largely indebted to theosophy—meaning "wisdom of the gods"—a self-styled mystical religion founded in New York in 1875.[128] The Theosophical Society, as it was called, was the brainchild of two contrasting figures: one was a mysterious, eccentric, and widely traveled Russian woman named Helena Petrovna Blavatsky (1831–91), and the other was an American scientist and lawyer named Henry Steel Olcott (1832–1907). Bringing together a group of devoted followers—primarily persons with an artistic and literary bent—the movement sought to explore "the expansion of human powers of mind and spirit" and Western occultist traditions and to integrate them with the ancient religions of the Indian subcontinent, especially Hinduism and Buddhism. The key precepts of theosophical teachings were evolution, man's constitution, karma and reincarnation, and after-death states.[129] The newly founded Theosophical Society's motto, "There is no religion higher than truth," qualified the group's working hypothesis that certain moral truths structure all worldly belief systems and their physical manifestations. The pursuit of truth, therefore, ought to be the primary mission of the enlightened theosophist, who generally should seek to reconcile all religions and traditionally opposed conceptions of past and present, science and religion, East and West, through a thread of eternally valid divine wisdom. With its antirationalist tenets and promise of spiritual healing through an Oriental interpretation of life, the movement appealed to those disillusioned with the alleged abstract, materialistic, and impersonal worldviews of Western societies. From its inception to the beginning of World War I, theosophy exerted a widespread influence on key thinkers and various artistic movements, such as art nouveau and expressionism.[130]

A central theme in theosophical thinking was the provocative idea of mahatma (a Sanskrit word meaning "the great soul"; *maha* is great, and *atma* is soul).[131] The mahatma was considered a "master" or a "teacher" who rose out of the ranks of ordinary men—unlike conventional prophets with provenance in celestial realms—and attained his greatness by ascending to a higher consciousness to be able to grasp and emit cosmic secrets. According to proponents of theosophy, the mahatma was not a disembodied entity, or a yogi, or an ascetic self-exile in a world of personal atonement, or a teacher in the conventional sense. Rather, all theosophical truths were derived from his experience of a journey to spiritual heights unattainable by ordinary mortals. The mahatma embodied a highly evolved mental prowess attained through personal efforts at "moral elevation and intellectual attainment."[132] The mahatma reached an astral consciousness through which he not only came to terms with his human limitations but also discovered the dormant divinity in the self. Blavatsky described this as "acquired individuality, first by natural impulse, and then by self-induced and self-devised efforts, thus ascending through all the degrees of intelligence, from the lowest to the highest Manas, from mineral and plant, up to the holiest archangel."[133]

Annie Besant who became the president of the Theosophical Society in 1908 (then headquartered in Adyar, India), described the conceptual significance of the mahatma to theosophy's very existence with this pithy statement: "Theosophy stands or falls on the existence of the Masters."[134] Theorized to revive the power of personal initiation at redemption and the ethos of self-knowledge, the ideation of mahatma bolstered the theosophists' cult of hero worship. It appealed to a diverse set of people across the world who wanted to examine alternative-reality theories, forming a so-called missing link between the empirical world of science and the unquantifiable realm of spiritualism. In all its mystical musings, theosophy revealed a fascination with ascension, be it the signifier of the mahatma's astral ambition or a higher being's superconsciousness.[135] Whether they were theosophists or just interested in spiritual phenomena, many authors in the late nineteenth and early twentieth centuries speculated on the nature of the mahatma-like figure. Henry Drummond's *The Ascent of Man* (1894), Richard

Maurice Bucke's *Cosmic Consciousness: A Study in the Evolution of the Human Mind* (1901), W. Tudor Jones's *The Spiritual Ascent of Man* (1917), Peter D. Ouspensky's *Tertium Organum: A Key to the Enigmas of the World* (1920), Alfred Machin's *The Ascent of Man by Means of Natural Selection* (1925), and Raymond B. Fosdick's *The Old Savage in the New Civilization* (1928), in their disparate ways, shared a collective preoccupation with the idealized development of man amid an anxiety of social decline. The leitmotif of these books was a mahatma character who, as W. Tudor Jones noted, "is not confined to the materialistic level: he is more than matter, and is able to soar to heights from which he may read the meaning of the universe, and may experience that meaning as a real element within his personality."[136]

The romance of the skyscraper's vertical isolationism and its illusion of removal from various modernist phobias provoked a range of spiritual reckoning that resonated with the ideation of the mahatma figure. It is not surprising that during the 1920s, many critics and architects described (and criticized) the vertiginous world of skyscrapers in mystical or religious terms. Lewis Mumford wrote, indignantly: "We are face to face with a religion, with a deep mystical impulse, a hierarchy, and a theology."[137] Herbert Croly, the author of *The Promise of American Life* (1909) and editor of *Architectural Record* and the *New Republic,* was more sanguine in his assessment of the skyscraper's ability to carry on the soaring spirituality of Christianity.[138] The novelist Theodore Dreiser, accompanied by Stieglitz, looked down at Manhattan from a tall building and philosophized about "the panorama of roofs and spires and jetting steam-pipes, and the narrow grottoes of streets, in the depths of which the turgid stream of humanity flowed noisily."[139] Dreiser was drawn to theosophy, and one wonders whether his highbrow self-distancing from the "turgid stream of humanity" below recalled the liberatory self-consciousness of the mahatma.[140] As Thomas van Leeuwen has suggested, there was a mutually informing relationship between Bragdon's theosophical orientation and his writings on the philosophical energies unleashed by the skyscraper and its height.[141]

In the early 1920s, Ferriss began to prophesy on the future city, and at the same time theosophical exchanges appeared between Ferriss and Bragdon in New York City. Later, in 1927, Bragdon solicited Ferriss to produce a characteristically atmospheric drawing for the architect's gargantuan design for a "Theater and Temple of the Dance." Ferriss's first published prediction of the metropolis was "Imaginary View of New York in 1942—Showing Effect of the Zoning Laws." This 1922 drawing suggested a journey into the future from a historicist perch framed by lion sculptures and Egyptian columns.[142] By 1924, Ferriss's drawings, frequently published in such magazines as *The Arts* and *Century Magazine* and in newspapers such as the *Christian Science Monitor, New York Times,* and *Chicago Evening Post,* catapulted him to the forefront of city experts.[143] In a 1928 article titled "Prophet of Sky Lines," an author romanticized Ferriss's stature in terms that abstracted theosophical depictions of the mahatma, as if probing the nature of the modern metropolis:

> [When] I entered [his] studio, I found the person I had come to see perched nonchalantly upon the railing of the balcony, some two hundred feet above the ground, a striking and fitting silhouette against the sky. I say fitting, for the qualities of bird and mystic are in this man. And the sky, especially the sky of Manhattan, with its miles of steel towers, is peculiarly his province. He is Hugh Ferriss, poet and prophet of architecture.[144]

Here was a sentimental portrayal of an "avian" mystic whose astral vision searched for the truth against the foil of New York City's escalating urban form.

In the 1920s, it was not Blavatsky-esque theosophy, per se, but the cultic provocateur Gurdjieff's esoteric teaching that appealed to the fractious postwar attitude of New York's intelligentsia (Figure 1.20).[145] Characterizing this attitude were, on one hand, the Progressive movement's hopefulness, "urban optimism," and the bourgeois excesses of a Fitzgeraldian "Jazz Age mood" and, on the other, the nihilism of what Ezra Pound called "a botched civilization," the Spenglerian specter of civilizational decline, and the spiritual hollowness of the machine man.[146] The author Waldo Frank, an occasional member of the Gurdjieff circle, wrote ominously in *The Re-discovery of America* (1929): "The old spiritual body is breaking up. Ere we can be whole and hale again, we must create a new spiritual body. And that means birth."[147] Gurdjieff could not have come to America at a more opportune moment, for his pedagogy promised this "birth," firing up the spiritualist imagination of many members of New York's avant-garde.

As a key affiliate of the Gurdjieffite society, Ferriss audited lectures and attended demonstrations of the trancelike "sacred dance" based on the Russian mystic's teaching. Alfred Orage, Gurdjieff's trusted pupil, came to New York in December 1923 with the mission of disseminating the Gurdjieffite tenet: to know oneself through a process of intensely focused self-observation in order to perfect the balance of mind, body, and feeling (Figure 1.21).[148] Upon arrival in America, Orage found a receptive audience for Gurdjieff's mantras. Three years earlier, in 1920, the mystic's famous acolyte Russian philosopher Ouspensky had published (in the United States) his magnum opus on the subject of higher consciousness, *Tertium Organum,* translated from Russian and with an introduction written by Bragdon.[149] In 1922, at the height of his literary career and celebrity, Orage relinquished his scholarly occupation to move to the seventeenth-century Chateau Le Prieuré at Fontainebleau near Paris, where in the same year Gurdjieff established his Institute for the Harmonious Development of Man.[150] Although mired in controversy and notorious for Gurdjieff's

Figure 1.20. George Ivanovich Gurdjieff. Library of Congress, Prints and Photographs Division, Janet Flanner-Solita Solano Papers (LC-USZ62-112902).

Figure 1.21. Alfred Orage (August 1928). Photograph by Ansel Adams. Copyright 2011 The Ansel Adams Publishing Rights Trust.

extreme daily routine of menial work, communal eating, and "sacred dance," the institute became a pilgrimage destination for many people, the devoted and curious alike, from America and abroad: Waldo Frank, Jean Toomer, Gorham Munson, Katherine Mansfield (who reportedly died there of tuberculosis in 1923), Frank Lloyd Wright's soon-to-be wife Olgivanna Milanoff, and many others. As Ferriss's daughter Jean Ferriss Leich disclosed to Carol Willis, in the mid-1920s Ferriss and his wife Dorothy visited the Prieuré at Fontainebleau, although they reportedly left the premises more confused than enlightened by the sect-like, cultic environment of the institute (Figure 1.22).[151]

Gurdjieff was an enigmatic and divisive figure, "the strange head of a strange practical religion" or the "harmonious developer," as *Time* magazine dubbed him in 1930. Part of his mystique was created by the wondrous stories his followers intentionally wove into his biography while retaining his lectures and activities in total secrecy.[152] The first four decades of Gurdjieff's life were fraught with near-mythical, hyperbolic stories that contributed to his legend as a man with supernatural power: from "miraculous" healing cultures in his birthplace of Alexandropol on the Persian frontier of Russia to his soul-searching, Odysseus-like travels through the Middle East, Russia, Armenia, Greece, and

Figure 1.22. Chateau Le Prieuré at Fontainebleau, location of Gurdjieff's Institute for the Harmonious Development of Man; reprinted in James Webb, The Harmonious Circle *(Boston: Shambhala Publications, Inc., 1980), insert.*

Tibet in search of "universal knowledge" and to meet "remarkable men."[153] The *New York City America* contributed to his mythmaking in its coverage of Gurdjieff's arrival in America with a sensationalist report alluding to Fontainebleau's cultish daily rituals: "'Dr.' Gurdjieff and His Magical Secret Life: How to Be a Super-Man or Super-Woman by Feeding Pigs, Dancing Weird Dances All Night and Other Fantastic Antics. From time to time the originators and high priests of various strange creeds of health, philosophy or spiritual conduct float over to America from Europe, and the latest strange bird to arrive here (where dollars are easy to pick up) is 'Dr.' George Gurdjieff."[154] For American writers, artists, intellectuals, and spiritual seekers drawn to Gurdjieff, his 1924 arrival in America, following epic sagas of escape from the Russian Revolution in 1917, had the aura of a modern-day messiah's descent into a chaotic world in need of redemption. Even the archcurmudgeon Frank Lloyd Wright seemed swayed by the Russian's machismo:

> Real men who are real forces for an organic culture of the individual today are rare. I venture to say one might count them on the fingers of one hand with the thumb to spare—unless the thumb were to go to George Gurdjief *[sic]* of the Prieure at Fontainebleau, France. There is only one Gurdjief. His career is as unique as is the man himself. Gurdjief seems to have the stuff in him of which our genuine prophets have been made. And when prejudice against him has cleared away his vision of truth will be recognized as fundamental to the man men need.[155]

Gurdjieff's pedagogy negotiated the polarities of crisis and a reformatory mind-set that defined America's social environment of the 1920s. His working hypothesis behind the Institute for the Harmonious Development of Man could be understood, if partially, from the name of the institute itself: that the modern man—the tragic figure of a haphazardly modernizing civilization—was incapable of thinking to the full capacity of his consciousness because he was subservient to a constant inner warfare of multiple, mutually exclusive, and hostile personalities, namely, the "Is" (ego, identity, self).[156] A litany of external stimuli, such as warmth, sunshine, education, religion, and tradition, created in the modern man were of various hierarchies and values, a disruptive psychomechanical process of influence beyond the control of an ordinary-thinking person. The mental life of that person was, therefore, subjugated by one type of self-awareness or ego at a given time, depending on specific stimuli, exacting the fragmentation of his consciousness.

Gurdjieff's remedy to this psychic imbalance was "the Fourth Way," which transcended the common ways of the physical, the emotional, and the intellectual—the three so-called operative centers of the brain.[157] The methods of achieving the Fourth Way's all-embracing self-consciousness were "intentional suffering," or "the Work," and "the Sacred Dance," or "the Movement." These two key daily rituals at the Prieuré required all institutees, celebrities and common folk alike, to submit to rigorous physical work during the day and trance-like dances in the evening.[158] Gurdjieff suggested that the cosmos is fueled by a constant supply of energy that is produced by various organisms of nature, including human beings, by means of conscious experience.[159] It was, therefore, a cosmological imperative that human beings become conscious, wake up from "sleep." But man was not at the top of the energy-producing chart, nor was the energy he provided for the cosmic wheel the most important. Therefore, the implicit goal was to constantly attain higher consciousness and scale the graph of energy production. While the esotericism and mesmerism of Gurdjieff's teachings appeared cultic—further complicated by his pupils' often bizarre and hypnotic loyalty—his goal was to examine man's psychic equilibrium through inquiries into the subconscious that echoed the theosophical imagination of the mahatma and eventually see a new breed of highly conscious men or perhaps, hauntingly, supermen. These "bigger and better men," to quote the psychologist C. Daly King, one of Gurdjieff's pupils, were the means by which the social decline hastened by industrial capitalism could be reversed.[160]

It was hardly surprising that the spiritual quandary of the modern man and its remedy through his "harmonious development" would intersect—and even draw ideological nourishment from—evolutionary politics that had already made inroads during the interwar period. Anxious about the detrimental effects of immigration, urbanization, racial decay, and a Malthusian fear of population growth disproportionate to food production, social reformers were receptive to provocative ideas of salvation such as eugenics.[161] Madison Grant's *The Passing of the Great Race* (1916; revised editions appeared in the early 1920s), Ouspensky's *Tertium Organum* (1920), Edward M. East's *Mankind at the Cross Roads* (1923), Raymond B. Fosdick's *The Old Savage in the New Civilization* (1928), H. G. Wells's *What Are We to Do with Our Lives* (1931), and Alexis Carrel's *Man, the Unknown* (1935; Ferriss read and cited from this book upon its publication) are examples of an intellectual trend that shared the idea of social degeneration crippling the West and advocated its cure by the propagation of men with higher intelligence, or, essentially, supermen. In *The Old Savage in the New Civilization,* Fosdick, a New York lawyer during the 1920s and later president of the Rockefeller Foundation, implicitly advanced the idea of a philosophic oligarchy ruling the world.[162] Writing his introduction to Ouspensky's *Tertium Organum,* Bragdon had already anticipated this oligarchic machination of the world:

> Ouspensky's "superman" . . . in developing higher-dimensional, or "cosmic" consciousness will indeed inherit—will control and

> regulate human affairs by reason of [his] superior wisdom and power. In this, and in this alone, dwells the "salvation" of the world. His superman is the "just man made perfect" of the Evangelist. The struggle for mastery between the blind and unconscious forces of materialism on the one hand, and the spiritually illumined on the other, is already upon us, and all conflicts between nations, peoples, and classes must now be interpreted in terms of this greater warfare between "two races" of men, in which the superior minority will either conquer or disappear. Like birds of the air, their fitting symbol, they are at home in realms which others cannot enter. Nor are these heavenly eagles confined to the narrow prison of the breast. Their bodies are as tools which they may take up or lay aside at will. This phenomenal world, which seems so real, is to them as insubstantial as the image of a landscape in a lake. Such is the Ouspenskian superman.[163]

These views of a hierarchical society and an oligarchy of supermen or mahatmas to rightfully rule the world affirmed Gurdjieff's position. While all humans were by the nature of their psychic constitution equally capable of self-knowledge, an equitable evolution of large masses of humanity was ultimately impossible. At the center of Gurdjieff's whole spectrum of the "objective" harmonious development of man lay a contradictory pedagogy that "earthly human life is a vast self-fertilizing garden in which only a few self-perfecting individual trees may grow."[164] Orage had already expressed similar ideas in his book *Consciousness: Animal, Human and Superhuman* (1907), published during the heyday of his theosophical affiliation in London.[165]

Orage disseminated Gurdjieff's ideas in America. Upon arrival in New York in December 1923, Orage faithfully presented to an American audience Gurdjieff's message of "a new quality of concentration and attention and a new direction of the mind" through synchronized body movement, the so-called sacred dance, believed to embody religio-scientific significance.[166] The reception of the dance demonstrations at Lesley Hall, Neighborhood Playhouse, Carnegie Hall, and Rosetta O'Neill Studio in Manhattan was mixed, ranging from admiration and curiosity to ridicule and confusion. However, the erudite orator Orage—with his smooth marshaling of Gurdjieff's concepts of knowing the self—became immensely popular in the United States between 1924 and 1931.[167]

More accessible, democratic, and personable than Gurdjieff, and already well respected in transatlantic literary circles, Orage struck a popular chord with the New York intelligentsia, many of whom met regularly at weekly séances. According to a number of critics, it was Orage, not Gurdjieff himself, who made

Gurdjieff's ideas float in America.[168] The literary critic and 1920s Greenwich Village writer Gorham Munson proclaimed, "Orage was the best talker I have ever listened to, a man of beautiful lucid speech such as I imagine Plato was in the Garden of Academe."[169] Bragdon, too, gave much credit to Orage: "It was Orage, the perfect disciple, the Plato to this Socrates, who was responsible for most of the success which attended the movement in America. His charming manner and brilliant mind did much to counteract the bewilderment in which Gurdjieff so often left his auditors."[170] Even if these estimations were exaggerated, the devotion and the number of people who gravitated to Orage lent some credence to the hyperbole. After Orage's first lecture on Gurdjieff's ideas on January 9, 1924, the *New York Times* Sunday art supplement (February 10, 1924) offered a glowing tribute to Gurdjieff's institute.

The magnetic personality that earlier in England had attracted T. S. Eliot, W. D. Howells, Ezra Pound, and Bernard Shaw to the progressive journal *New Age* now intrigued Americans across the social spectrum and drew them to Orage's lectures.[171] As fellow theosophists, Orage and Bragdon had known each other prior to Orage's arrival in New York. Before his American sojourn, Orage wrote to Bragdon: "I expect to be sailing for New York on December 15 to spend a few weeks in preparing the way for Mr. Gurdjieff's visit in January. Naturally I should come with even more timidity if I did not expect to find you there. I should be most grateful if you would collect such material as might be useful for my mission. Your friends, I feel, are bound to be friends of the Institute."[172] Orage expected Bragdon to find him American listeners for Gurdjieff's mystic mantra.

As a man who believed that the goal of a "self-conscious Architecture" would be to contribute to "the harmonious development of man," and who implicitly but consistently employed the notion of heights as a metaphor for some kind of spiritual liberation and future-gazing, Ferriss was an easy and willing recruit for Oragean séances. Ferriss wrote in *The Metropolis of Tomorrow*:

> Architecture influences the lives of human beings. City dwellers react to the architectural forms and spaces which they encounter: specific consequences may be looked for in their thoughts, feelings and actions. Their response to Architecture is usually subconscious. Designers themselves are usually subconscious of the effects which their creations will produce. Nevertheless, we may look forward to some stirring of thought—perhaps even to some specific training—which will put a considerable body of students in command of the architectural influence. Our criterion for judging this self-conscious Architecture will be its effect on human values: its net contribution to the harmonious development of man.[173]

The stated goals of Ferriss's "self-conscious Architecture" were in line with those of Gurdjieff's Institute for the Harmonious Development of Man. At theaters across Manhattan, Ferriss attended Orage's Gurdjieff-inspired lectures and demonstrations of "various movements of the human body taken from the art of the Ancient East—examples of sacred gymnastics, sacred dances and religious ceremonies" that Gurdjieff and Orage championed as having "scientific" healing power through securing the balance of mind-body composition.[174] Gorham Munson vividly described Orage's weekly lectures, which often took place at the bohemian socialite Muriel Draper's salon at 24 East 40th Street in Manhattan, and identified Ferriss as one of the regular attendees: "In the center of the room could usually be found . . . the brooding architect Hugh Ferriss."[175] As revealed by a solicitation letter to patrons requesting money to support Gurdjieff's cause in America, Ferriss's affiliation with the Oragean circle continued at least through 1930 (Figure 1.23).[176]

Louise Welch, a devoted pupil of Orage, wrote about a postdemonstration gathering of a number of inner-circle members at Ferriss's skyscraper apartment:

> After the demonstration we went with Hugh Ferriss, whose architectural drawings I admired enormously, and his painter wife, Dorothy, to their new apartment at the very top of a high building looming magnificently among its neighboring skyscrapers above chiaroscuro streets. The room was bare except for one armchair and a few unpacked wooden crates. The armchair was presented to Orage. The rest of us sat on the floor.[177]

In Ferriss's apartment overlooking the city, Orage discussed how he first met Ouspensky and, later, Gurdjieff, the accidental encounters that sparked his research into the issues of spiritualism, self-knowledge, and the salvation of man. Two persistent themes cutting through Orage's conversation were the psychic imbalance allegedly caused by scientific determinism and attainment of higher consciousness as a means to neutralize the premonitions of disaster swirling around the materialistic industrial society. Orage's self-development doctrines went so far as to suggest that anguish was a necessary step toward nirvana: all of the fragmentary blocks of life had to be conjoined in order for the enlightened man to visualize the whole picture, recalling the theosophist's penchant for spiritual ascension over the divisive stimuli of terrestrial life. As Orage's séance continued, at a captivating elevation above the city, his prescription for combating the pitfalls of psychic disharmony seemed to echo in the group's parallel rumination on the sprawling city, a total spectacle at least momentarily devoid of all fissures and imbalance.

Orage's listeners dutifully took notes. In one instance, on December 16, 1925, Ferriss produced four pages of notes while attending his lecture.[178] The notes

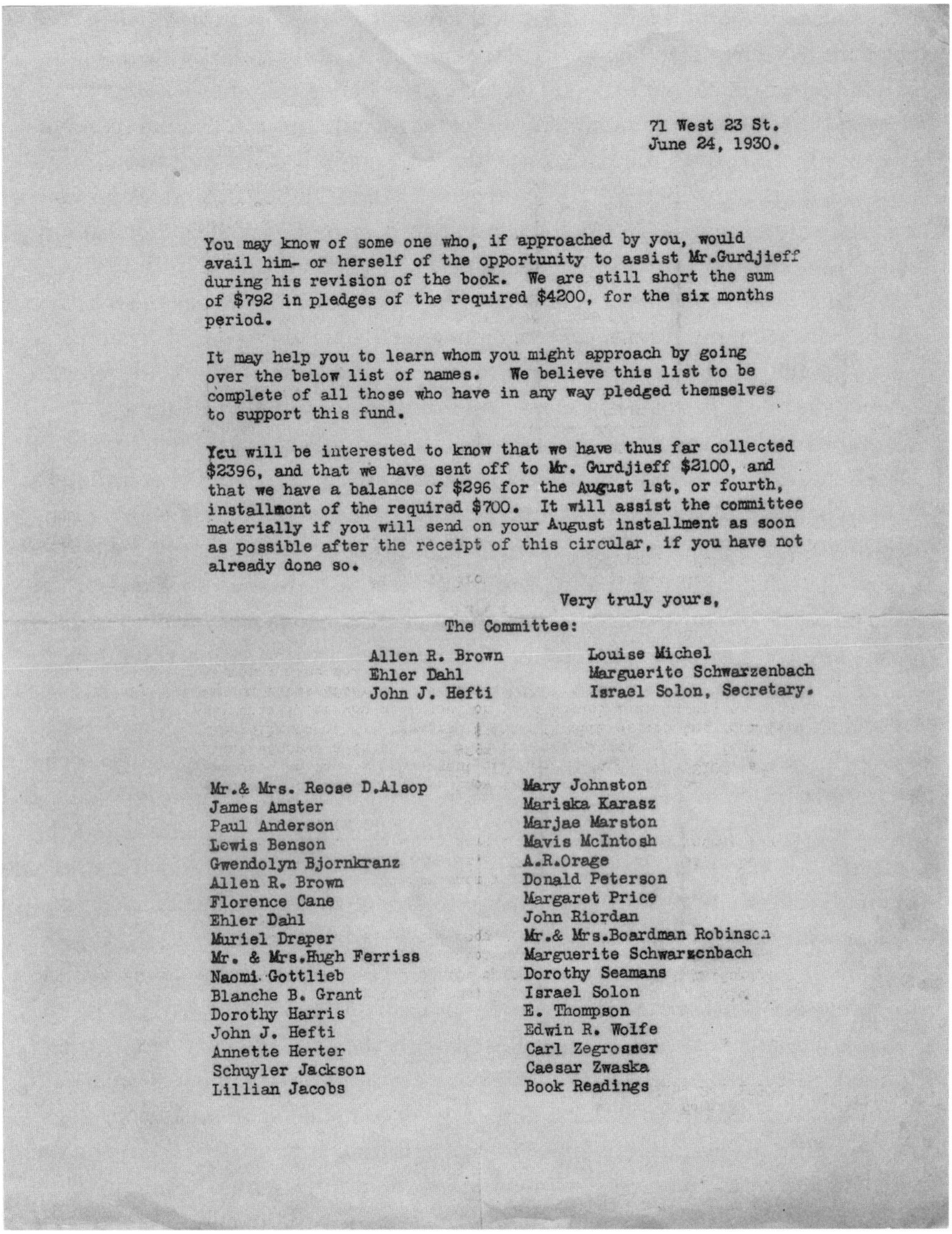

71 West 23 St.
June 24, 1930.

You may know of some one who, if approached by you, would avail him- or herself of the opportunity to assist Mr.Gurdjieff during his revision of the book. We are still short the sum of $792 in pledges of the required $4200, for the six months period.

It may help you to learn whom you might approach by going over the below list of names. We believe this list to be complete of all those who have in any way pledged themselves to support this fund.

You will be interested to know that we have thus far collected $2396, and that we have sent off to Mr. Gurdjieff $2100, and that we have a balance of $296 for the August 1st, or fourth, installment of the required $700. It will assist the committee materially if you will send on your August installment as soon as possible after the receipt of this circular, if you have not already done so.

Very truly yours,

The Committee:

Allen R. Brown
Ehler Dahl
John J. Hefti
Louise Michel
Marguerite Schwarzenbach
Israel Solon, Secretary.

Mr.& Mrs. Reese D.Alsop
James Amster
Paul Anderson
Lewis Benson
Gwendolyn Bjornkranz
Allen R. Brown
Florence Cane
Ehler Dahl
Muriel Draper
Mr. & Mrs.Hugh Ferriss
Naomi Gottlieb
Blanche B. Grant
Dorothy Harris
John J. Hefti
Annette Herter
Schuyler Jackson
Lillian Jacobs
Mary Johnston
Mariska Karasz
Marjae Marston
Mavis McIntosh
A.R.Orage
Donald Peterson
Margaret Price
John Riordan
Mr.& Mrs.Boardman Robinson
Marguerite Schwarzenbach
Dorothy Seamans
Israel Solon
E. Thompson
Edwin R. Wolfe
Carl Zegrosser
Caesar Zwaska
Book Readings

Figure 1.23. A committee of Gurdjieff supporters sent out this letter (dated June 24, 1930) soliciting money to support the mystic's mission in America. The patrons, whose names are listed at the bottom of the letter, included "Mr. and Mrs. Hugh Ferriss." Muriel Draper Papers, Yale Collection of American Literature, Box 20, Folder 625, Beinecke Rare Book and Manuscript Library, Yale University.

are cryptic, but they demonstrate Ferriss's entrenched interest in the psychological constitution of man. As his notes suggest, Orage identified two sets of elements in each human being: six potential centers at the biological conception and four stages of consciousness. The three basic biological centers were instinctive (muscular system), emotional (visceral system), and intellectual (cerebral system). Distinct from the basic ones, there were three other "advanced" centers: will (higher instinctive), consciousness (higher emotional), and individuality (higher intellectual). The four stages of consciousness were sleeping, waking, self-consciousness, and cosmic consciousness. When equipped only with the three basic centers, a person remained a mere "creature," whereas the mastery of the advanced, but latent, centers empowered a person to see his inner universe, or the I, an essence that existed in each person as an "undeveloped germ." The two types of centers and consciousnesses were aligned along a vertical graph in which the attainment of an advanced center meant a higher position of self-awareness. According to Ferriss's notes, Orage lamented that although human beings had six potential centers and four stages of consciousness in the original biological self, the collective societal framework, or what he called the "crowd psychology," eventually conditioned mental growth in such a way that human beings were unable to mine these centers and consciousnesses so as to explore their full human capacity. Therefore, "our first effort should be to abandon this sociological status and return to our native biological status."[179] In lockstep with Gurdjieff, Orage believed that the psychological prowess to exact this self-cleansing process eluded the majority of humanity. It was the sacrosanct province of a self-selected few. Thus, following Gurdjieff's desire to create a new breed of hyperconscious human beings, Orage advocated the "rebirth" of man through the unlearning of received wisdom and the subsequent development of a focused strategy to engage the full array of centers and consciousnesses.

Early in his literary career, Orage was an avid reader of Nietzsche, who became the philosopher à la mode in England between 1909 and 1913, as W. B. Yeats, George Bernard Shaw, and Orage flirted with the idea of the *Übermensch*.[180] Influenced by Nietzsche, Orage's 1907 book *Consciousness: Animal, Human and Superman* trekked a theosophical path to what he called "superman consciousness," a kind of "ecstasis" or super mental ability to situate the self outside the cerebral realms of human consciousness.[181] The ordinary human mind, Orage argued, was delimited by the body; in contrast, the ecstatic mind of the superman interiorized the body itself. In this rather cryptic argument, Orage consistently attempted to articulate the superman as a lofty creature: "Certainly the altitude at which they must have stood above the contemporary humans, the influence they exerted apparently with so little effort, and the genius of their inventions, all point to a difference of kind between such beings and men."[182] Just as Nietzsche sang the virtues of his prophet Zarathustra's "winged vision,"

Orage imagined the superman's mental world as liberated from ordinary human consciousness, which was tragically conditioned by the body on the ground: "In relation to pedestrian feeling," Orage opined, "superconscious feeling is, as it were, winged."[183] Although Orage's Nietzschean orientation waned somewhat after he came in contact with Gurdjieff in 1922, his core interest in higher consciousness as a prerequisite of the modern man, perched atop a high plateau of evolution, remained. To attain superman consciousness was to ascend, with both body and mind. This was indeed a leitmotif in the Gurdjieffite circle. Jeanne de Salzmann, a Gurdjieff disciple, noted that to see and purify oneself, one must rise above the associations that hold consciousness captive: "Objective thought is the look from above."[184]

Ferriss shared Orage's Spenglerian conviction that there was a crisis in Western civilization, which was set on a precipitous path of social degeneration and neurosis.[185] Ferriss believed that the modern world's technological advancement had not been paralleled by a "corresponding and compensating evolution in the psychological world—in personality traits, understanding of the deeper needs of society, human relationships, spiritual aptitude or esthetic development. The discrepancy creates a disturbing and dangerous situation."[186] Reduced to one dimension, man's inner world had been fractured, Ferriss believed, as a result of the mutual exclusion of science and arts, inner knowledge and outer influences. The fractured self was now reflected in the uncoordinated, unhygienic, and clotted body of the city: "As the avenues and streets of a city are nothing less than its arteries and veins, we may well ask what doctor would venture to promise bodily health if he knew that the blood circulation was steadily growing more congested!"[187] The disintegration of the self, along with its negative impacts on the city, was the source of a classic modernist angst that provoked haunting speculations on the expedited evolution of the human race, for which Darwin, Spencer, and archprovocateur Nietzsche provided sustained intellectual frameworks.

The publication of Ferriss's *Metropolis of Tomorrow* in 1929 in the backdrop of Oragean activities in the 1920s reveals how the two men came together with a common cause. To generate funds for Gurdjieff's "maintenance" expenses, Orage undertook a series of paid lectures on literature that developed into a workshop for professional and amateur writers. From 1928 through 1930, about seventy-five writers or prospective writers—including Ferriss, Draper, Toomer, publicist Amos Pinchot (writing a rebuttal of Walter Lippmann), poet Melville Cane, novelist Isa Glenn, editorial staff at the *New Republic,* and Princeton alumnus Thomas Stanley Matthews—participated in the so-called Orage course.[188] Orage asked participants to bring their ongoing writing samples and provided them with constructive criticism regarding problems in rhythm of thought, taste, structure, and their ability to engage the reader. His encouragement of participants to consider publishing their work reportedly resulted in a number of literary works dedicated to the acclaimed editor, their *cher maitre*: among others, Muriel Draper's European memoir *Music at Midnight* (1929), Isa Glenn's human story of sea voyage *Transport* (1929), and T. S. Matthews's *To the Gallows I Must Go* (1931).[189] Published around the same time, Waldo Frank's *The Re-discovery of America* (1929) and Harlem Renaissance leader Jean Toomer's essay on racial politics in modern society in *Problems of Civilization* (1929) carried on the Oragean discourse of sociospiritual decline in the modern era. Louise Welch claimed that Ferriss's *Metropolis of Tomorrow* was also a product of the Oragean literary workshop:

> [A] book that emerged from the New York vow came from Hugh Ferriss, architectural artist, whose drawings of New York were a haunting combination of prophecy and dream. *The Metropolis of Tomorrow,* generally illustrated with Ferriss's drawings, suggested an ideal city on the island of Manhattan, washed by the tidal waters of three rivers and the sea. To leaf now through Hugh's shimmering vision of a city designed for human beings is to shudder once more at the reality of New York's Gadarene descent into real estate speculation.[190]

Welch might have overestimated Orage's influence on *The Metropolis of Tomorrow*. This is especially problematic since Ferriss began his futurist predictions from the early 1920s. Yet it is entirely possible that Ferriss's decision to put all of the drawings together into a philosophical, if somewhat hackneyed, story line might have dawned on him after he attended Orage's literary workshop. The theme of Ferriss's book—that a city (and its inhabitants) goes through a crisis-ridden "today," followed by an exploratory trial-and-error period, and finally arrives at the hoped-for utopia (designed by Ferriss)—reverberated with Orage's elaboration of a tortured path to spiritual redemption. And the captions that accompanied the drawings in *Metropolis*—a lyrical-expressionist fusion of practicality, spirituality, and intellectualism, all ultimately aimed at rectifying a decaying civilization—contained sparks of mystical sensationalism common to the Gurdjieff–Orage séances. Skeptical and committed Gurdjieffites alike agreed that Orage was a galvanizing speaker who combined Gurdjieff's ideas with his own erudite explanations of current philosophical trends. In all likelihood, Orage inspired Ferriss to see the promise of his own futurist visions in a book format. Orage provoked attendees at the workshops to venture beyond the narrow limits of their respective fields, to ask fundamental questions about the self and civilization.[191] Ferriss's *Metropolis*—as much as it was an extravagant prognostication on future urbanization—was an anthropological study of social degeneration and its remedy through man's spiritual rebirth, for which the planned city provided an apt metaphor.

If the occasionally calamitous tone in some of his *Metropolis* captions was any indication, Ferriss debated the concept of crisis as a central theme in a discourse of regeneration. In a quartet of previously unpublished drawings, Ferriss revealed how entrenched he remained in the existential anxiety of the modern man and his search for salvation.[192] Marked by a lingering parable of solitude, all four drawings showed in draconian settings a lonely figure, tortured yet resilient in his quest for enlightenment. In one poignant, nocturnal, and Piranesi-like pastel drawing, Ferriss depicted the primal figure of an agonized man—head bent downward in pain and defeat—in a subterranean chamber. Next to the figure stood his triumphant alter ego, head upward and hands outstretched, jubilant in his self-discovery (Figure 1.24). Without an entablature binding them into one structure, six aboveground Doric columns may have alluded to—among other possibilities—what Orage called the six centers of the mind–body system. In another mystical drawing, Ferriss centrally positioned a solitary human figure, face downward in rumination of his own self, represented by his mirror

Figure 1.24. Hugh Ferriss's pastel rendering of a man and his alter ego in alternate states of despair and ecstasy. This is one of four unpublished drawings cataloged in the Hugh Ferriss Papers, Section III: Drawing, Library, Center for Advanced Study in the Visual Arts, East Gallery, National Gallery of Art, Washington, D.C. Courtesy, National Gallery of Art Library, David K. E. Bruce Fund.

image in water (Figure 1.25). Ferriss's enigmatic caption suggested a "healing" man within the triadic cosmology of ebb tides, sands, and stars. The lost soul in the canyons of a vertical city in the third sketch could not have offered a more tragicomic representation of the modern man (Figure 1.26). In the fourth sketch, the question "Is my true love hiding?" was more a rhetorical suggestion of man's struggle to reach an elusive destiny than a personal lament (Figure 1.27).

The experience of psychic disharmony, as Ferriss sketched out, was no less a precondition for a redemptive "tomorrow." Thus, one could draw parallels between the Gurdjieffite ambition of engendering an exclusive fraternity of harmonious supermen—as evidenced by the name of Gurdjieff's mystical enclave, the Institute for the Harmonious Development of Man—and Ferriss's concern for the psychic reconstruction of man through the language of architecture. Orage's calling for a rebirth of the human race uncannily reverberated in Ferriss's own dreaming of a metropolis as a vehicle for the harmonious creation of a new people: "Our criterion for judging [a] self-conscious Architecture will be its effect on human values: its net contribution to the harmonious development of man."[193] Ferriss's program comprised a seductive trinity: first, the purported spirituality embodied in the upper domain of the metropolis; second, the attainment of higher consciousness as a panacea for the modern man's existential angst; and third, the anticipated advent of a master class. At the center of this trinity remained Ferriss's belief that the mahatma-like figure perched at the peak of spiritual consciousness found its most persuasive symbolization in the seer atop the skyscraper, who, in turn, could most convincingly present the early twentieth-century image of the master planner, poised, omniscient, and intent on creating the city of tomorrow from godlike heights.

Orage's influence on Ferriss resided in the epistemic relationship between his concept of a "winged" superconsciousness and Ferriss's trope of a lonely spectator perched on the skyscraper parapet, gazing through the morning mist. It was a curious coincidence that a reviewer of *The Metropolis of Tomorrow* wrote in the *Christian Science Monitor* that Ferriss had converted the future American city into "things of winged beauty."[194] Did Ferriss's association with such Gurdjieffites as Orage, Bragdon, Jane Heap, Muriel Draper, Jean Toomer, and others account for his own search for the "harmonious development of man" against the foil of the modern metropolis? Ferriss argued that the metropolis was, after all, a "human drama" and that the "vast architectural forms are only a stage set. It is those specks of figures down there below who are, in reality, the principals of the play . . . it is indeed true that the human values are here the principal values."[195] The solitary figure that appeared in "The Lure of the City" or "The City at Night" hardly seemed like a coincidental graphic insertion; rather, it was a symbol of a protracted investigation into the psychological nature of the modern man, navigating the labyrinth of the metropolis. Given Ferriss's interest in creating an ideal city, it was hardly surprising that

Figure 1.25. Hugh Ferriss's drawing of a solitary figure on the beach with the caption "Ebb sands and stars—these be the healing mutes . . ." Hugh Ferriss Papers, Section III: Drawing, Library, Center for Advanced Study in the Visual Arts, East Gallery, National Gallery of Art, Washington, D.C. Courtesy, National Gallery of Art Library, David K. E. Bruce Fund.

Figure 1.26. Hugh Ferriss's drawing of a man dwarfed by skyscrapers. Hugh Ferriss Papers, Section III: Drawing, Library, Center for Advanced Study in the Visual Arts, East Gallery, National Gallery of Art, Washington, D.C. Courtesy, National Gallery of Art Library, David K. E. Bruce Fund.

Figure 1.27. Hugh Ferriss's drawing of a man despairing in the foreground while his putative companion stands in the background with the caption, ". . . is my true love hiding? Ah, where is she hiding?" Hugh Ferriss Papers, Section III: Drawing, Library, Center for Advanced Study in the Visual Arts, East Gallery, National Gallery of Art, Washington, D.C. Courtesy, National Gallery of Art Library, David K. E. Bruce Fund.

the modern man's path to spiritual redemption symbolically climaxed at the top of the skyscraper, recalling the heroic posture of the master builder poised to create a great metropolis of tomorrow.

The final illustration in *The Metropolis of Tomorrow* was a curious representation of Ferriss's view of the master builder as the consummate Man (Figure 1.28).[196] Fraught with Oragean vestiges—namely, the Ferrissian city's triadic divisions of science and business as corresponding to the Oragean body's three centers: thoughts, feelings, and senses—Ferriss's drawing revealed how consistently interested he was in probing the psychological conditions of man through the prism of the city. As the creator of an ideal environment, the architect, Ferriss argued, must fashion himself in the image of the Creator, not unlike the medieval articulation of God as the architect of the universe.[197] Ricocheting between being a curio and a coded narrative, his final drawing summed up his

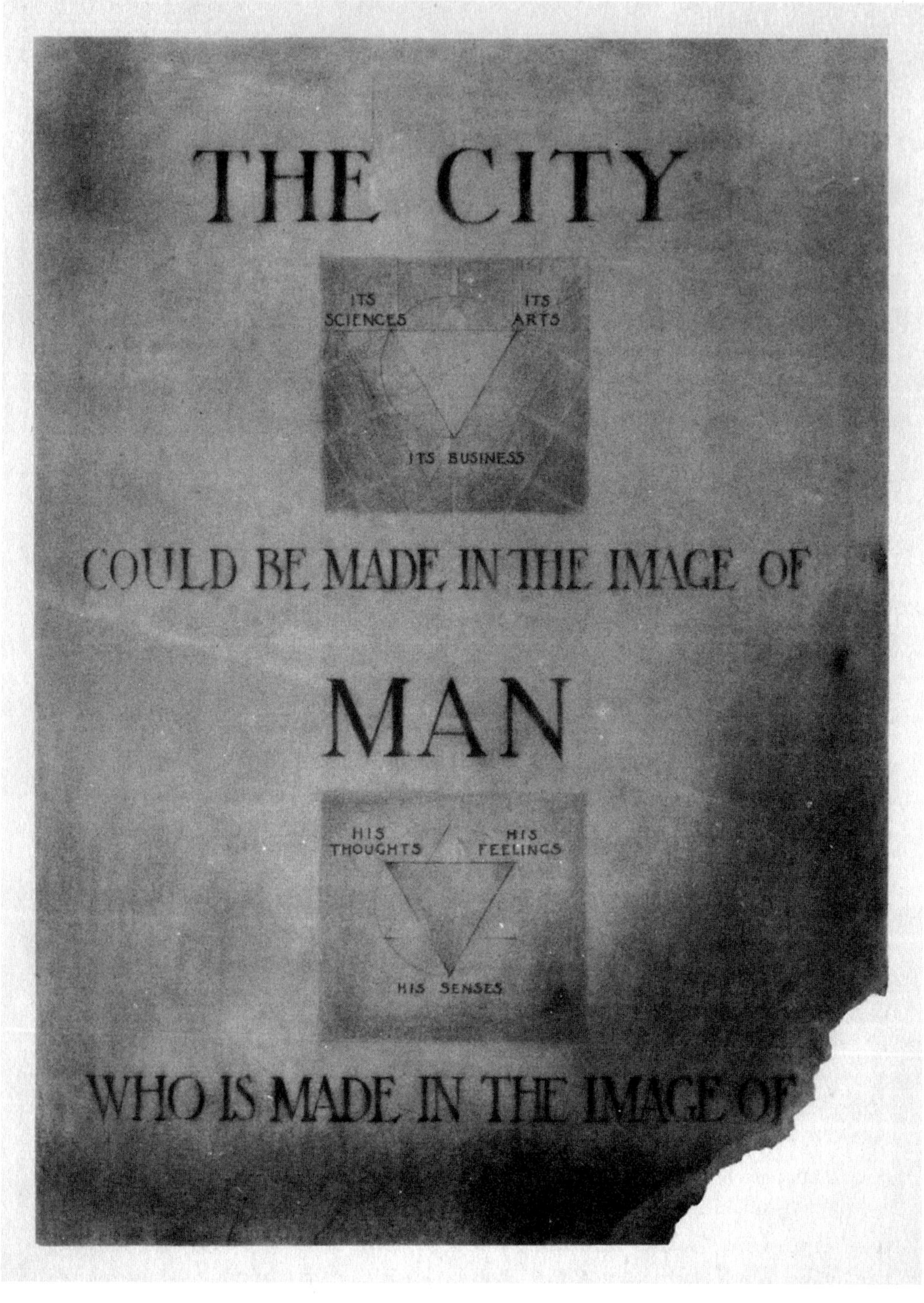

Figure 1.28. Hugh Ferriss "The City in the Image of Man," final illustration in The Metropolis of Tomorrow *(New York: Ives Washburn, 1929), 143.*

whole *Metropolis* project: "The city could be made in the image of Man who is made in the image of . . ." That godlike perfection was the final destiny in the Man-Architect's evolutionary march was unmistakable in the drawing's deliberately truncated statement. The textual ambiguity here could be read as a euphemism for what Ferriss called the "curtain of mist," the tortured path to self-discovery or harmonious self-development. In a set design for the play *New Year's Eve in New York* in 1930, Ferriss's modern man—the creator of the metropolis of tomorrow, hand raised triumphantly heavenward—was exalted as a dramatic index of the towering city, while the huddled masses, engaged in hero worship, merely formed a pedestal for their savior, a mahatma figure, a superman, or a mortal god (Figure 1.29).[198] Architecture of the metropolis became the consummate stage set for the man to become Man. The juxtaposition of the spectacle of the glittering skyscraper and the perfected Man represented the eventual convergence of two evolutionary paths.

Figure 1.29. Hugh Ferriss's sketch for the set design of New Year's Eve in New York, *Neighborhood Playhouse; from* Theater Guild Magazine, *March 1930.*

TWO

Ascension as Autobiography

Buckminster Fuller and His "Land to Sky, Outward Progression"

FULLER'S ASCENSION

Richard Buckminster Fuller presented his Dymaxion House before the Architectural League of New York on July 9, 1929.[1] A staged photo, taken for promotional purposes shortly after this presentation, offers a glimpse into the mind of the New England native and emerging social provocateur (Figure 2.1).[2] Looking every bit the suave salesman in his crisp suit, the thirty-four-year-old Fuller poses with a scale model of his proposed single-family house, ready to insert a triangular duralumin panel on its second floor. A teardrop-shaped car is parked in front of the model. Strikingly, a die-cast miniature airplane sits in the ground-floor garage. This photo of the "house of tomorrow," with its provocative inclusion of a family "flying machine," provides a vital clue to how modernist designers embraced the theme of flight to showcase a range of futurist aspirations during the interwar period.

As observers in contemporaneous popular magazines suggested, the technology of Dymaxion House was radical.[3] The mass-producible and transportable house consisted of a lightweight central mast bolted to a concrete caisson that contained a septic tank and oil storage facility below grade (Figures 2.2 and 2.3). Containing an elevator and the building's utilities, the mast was surmounted by a conical climate-control device that supplied light and air to the main rooms on the second floor, maintaining an optimal interior microclimate year-round. Metal tubes functioned as load-carrying beams, radiating out from multipoint sockets in the mast and joined together by peripheral tubes. A mesh of tension wire above the base floor, supported by metal beams and covered with pneumatic materials, created the main floor of the house. As if to simulate an airplane in flight, the house could be hung by high-strength tensile wire at a convenient elevation, depending on geographic necessities. Fuller articulated the central concept of his house: "The basic idea of the construction is that all

Figure 2.1. A staged photograph of Buckminster Fuller presenting Dymaxion House. The Special Collections of Green Library, Stanford University. Courtesy of The Estate of R. Buckminster Fuller.

elements shall be suspended from above rather than rest upon supports from below."[4]

Dymaxion House presented an iconoclastic, technology-driven domestic environment, inspiring both admiration and ridicule from the media and professional circles. But as the promotional picture illustrates, Fuller narrated an equally intriguing personal story through the paraphernalia of the model. The placement of a model airplane appears neither coincidental nor trivial. Horizontally corrugated duralumin walls at the ground level create a garage from

Figure 2.2. Buckminster Fuller's model of Dymaxion House. Joachim Krausse and Claude Lichtenstein, eds., Your Private Sky: R. Buckminster Fuller, the Art of Design Science *(Lars Müller Publishers, 1999), 130. Courtesy of The Estate of R. Buckminster Fuller.*

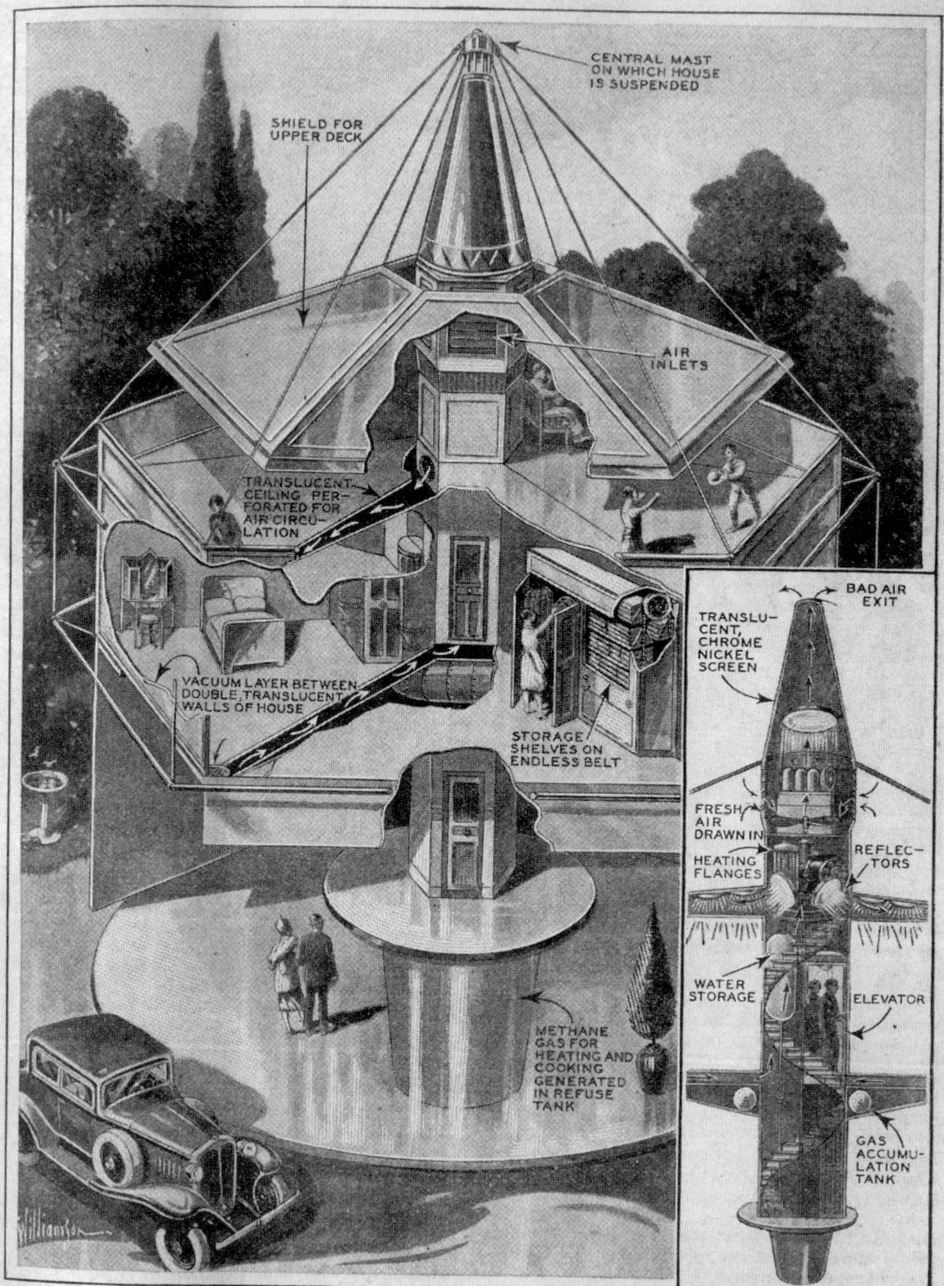
a POLE!

"Toad Stool" Houses for $2000
by KNIGHT DEACON

Essential details of "toad stool" house are shown here. Note that central mast from which it is hung contains all operating mechanism—air conditioners, heating units, reflectors for lighting rooms, and generator for creating gas from refuse.

Figure 2.3. An article featuring Dymaxion house in Modern Mechanix and Inventions *8, no. 5. (Minneapolis: Fawcett Publications, Inc., September 1932), 41.*

which the airplane protrudes, as if it has just returned from a flight to a neighboring town. The idea may not have been too far-fetched for Fuller. Seven years earlier, he flew in an airplane when most people had never been in one and could only imagine the sensation of flight. In placing the airplane in Dymaxion House, Fuller seemingly recounts the exhilarating moments he experienced flying along the East Coast between New York and Maine in 1922 after borrowing his wealthy U.S. Navy colleague Vincent Astor's seaplane for two weeks.[5]

Modernist designers demonstrated their futurist commitments by drawing on the most advanced transportation machine. The idea of having an airplane for family usage, "an airplane in every garage," just like an automobile, became a common iconographic trope of the interwar period.[6] In a salubrious representation of the "home of tomorrow" on a 1931 *Popular Aviation* magazine cover, the family airplane flies away just as a car would leave the driveway (Figure 2.4). George Fred Keck's House of Tomorrow, designed for the Century of Progress Exposition in Chicago in 1933, and William Lescaze's 1937 design for the House of 2089 continued this trend through the 1930s as a way to reinforce the futurism of their architecture.[7] Science fiction pulp, popular journals, and advertisements demonstrate the internalization of the flying machine within the American domestic vernacular. As a much-hyped symbol of progress, the airplane—either in flight or nestled at the house—predictably yet effectively created a popular image of tomorrow.

Fuller's placement of an airplane in the garage of Dymaxion House was more than a semiological tactic for suggesting the project's futurism. The inclusion of an airplane in domestic architecture raised questions about what aviation meant to America socially and culturally in the 1920s, and to Fuller personally, especially in the wake of Charles Lindbergh's sensational transatlantic flight in 1927. Fuller's reaction to Lindbergh's flight illuminates how he cultivated an aesthetic of ascension in developing his architectural manifesto, as well as demonstrated his understanding of the world and his anticipated role in it. Like Le Corbusier, whose manifestos he had studied upon their publication in the 1920s, Fuller took a sustained interest in Lindbergh's flight. He published *4D Time Lock,* a mimeographed manifesto outlining the philosophical basis of his single-family house, on the first anniversary of Lindbergh's epoch-making flight.[8] In the booklet's cover letter, dated May 21, 1928, he established the technological advancement of the *Spirit of St. Louis* as an inspirational trajectory for Dymaxion House: "Was it not a product of complete segregation of the SPIRIT and the MATERIAL, with truthful standardization of the latter, that so immortalized the name of St. Louis, but a year ago today[?]"

Many advocates of modernism celebrated Lindbergh's flight as the harbinger of a new age of heroes (before the aviator earned notoriety for his extremist politics during the 1930s). "Paris hastens by all roads towards this wonder man," wrote Le Corbusier of Lindbergh's spectacular descent into the French capital.[9]

Figure 2.4. A Vacuplane flying over a house on the cover of Popular Aviation, *November 1931.*

Fuller's reference to Lindbergh is indeed instructive of his fledgling worldview, as he retained his youthful excitement more than a decade later in his first published book, *Nine Chains to the Moon* (1938): "The significance of the REALIZATION OF FLYING is so far-reaching as to be beyond calculation, though it may be dimly appraised in the world-wide, nation and breed surmounting, exultation and 'god' making of Lindbergh's achievement."[10] What was this "'god' making" of Lindbergh all about? Did figures like Le Corbusier and Fuller celebrate Lindbergh's accomplishment to advance their own philosophical positions?

To answer these questions, it is useful to look more closely at Lindbergh's flight. On May 21, 1927, Lindbergh, a lanky twenty-five-year-old former Minnesota farm boy, touched down on the tarmac of the Le Bourget airfield on the outskirts of Paris thirty-three and a half hours after ascending to the sky from Roosevelt Field on Long Island, New York (Figure 2.5). The event caused an unprecedented popular delirium on both sides of the Atlantic and spawned a frenzied spell of hero worship in "the year the world went mad."[11] Four million people converged in Manhattan to greet "Lucky Lindy" upon his triumphal return from Paris on the American cruiser USS *Memphis,* which had been dispatched to bring back the aviator on the orders of President Coolidge himself (Figure 2.6). Other aviators had crossed the Atlantic by air before Lindbergh. British aviators John Alcock and Arthur Whitten Brown had flown nonstop from Newfoundland to Ireland in 1919; British and German dirigibles and American army planes accomplished similar feats. What, then, was the novelty of Lindbergh's flight? For the American cultural historian Frederick Lewis Allen,

> the explanation is simple. A disillusioned nation fed on cheap heroics and scandal and crime was revolting against the low estimate of human nature which it had allowed itself to entertain. For years the American people had been spiritually starved. They had seen their early ideals and illusions and hopes one by one worn away by the corrosive influence of events and ideas—by the disappointing aftermath of the war, by scientific doctrines and psychological theories which undermined their religion and ridiculed their sentimental notions, by the spectacle of graft in politics and crime on the city streets, and finally by their recent newspaper diet of smut and murder. Romance, chivalry and self-dedication had been debunked; the heroes of history had been shown to have feet of clay, and the saints of history had been revealed as people with queer complexes.[12]

Anxious Americans now needed, Allen reasoned, "real" heroes. The hyped image of Lindbergh—indomitable, solitary, daredevil, and self-reliant—provided a fitting figure for the romanticization and rediscovery of allegedly lost American ideals. As the American historian John Ward explained the public enthusiasm:

> The wild medley of public acclaim . . . make[s] one realize that response to Lindbergh involved a mass ritual in which America celebrated itself more than it celebrated Lindbergh. Lindbergh's flight was the occasion of a public act of regeneration in which the nation momentarily rededicated itself to something, the loss of which was keenly felt. It was said again and again that "Lindy" taught America "to lift its eyes up to Heaven."[13]

Not all opinions of his flight were celebratory, however. One author asserted that the brouhaha surrounding the aviator was best understood as the mass media's eagerness for sensational stories befitting the prevailing culture of celebrity.[14] In another instance, contrary to the media's portrayal of Lindbergh as demure, opposed to self-promotion, and reluctant to cash in on his success, a searing article in the *New Yorker* painted a starkly different portrait of an insolent exhibitionist and manipulator of public opinion.[15] The reception of Lindbergh's flight was full of contradictions.

Like many others of the airplane generation who came of age in the 1920s, Fuller embraced an idealized version of the Lindbergh story. Enshrined in myths and fables, the dream of the flying man became a reality in the early twentieth century, and cultural reactions to it were predictably euphoric. The cultural historian Robert Wohl has observed that the valorization of the aviator captured the popular imagination because it drew on a long masculinist tradition of mythological heroes "who had haunted (and delighted) the Western imagination for hundreds of years."[16] A staple in popular technological utopias, science fiction, and a range of spiritualist and avant-gardist reckonings during the 1920s, the aviator was viewed by many as a modern hero, a lofty symbol of the machine age, a godlike seer of the world, and, no less, a Darwinist emblem of highest evolution. Three months after Lindbergh flew across the Atlantic, an author in *Science News-Letter* invoked a characteristic set of evolutionary qualifications, a sort of "aviation hygiene," for prospective aviators to meet (Figure 2.7).[17] The underlying message of the essay was that the aviators must be a new breed, endowed with well-developed physical attributes to withstand the travails of altitude and mental alacrity to fulfill the sensitive demands of controlling a flying machine. "Just what per cent of the youth of this country is potentially Lindberghian is a question of intelligence tests and physical tests," mused the author. Lindbergh became a new measure of manhood and heroics.

For many observers, the Minnesota pilot's aeronautical voyage also rekindled the myth of the pioneer and the western frontier. Lindbergh was, as one author suggested in the wake of the 1927 flight, "a Daniel Boone or Davy Crockett of

Figure 2.5. Charles Lindbergh standing in front of his plane, the Spirit of St. Louis, *at Roosevelt Field for a publicity shot promoting his transatlantic flight in 1927. Library of Congress, Prints and Photographs Division (LC-USZ62-13140).*

Figure 2.6. Charles Lindbergh's reception in New York City on June 13, 1927. Lindbergh Picture Collection (MS325B). Manuscripts and Archives, Yale University Library.

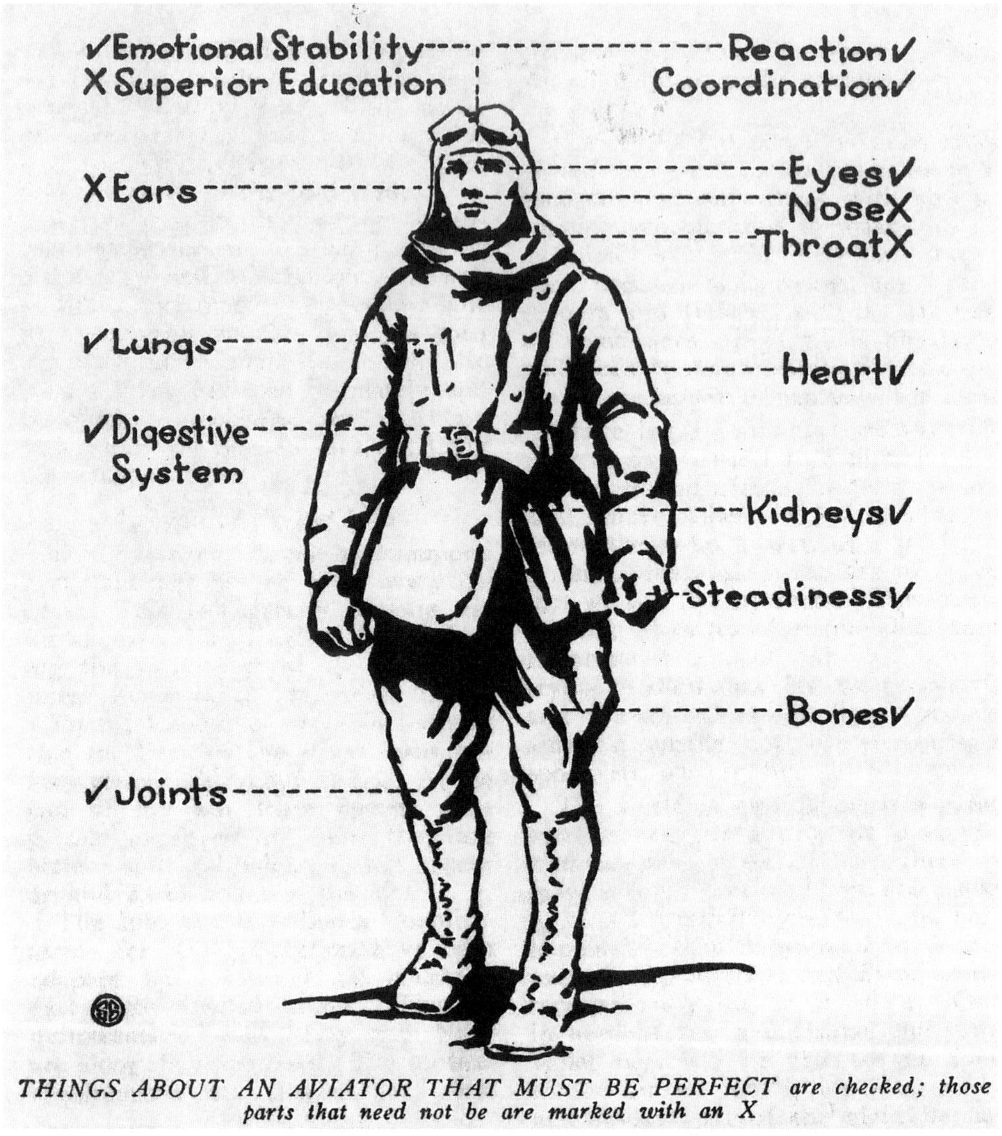

Figure 2.7. An image from Science News-Letter *depicting the required physiological characteristics of an aviator.* Science News-Letter *12, no. 332 (August 20, 1927), 119.*

the air," a twentieth-century reincarnation of the frontiersman of the Wild West.[18] Yet in this new, aerial romance of the frontier, the man–machine relationship warranted new reflections. As the American author James Oliver Robertson has argued, there was a structural difference between a "classic" frontiersman like Davy Crockett and an aviator like Lindbergh, for the two represented very different conceptualizations of the frontier idea.[19] Unlike the physical space of the American West, which could be occupied, developed, and internalized within a lived experience, the "frontier of the air" remained primarily a conceptual space whose conquest demanded a new type of symbiotic relationship between man and machine. This relationship would be based on a balance of mutual dependence in which "the complex machinery of impersonal industrial society was at the disposal of, and dependent on, the virtues of the lonely, independent, free American." In other words, without the prosthetic wings, Lindbergh could not be a pioneer, whereas the prosthetic wings entailed the resolve and perseverance of an individual. Robertson explains the man–machine network in the context of an enduring frontier myth:

> The frontier sought by Americans in the twentieth century is no longer the frontier of settlement; their ambition is no longer to carve out agrarian empires from the wilderness. The "new frontier" is to be found in the modern urban and industrial world. It was explored and tamed by invention, by production, by men with machines, like Edison and Ford and Lindbergh. Lindbergh's frontier was a combination of new frontiers: it was a test of human endurance, an exploitation of industrial genius, and a pushing of the limits to set a new record. His frontier was the wilderness of the air. Lindbergh, the self-reliant American individual, pioneered, but only in a complex machine produced by the engineering and mechanical skills of a team of hard-working experts funded by progressive businessmen. Man and machine and industrial organization were one. The Lindbergh myth triumphantly celebrated the inseparability of the individual and industrial society. The acceptance of Lindbergh as a national hero indicated that there were roles available for Americans—the spiritual if not physical descendants of the frontiersmen, the pioneers, and the cowboys—in the modern world of airplanes, automobiles, vast industries, and sprawling cities.[20]

The polemical suggestion that the aviator and airplane—the man and machine—together formed a dialectic to create a new aerial frontier helps explain how Fuller attributed a particular significance to the role of machines in the development of human capital. His relationship with technology was different, as Reyner Banham argued, from that of the European avant-garde architects identified with the "International Style," such as Le Corbusier.[21] If Le Corbusier promoted a house that worked like a machine, Fuller proposed a house that was a machine. Le Corbusier's house represented an avant-garde formalism that assimilated the image of technology as a foil for futurist aspirations, whereas Fuller sought a radically direct engagement with technology, one that was more operative than semiological in his avowed goal to exploit machines for developing human capacity. Thus, the shared modernist fascination with technology found different expressions in contemporaneous projects such as Le Corbusier's Villa Savoye and Fuller's Dymaxion House. Villa Savoye's architectural body, suspended by means of a series of slender *pilotis,* invoked the illusion of an architectural counterpart of the Wright brothers' airplane.[22] Dymaxion House made its architectural suspension the very basis of its rapid constructability, freeing up construction labor and time for harnessing what Fuller frequently referred to as the intellectual capital of the "thinking type."

Providing the metaphor of the airplane, the American art critic Sheldon Cheney encapsulated Fuller's man–machine dialectic in the anticipation of a technology-driven house: "Here is a clean athletic transportation-machine for the modern clean athletic body—and we should have houses to match."[23] Man, machine, and dwelling formed a tripartite modernist ambition for Fuller. He believed that the single-family house, an icon of the American vernacular, needed to undergo an evolutionary process to reach perfection, just as modernist design theorists deemed the airplane the consummate representation of human thinking.[24] Fuller reasoned: "Our airplanes span oceans, while multimillions of dollars worth of homes burn to the ground . . . 24 hour service is planned by airway between London and New York, while still we take from six months to a year to build a simple dwelling."[25] Fuller's view of the dwelling machine hardly concealed a broader humanist concern. If the *Spirit of St. Louis* "created" Lindbergh, the same spirit behind the single-family house would create ideal citizens like Lindbergh. This mutual dependency of man and machine was central to Fuller's design philosophy, one in which the theme of ascension suggested a provocative relationship between the "conquest" of gravity and the evolution of the human mind.

Fuller started to intellectualize an aesthetic of ascension from 1927, and it would continue to manifest itself with a range of phantasmagoric ideas, including the air-deliverable houses of 1928 and, later, the astronaut of Spaceship Earth (1951) and floating cities called Cloud Nine (circa 1960). For one's body to defy gravity, Fuller posited, one must ultimately master a heightened awareness of the self. A body that remained horizontal was entrapped in a false sense of security, dulled by a lack of creative initiative. Decades later Fuller wrote: "Babies live horizontally for months before they spontaneously take the initiative and, coordinating their own complex of control facilities, stand vertically. Vertical is objective. Horizontal is subjective, yielding. In extreme, the vertical characterizes life and the horizontal characterizes death."[26] The technological conquest of gravity was, then, reflective of the most perceptive mind. This enduring Fullerian philosophy harked back to the aviation culture of the early twentieth century.

When he designed a futuristic flying machine described as a "triangular framed auto-airplane with collapsible wings" in 1927, its fuselage was emblazoned with Fuller's logo, "4D," which illustrated not only his fascination with technological efficiency, mobility, and lightness but also what could be called his vertically thinking mind (Figure 2.8).[27] As Linda Henderson has demonstrated, by the 1920s, the concept of the "fourth dimension" had become infused with both mystical and popular scientific appeal, thanks to regular publication on the topic in such magazines as *Popular Science Monthly, Science, New Science Review, Harper's Weekly, McClure's,* and *Current Literature*.[28] Going beyond the Euclidean world, the notion of the fourth dimension assumed numinous overtones, inspiring modernist artists to seek philosophical space outside positivism and materialism. Introduced to the concept in 1914 at the Harvard mathematics department, Fuller furthered his understanding of the fourth dimension through his reading of the works of, among others, Claude Bragdon, H. G. Wells (who used the concept of the fourth dimension in various guises in his novels), and Albert Einstein.[29]

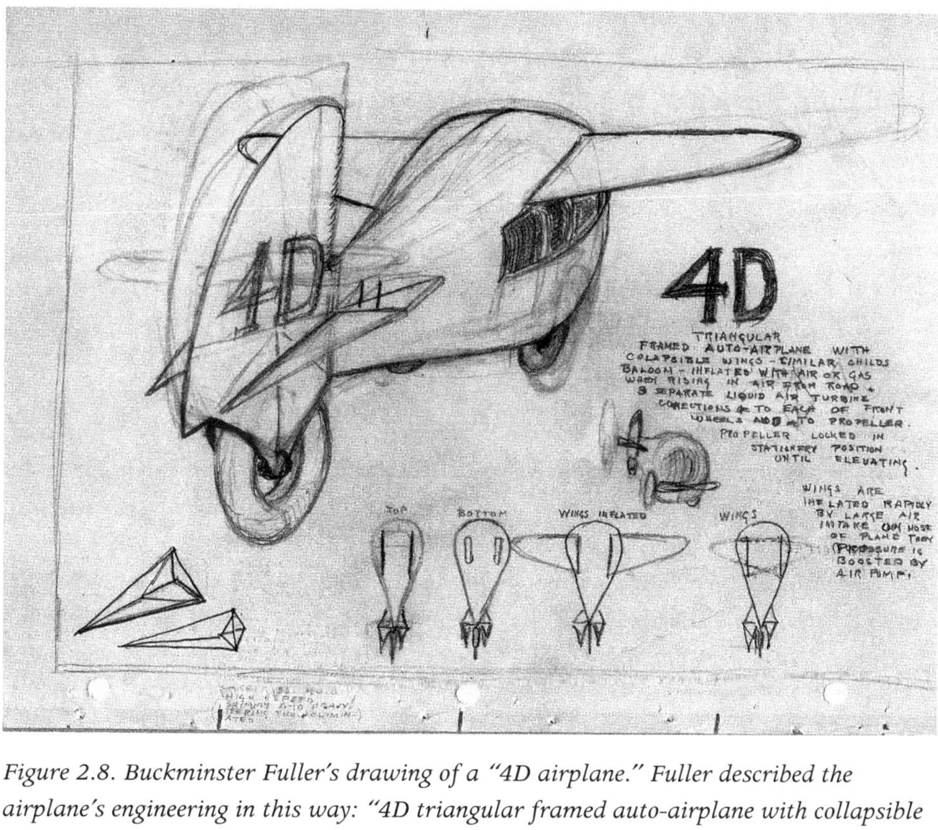

Figure 2.8. Buckminster Fuller's drawing of a "4D airplane." Fuller described the airplane's engineering in this way: "4D triangular framed auto-airplane with collapsible wings . . . inflated with air or gas when rising in air from road. 3 separate liquid air turbine connections to each of front wheels and to propeller. Propeller locked in stationary position until elevating. Wings are inflated rapidly by large air intake on hose of plane then pressure is boosted by air pump." Printed in R. Buckminster Fuller, 4D Time Lock *(Albuquerque: Lama Foundation, Biotechnic Press, 1929/1972), 107. Courtesy of The Estate of R. Buckminster Fuller.*

Fuller's "4D airplane" implied a radical dialectic between a conscious defiance of gravity (a term that many avant-garde designers at the time deployed to suggest social conformism and servitude) and a higher state of mind. Henderson has drawn on Fuller's manifesto *4D Time Lock* to suggest how Dymaxion House's "from the inside out" and suspended spatial design was inspired by early-twentieth-century avant-garde discussions on the fourth dimension and its promise of a nongravitational, nondimensional space.[30] Fuller's pronouncement that "vertical is objective" alluded to the nongravitational spatial politics that underpinned the avant-garde perceptions of both the fourth dimension and aviation during the 1920s.[31] By inhabiting a nongravitational space, the flying man offered an alluring symbol of the avant-garde imagination of a new human kind. As we shall see in the following sections of this chapter, it is plausible to view Fuller's "4D airplane" and Dymaxion House's suspended morphology as an allegory of the flying man and his evolved mental power.

Fuller's philosophical interpretation of human flight—and its associated notions of efficiency, weightlessness, and planetary mobility—provided an illuminating window on his industrially reproducible house, which he considered an ideal platform to extend the evolutionary arc of humanity. Fuller began to cultivate an aerial dimension in his ideation of the most evolved man, a sort of a genius for whom flight was a necessary epistemic vantage for a comprehensive understanding of the world.[32] The genius, for him, was a figure who remained suspended from the hegemonic ideologies of his era or who was even somewhat schizophrenic: "Genius's dual or multiple personalities may be said to be representative of a breadth of viewpoint, more-than-average, high worldly, and having an exquisite sense of Timeliness."[33] If society could foster more geniuses, then it would be in a more advantageous position to reexamine civilization's political and economic foundations. During the decade between *4D Time Lock* and *Nine Chains to the Moon,* Fuller presented a body of ideas about the development of human capital and its bearing on the progressive evolution of civilization. He proposed analogical relationships, if at times audaciously, between his Dymaxion House as a weight-defying, tensile, and air-deliverable mobile housing unit and the modern genius, a highly evolved philosopher and cosmic pilot of Spaceship Earth.

From archival sources it is hard to pinpoint when Fuller began to be influenced by evolutionary ideologies and various persuasions of social Darwinism. But it is not difficult to assess how he absorbed their rhetoric and tenets, which were intensely debated in 1920s American public life, so much so that Fuller considered it normative to apply them to a much wider discussion of social reform, aesthetics, and technology.[34] Since the 1920s he had collected books on human evolution, advancement of civilization, and themes related to the nature of highly intelligent men.[35] For instance, he read the American author and journalist William Harlan Hale's 1931 book *Challenge to Defeat: Modern*

Man in Goethe's World and Spengler's Century, as well as the review of the book in *The Nation* the following year.[36] Hale contended that Goethe's "Faustian" man, as an embodiment of the modern age's conflicts and contradictions, hopes and estrangements, might provide an alleviating answer to the so-called crisis of "Spengler's century." Hale's romantic portrayal of the modern man must have appealed to Fuller's own imagination of a savior, or what Hale called "the soul of man awakening from the surrounding chaos." It is the same romantic lens through which Fuller himself studied the lives of Leonardo da Vinci, Henry Ford, Lindbergh, Einstein, and other iconic figures.

As we shall see later in this chapter, Fuller's appreciation of trailblazers who rose above their peers with pioneering ideas was also influenced by mystically oriented teachers whom he met in New York City at the beginning of the Great Depression. He gravitated to esoteric teachings in spiritual development by the Russian mystic George Ivanovich Gurdjieff (whom Fuller met in Greenwich Village in the late 1920s) and his pupils P. D. Ouspensky and Claude Bragdon, and, in the 1930s, the controversial French American scientist Alexis Carrel. Drawing on these various sources, Fuller merged the design conceptions of his single-family house with mystical iterations on self-propelled evolution.

Fuller's evolutionary design philosophy revealed a crucial autobiographical connection. His romance of the aviator as a transcendental figure worked as a prop for what he sensationalized as a turning point in his life in 1927, in the wake of an entrenched feeling of personal and professional failure. He blamed himself for not being able to provide his family with a healthy domestic environment in Chicago, a failing he believed caused his daughter's untimely death in 1922 at the age of four. The melancholy spell continued. Fuller claimed to have reached his lowest point when the Stockade Building System, where he had worked for five years, ousted him for alleged financial mismanagement.[37] His firing led to the much-mythologized nadir in Fuller's life. Contemplating suicide on the shores of Lake Michigan, Fuller was overcome by an epiphany that he must dedicate his life to the cause of humanity, rather than wasting it. In 1939, Fuller retrospectively offered a dramatized account of his much-hyped suicide attempt:

> Now came the great crisis in his life. No job, no money, infant daughter, betrayed by people he had trusted. He walked over to the lake and thought about suicide. Should he call his life a bad job and throw it away? Or should he try to figure out some way to make all the experiences of it bitter or happy, useful? He took stock of himself, and realized that he had had a full life, and that he had acute understanding.[38]

The critical question for Fuller was how to overcome his overwhelming sense of guilt and failure. What kind of *hero* would fit convincingly into his narrative of self-rediscovery? Would the figure of the aviator, "seeing the world from the air," provide a befitting model for his own transcendence?[39]

If Fuller's writings and drawings are any indication, the mythology of the aviator's redemption and glorified perspective was Fuller's ruse in creating a failure-leads-to-success account. In the 1920s, as the Lindbergh phenomenon attested, none personified cultural veneration more profoundly than the winged protagonist in the sky. Fuller's idealistic approach to man's newfound aerial mobility and the planetary perspective it enabled neatly served his millenarian aspiration to be mankind's servant while erasing from his "cosmic" vista the troubling details of personal anxiety and sorrow.

A VERTICAL FRONTIER

Fuller was born at the start of the Progressive Era into a pedigreed New England family that produced eight generations of Boston clergymen and lawyers.[40] Fuller's great-great-grandfather, the Reverend Timothy Fuller, was a Massachusetts delegate to the Federal Constitutional Assembly and opposed the Constitution's sanctioning of slavery and its ratification.[41] Fuller's great-aunt Margaret Fuller-Ossoli (1810–50), an active member in the Transcendentalist circle and friend of Emerson and Thoreau, was a pioneer of nineteenth-century American feminism.[42] Nonconformity seemed to have been the Fuller clan's defining myth. Fuller was no exception to this family "legacy." Impatient with the rigors and discipline of college life, he was expelled from Harvard in 1914 for frolicking in Manhattan instead of taking his freshman midyear tests.

In 1915, Fuller moved to New York City with his mother. Thereafter, he had a range of hands-on job experiences that steadily shaped his view of machines as integral parts of a larger social mission. From installing textile machines in Sherbrooke, Quebec, to lugging meat at Armour & Company, from learning mechanics at the Naval Academy in Annapolis during World War I to erecting the walls of the Stockade building in the 1920s, Fuller developed a particular empathy for machines and their relationship to humanity.

Fuller's childhood was touched by the innocent excitement pouring out of Kitty Hawk and the Promethean promise of Orville Wright's groundbreaking flight. Fuller grew up as part of an American generation of boys enthralled by the wondrous, if sometimes wobbly and clumsy, flying machines. His childhood coincided with the beginning of the aviation age, when the older generation encouraged the youth to make model airplanes as part of an educational system that charitably conflated the notion of social progress with the development of technology.[43] Aeronautical interest, proponents of the mushrooming aero clubs of America argued, would create the "winged superchildren of tomorrow."[44] Much later in the 1960s, Fuller reminisced:

Plate 1. Frank Paul, "Flying Man," on the cover of Amazing Stories *3, no. 5 (August 1928).*

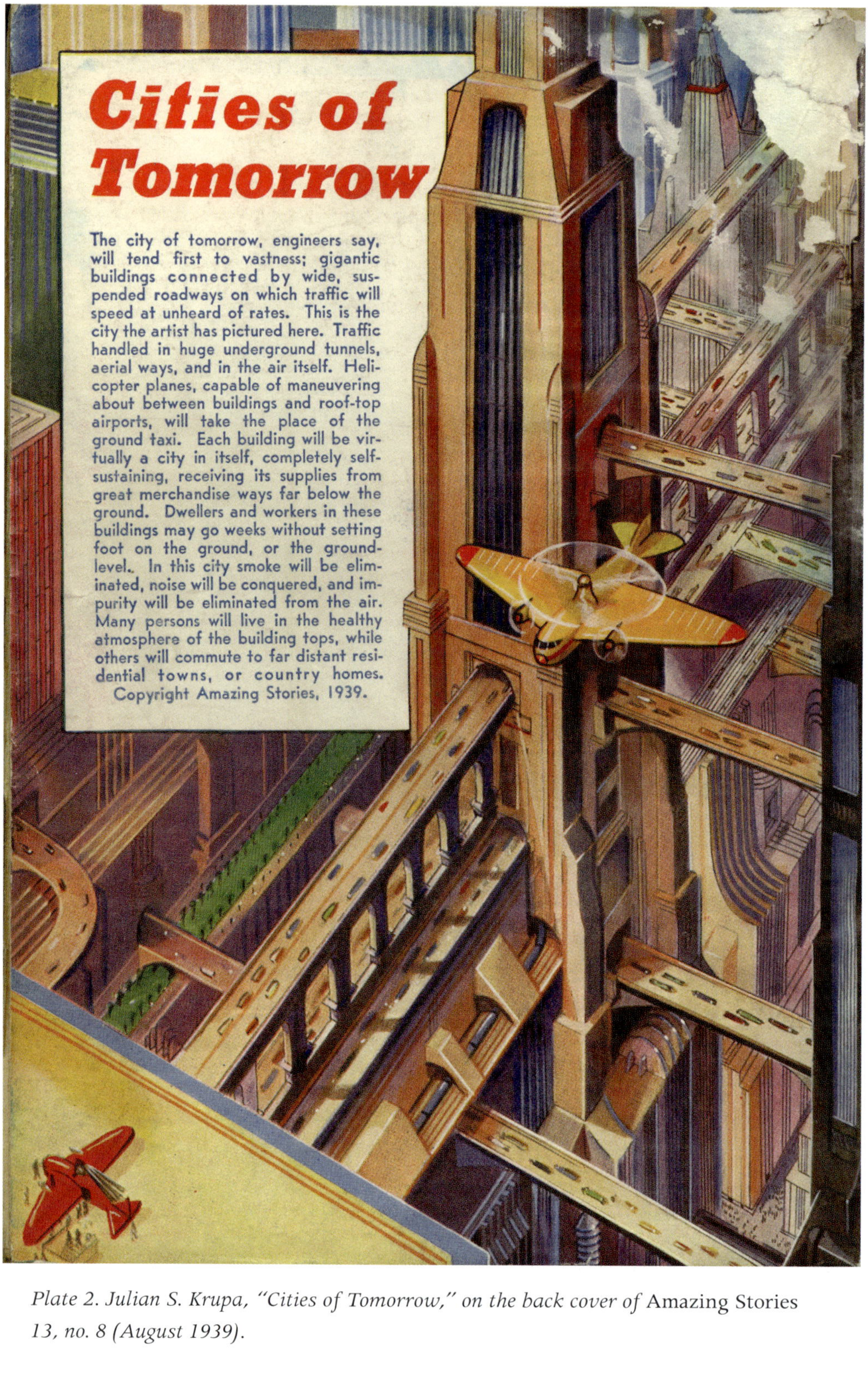

Plate 2. Julian S. Krupa, "Cities of Tomorrow," on the back cover of Amazing Stories *13, no. 8 (August 1939).*

Plate 3. George Bellows, New York, 1911. *Oil on canvas, 106.7 × 152.4 cm (42 × 60 inches). Copyright 1995 Board of Trustees, National Gallery of Art, Washington, D.C. Collection of Mr. and Mrs. Paul Mellon.*

Plate 4. Thomas Hovenden, Breaking Home Ties, *1890. Reprinted in Anne Gregory Terhune,* Thomas Hovenden: His Life and Art *(Philadelphia: University of Pennsylvania Press, 2006), 155.*

Plate 5. Georgia O'Keeffe, Radiator Building—Night, New York, *1927. Oil on canvas. Courtesy of Alfred Stieglitz Collection, Fisk University Galleries and Crystal Bridges Museum of American Art.*

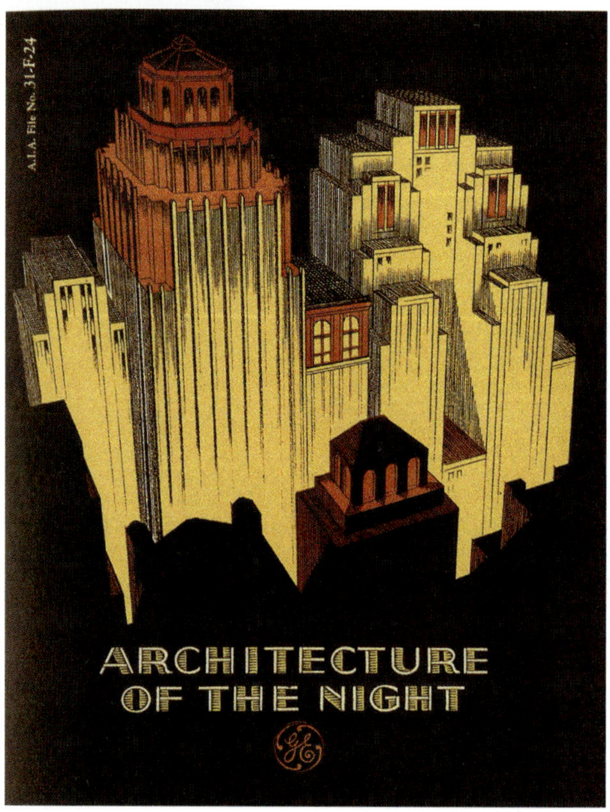

Plate 6. "Architecture of the Night," General Electric Company Bulletin GED 375, February 1930. Photograph courtesy of the Burndy Library at the Massachusetts Institute of Technology; Huntington Library and Collections, Santa Monica, California. Fagan Trade Catalogs, box 4, GE folder, GED 375.

Plate 7. Joseph Binder's winning poster on the cover of Official Guide Book: New York World's Fair, The World of Tomorrow, 1939 *(New York: Exposition Publications, 1939).*

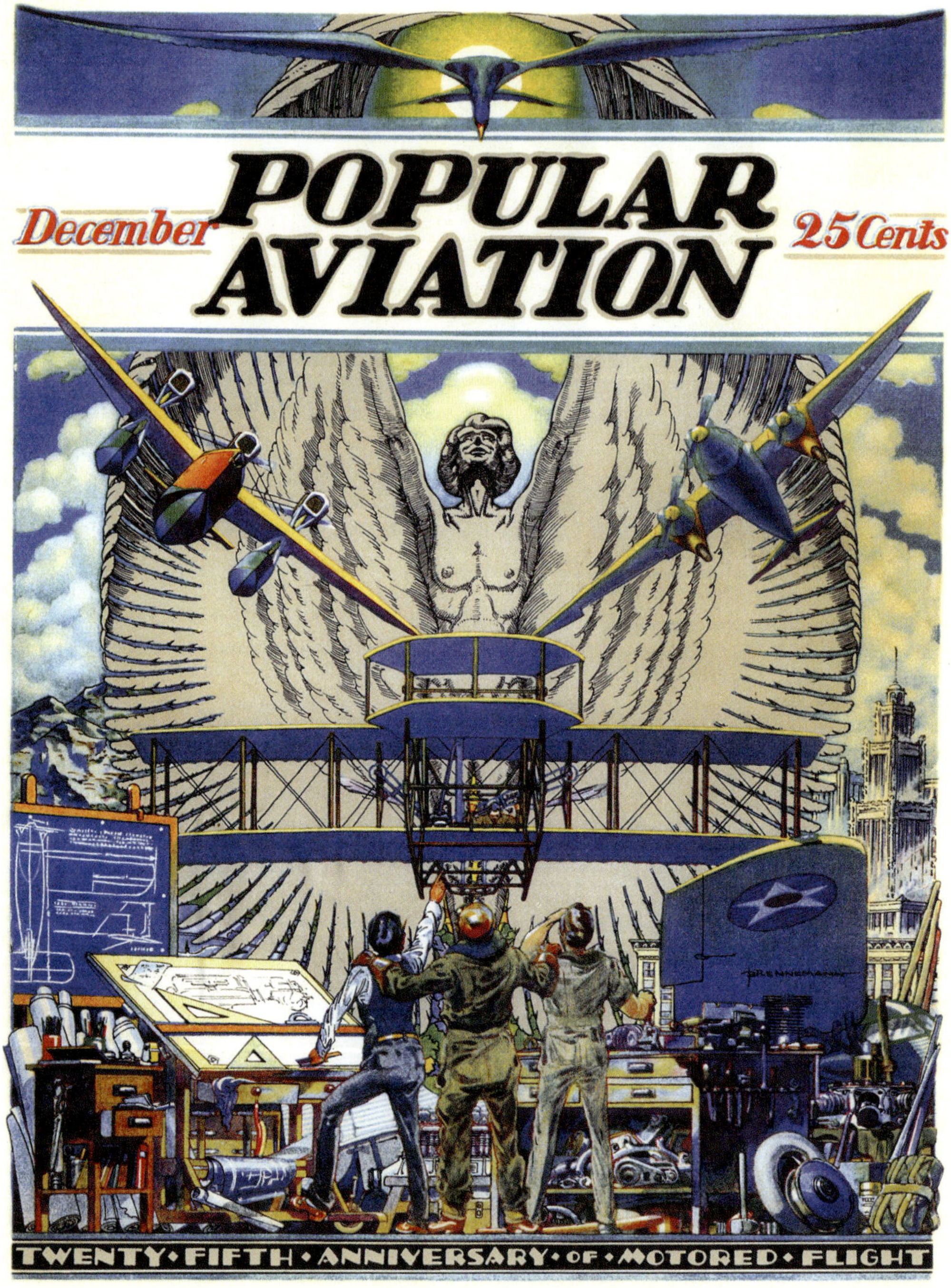

Plate 8. The cover of the twenty-fifth anniversary edition of Popular Aviation *3, no. 6 (December 1928).*

Plate 9. Norman Bel Geddes, *"Aerial Restaurant," on the cover of* Popular Mechanics, *July 1930.*

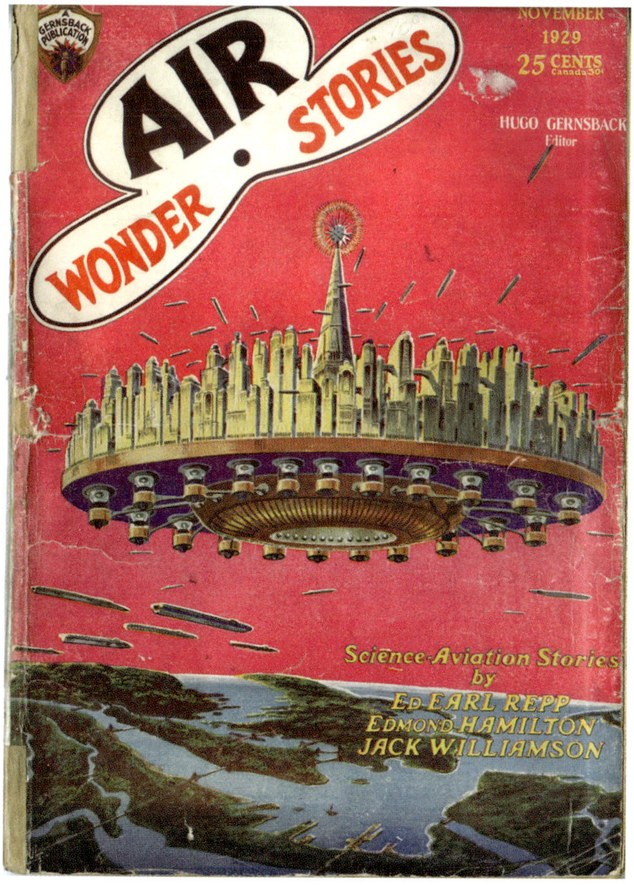

Plate 10. Frank R. Paul, "City in the Air," on the cover of Air Wonder Stories, *November 1929.*

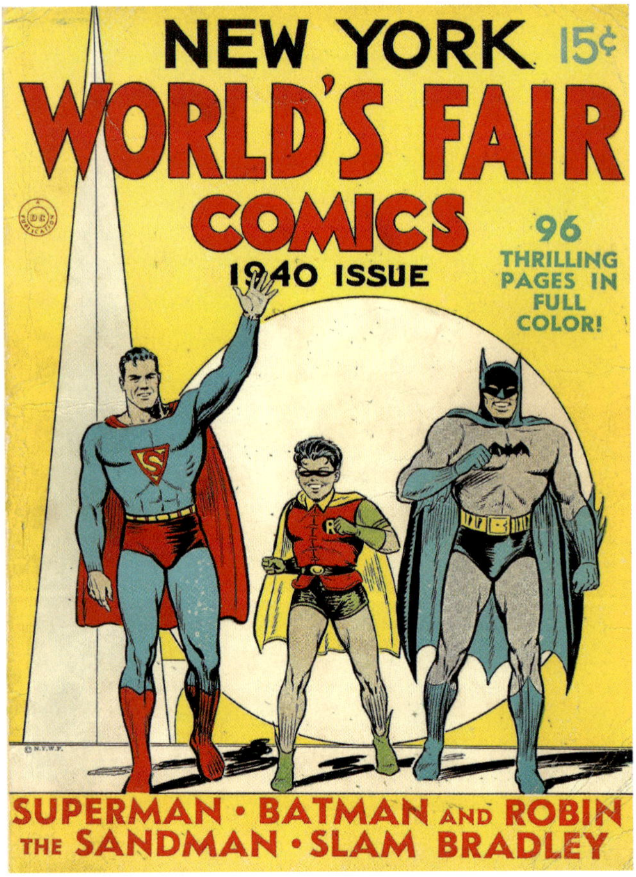

Plate 11. Jack Burnley, "Superman at the World's Fair," on the cover of New York World's Fair Comics, *depicting Superman, Batman, and Robin in front of the iconic 1939 New York World's Fair symbols, Perisphere and Trylon. All characters are trademarked and copyright DC Comics. Reprinted by permission.*

Plate 12. Will Cotton, a satire of Futurama on the cover of The New Yorker, *June 1, 1940. Copyright 2009 Conde Nast. All rights reserved. Reprinted by permission.*

> When I was nine years old the airplane was invented but I did not see one flying until I was fourteen and I did not fly one until I was twenty-three. Along with millions of other boys, I had been trying to invent that airplane, first with paper dart models and then with box-kite-like multi-planed gliders. Despite our elders' doubts and engineering's down-to-earth negatives, imminent invention of the "airplane" was everywhere present in the mind-wind of my pre–Wright Brothers knee-breeches years.[45]

When, in a letter written on May 11, 1928, Fuller told his mother, "I suggest getting rid of all rail holdings and putting the money in Curtis, Wright and other good airplane manufacturers who are on an approved business basis," it was just an instance of how the corporate promotion of the aviation industry permeated both the economy and everyday life in America.[46]

In the fall of 1922, Fuller's first major flying experience along the Northeastern Seaboard provided him with a new angle from which to view the earth.[47] Although commercial air service in America began as early as 1914, flying was still a luxury and a matter of social prestige in the early 1920s; only the privileged had access to air travel before World War II.[48] Fuller's friend Vincent Astor, whom he met when they both served on the crew of a navy patrol ship, provided Fuller with the privilege of flying. Younger than Fuller by one year and with a knack for a flashy lifestyle, Astor hailed from a wealthy New England family and had inherited such a vast fortune that he owned a private airplane in the early days of aviation. Fuller struck an enduring friendship with Astor, and they socialized together at sumptuous parties in and around Newport, Rhode Island, Astor's home territory.

Once, as he left for Paris on family business for a couple of weeks, Astor lent Fuller his flying boat, designed and built by Grover Loening (designer of the first amphibious vehicle). Capable of carrying four passengers and powered by a 400-horsepower engine, the airplane was equipped with all the new aviation technology perfected during World War I (Figure 2.9). With Astor's airplane, along with its pilot, Clifford Webster, at their disposal in the fall of 1922, Fuller and his wife took the opportunity to fly extensively between New York City and Bar Harbor, Maine. Creating a media sensation, they traveled to friends' parties and marriage ceremonies in Long Island, Newport, and Boston in the seaplane (Figure 2.10).[49] From his home base in Bear Island, Fuller made short flights and rediscovered familiar places from a radically different aerial angle.[50] His biographer Alden Hatch describes Fuller's experience thus: "In October Maine is fantastically beautiful, the air cold and crystalline, sky and sea and earth sharply seen to the farthest horizon. Bucky made short flights every day . . . getting new perspectives on the channels and shoals so familiar to him from the water level."[51] Fuller's wonderment at gazing down on the sprawling geography below was captured by the large number of aerial photographs he took during the flights (Figure 2.11).[52] As his expanded vistas of "Connecticut shore," "West end of Cape Cod Canal," and "Ships in Boston Harbor," among others, reveal, no longer was he looking at the minutiae of the earth's surface; rather, he was seeing its geographic architecture.

The ability to see a vast geography from hitherto impossible heights provoked a self-aggrandizing modernist ambition. Le Corbusier epitomized this ambition: "By means of the airplane, we now have proof, recorded on the photographic plate, of the rightness of our desire to alter methods of architecture and town-planning. . . . The airplane instills, above all, a new conscience, the modern conscience. Cities, with their misery, must be torn down. They must be largely destroyed and fresh cities built."[53] For Fuller too, his flight provided a moment of twin introspections. Being able to see the world from the sky, the joy of rediscovering the familiar from a new angle, meant a possibility for rediscovering himself anew. Alongside the aerial photos, he had himself photographed looking out meditatively at the world below, foreshadowing a lifelong fascination with the idea of ascension and its continued association with, and influence on,

Figure 2.9. Buckminster Fuller flew along the East Coast in Vincent Astor's airplane. The Special Collections of Green Library, Stanford University. Courtesy of The Estate of R. Buckminster Fuller.

FLIES TO WEDDINGS

Photo by International News Reel

LIEUT. FULLER AND ASTOR AIRPLANE

Vincent Astor's private mono-seaplane had to be called into service by Lieutenant and Mrs. R. Buckminster Fuller after they found themselves with so many wedding invitations that they couldn't attend all of them any other way. First they flew down to New York to go to a wedding. Then they returned to Boston to attend another. Today they will fly to Glen Cove for still another, and when they get through there they must fly to Lawrence and thence back to New York. The lieutenant is a well-known member of the Naval Reserve, and a close friend of Mr. Astor.

Naval Reserves Under Astor Complete Cruise

NEW YORK, Sept. 9—Commanded by Lieutenant Vincent Astor, New York Naval Reserves of the Sixth Battalion, today completed a two weeks' cruise of New England waters. Lieutenant Astor's squadron was reviewed while at Newport by Rear Admiral Sims, and following an inspection, Rear Admiral Cleaves came aboard, accompanying the ships to Block Island.

The two eagle boats, H 15 and 4, here in command of Lieutenant R. B. Fuller and Frank Cogan.

WISCASSET

Special Despatch to The PRESS HERALD

Wiscasset, Sept. 21.—Wiscasset received two visitors Wednesday afternoon in a way that is novel. Mr. Martin Andrews nephew, Mr. Fuller, and Mrs. Fuller, literally descended upon the harbor for they came by hydro-aeroplane, which was loaned to Mr. Fuller by Vincent Astor of New York. The Fullers are visiting at Mrs. Andrews' home on Parker street.

The funeral of Mrs. Leilla E.

Figure 2.10. Media coverage of Buckminster Fuller's airplane trips. The Special Collections of Green Library, Stanford University. Courtesy of The Estate of R. Buckminster Fuller.

his aesthetic views (Figure 2.12). The photograph foretold what would later become a characteristic Fullerian solipsistic wish for auto-psychoanalysis and self documentation that resulted in his massive archive called Dymaxion Chronofile.[54]

Fuller represents a generation that invested deep faith in technology as an agent of social transformation. His early adulthood coincided with a time when the narrow nineteenth-century social understanding of technology as identifiable machines (such as the steam engine) changed, as Leo Marx argues, to one in which technology represented a broad abstract system of production and social progress.[55] Prior to the twentieth century, many social theorists considered political will as *the* driver of the Enlightenment objective of a just and universal society. Marx observes that in the early twentieth century an all-encompassing

Figure 2.11. Buckminster Fuller's aerial photographs taken during his flight in 1922 in Vincent Astor's airplane. The Special Collections of Green Library, Stanford University. Courtesy of The Estate of R. Buckminster Fuller.

idea of technology replaced political will as the main catalyst of human progress. He notes: "The cultural modernism of the West in the early twentieth century was permeated by this technocratic spirit [which] was made manifest in the application of the principles of instrumental rationality, efficiency, order, and control of the behavior of industrial workers."[56] Fuller was a representative of this cultural modernism invested in a "technocratic spirit."

In the 1920s, American industry was influenced by Frederick Winslow Taylor's thesis of scientific management, which proposed a new model of efficient organization and standardization of both labor and tools for optimizing production.[57] As Mauro Guillen explains, "The overarching idea in scientific management was that of order, one that subsequently captivated the modernist architects because it enabled them to move away from the prevailing eclecticism and to present themselves as organizers, as technocrats who could ameliorate social

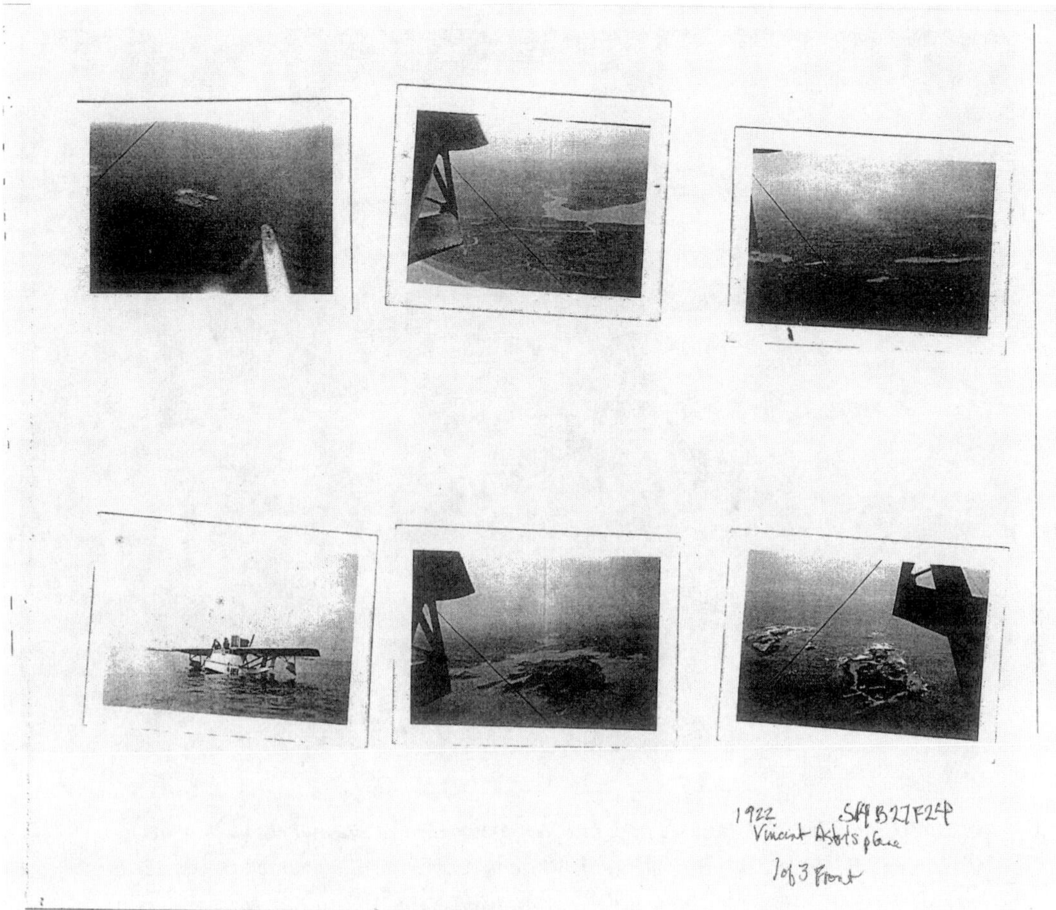

conflict and improve standards of living."[58] The ethos of scientific management was evident in Fuller's design of the 4D House—or the "4-D Utility Unit," as he called it—in which "all functions have been segregated as in [the] human body" so that they could be efficiently assembled on-site like automobiles. The spatial organization of domestic activities, such as working, eating, sleeping, and cleaning, hinged on a Taylorist formula of "logical arrangement of functions, just as much as there is in a factory." Fuller proposed the assembly-line production of the "utility units," which were to be "manufactured and delivered in toto, for modular, manifold hookup and no further adjustment."[59] For Fuller, the transportation revolution exemplified the promise of mass production, and hence the technological model for the house of tomorrow should naturally be inspired by transportation machines.

The technocratic relationship between the house and mechanical locomotion was all too common among the forward-looking theorists of the period. In a 1931 article titled "The House of the Future," the architecture critic Douglas Haskell argued that the construction processes used for modern buildings failed

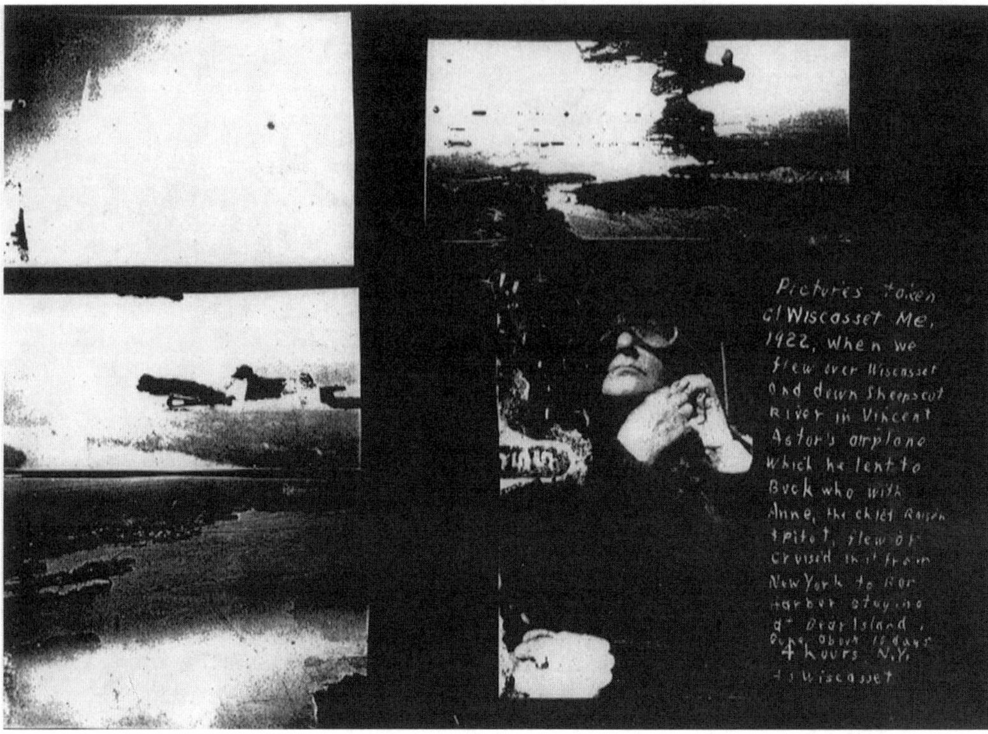

Figure 2.12. Buckminster Fuller observed geography from Vincent Astor's airplane in 1922. His notes read: "Pictures taken at Wiscasset, Me., 1922, when we flew over Wiscasset and down Sheepscot River in Vincent Astor's airplane which he lent to Buck who with Anne . . . & pilot, flew or cruised in from New York to Bar Harbor staying at Bear Island . . . about 10 days, 4 hours, N.Y. to Wiscasset." The Special Collections of Green Library, Stanford University. Courtesy of The Estate of R. Buckminster Fuller.

to incorporate the scientific principles that create the means of modern transportation, such as cars and airplanes—true representations of technological advancement.[60] Fuller expressed a similar attitude through a historic narrative, outlining the contribution of various modes of transportation in bringing geography under a favorable arrangement for humanity:

> The second great adventure of nature after detaching itself from the earth sphere, to which it had been rooted as are trees, floated away by the deluge, and finding its epitomy *[sic]* in man, was the conquest of the second or liquid sphere of the earth. Boats, and their river and sea commerce, opened up the second stage of progress. Such we found its zenith in England's control of the world. The third great era of man is his conquering of the air or vapor sphere, always expanding outward.[61]

Within the culture of technological utopianism and "expanding outward," transportation machines promised a kinesthetic future of human habitation, as "the airplane was but one variant of the technology that sundered protective frontiers and created new spatial dynamics."[62] Fuller sought to develop, within this kind of spatial description, an operative strategy wherein he could connect human flight not only with the issues of social advancement and a "popular revolt against [the] monopoly dominance" of the corporate–market nexus but also with his own pursuit of a worthy cause to live.[63]

If Fuller's own hand in his legend has been studied, what has not been examined are the discursive ways in which Fuller employed certain types of imagery to visualize his "transcendental" cause to live.[64] The mystery is not that Fuller sought to resuscitate his life out of the morass of personal failures and tragedies, but *how* he did that. Even if his fabled attempt to end his "miserable" life was a retroactive concoction on his part, he did indeed display a sudden burst of creative activity between the summers of 1927 and 1928. It explains less a failed entrepreneur still trying to make it than an embattled persona seeking to come to terms with himself. Whatever the reason for Fuller's sudden outpouring of creativity, an examination of his work itself helps us decode the psychology of his autoconstruction. Fuller consciously developed an aesthetic of ascension that served his cause to serve humanity. The image of the airplane and the aviator, the allegory of heights, and references to Lindbergh and other protagonists of history—all helped him develop such an aesthetic theory. In his attempt to build a cogent autobiographical narrative, Fuller tapped into the cultural veneration of the aviator as the so-called apotheosis of a new age, an evolutionary icon that would provide him with a model for his self-definition.

Fuller's drawings from the 1920s are particularly instructive. The *4D One Ocean World Town Plan* (circa 1927)—included in *4D Time Lock*—represents Fuller's prescient astronaut's gaze that scans "85% of all Earth's dry land" (Figure 2.13).[65] The drawing presents a parable of the future, tinged with humanist ideals, in which aerial transportation would enable the maximum usage of Earth's dry land by simultaneously decentralizing human habitations across the planet and connecting them. Fuller's omnidirectional "god's eye," as if perched on a space-ascending machine, would be capable of doing away with all political-national boundaries by dotting such improbable locations as the Arctic Circle, Alaska, the Amazon forest, the Sahara, Siberia, and China with what he called 4D towers. Each of these towers would consist of ten floors suspended from a central mast. Apt symbols of humanity's planetary mobility, Lindberghian monoplanes would swarm the skies of Earth and aerially connect the 4D towers with existing population centers. Fuller's biographer Lloyd Steven Sieden explains:

> [This drawing] portrayed not only radical housing concepts, but a revolutionary method of viewing the entire environment,

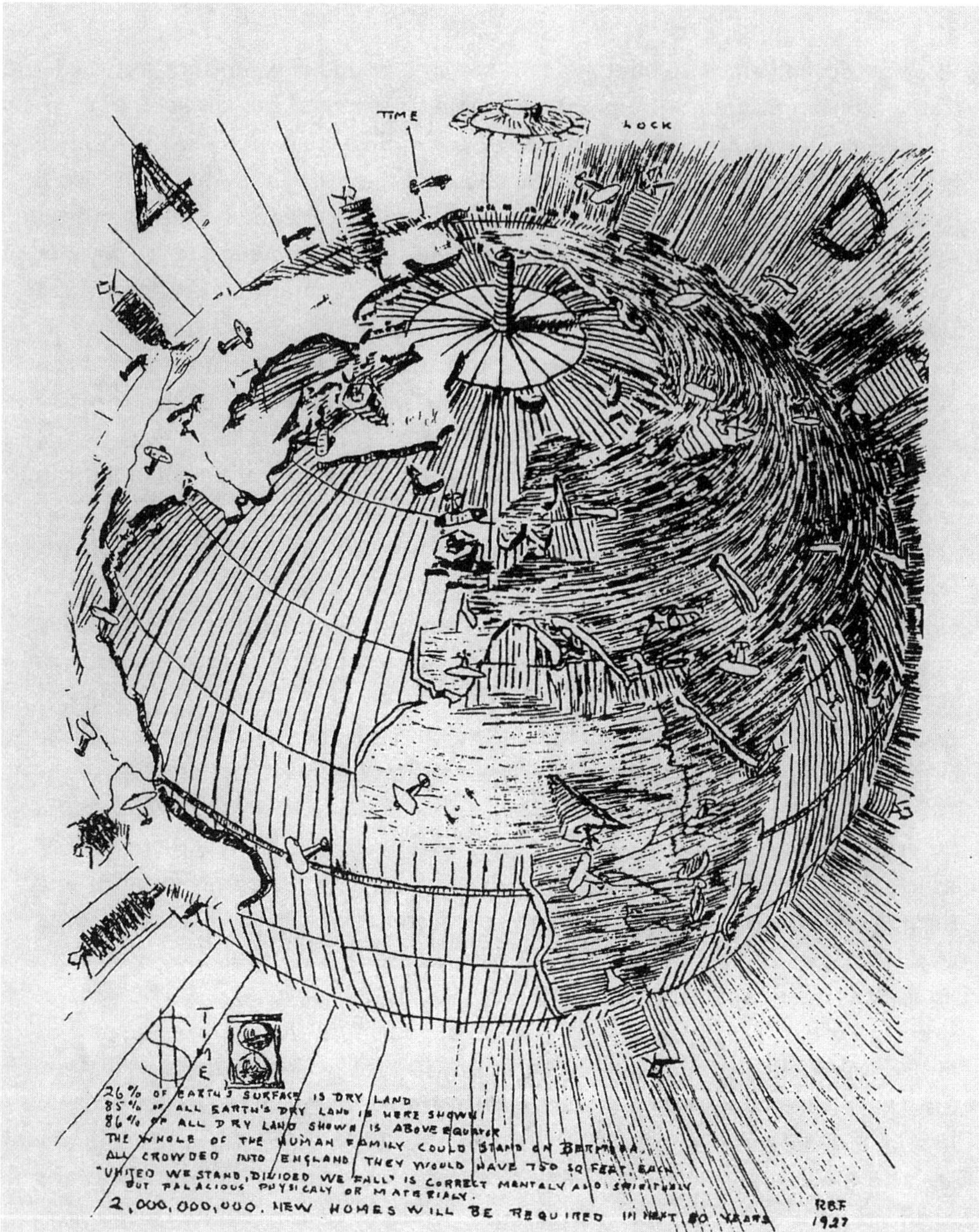

Figure 2.13. R. Buckminster Fuller, 4D One Ocean World Town Plan *(1927). His caption reads: "26% of Earth's surface is dry land. 85% of all Earth's dry land is here shown. 86% of all dry land shown is above equator. The whole of the human family could stand on Bermuda. All crowded into England they would have 750 sq feet each. 'United we stand, divided we fall' is correct mentaly and spiritually but falacious physically or materialy. 2,000,000,000 new homes will be required in next 80 years." From R. Buckminster Fuller,* 4D Time Lock *(Albuquerque: Lama Foundation, Biotechnic Press, 1929/1972), 111. Courtesy of The Estate of R. Buckminster Fuller.*

> especially Earth. Fuller felt that in order to operate effectively in the largest context, he should consider the entire Planet and search for large patterns operating throughout the Universe when examining any issue. He also believed that in order to work with the entire Earth, he had to be able to view as much of it as possible on a single sheet of paper with no visible distortion. . . .
>
> Because Bucky sought to show as much of the Earth's landmasses as possible in a single drawing of a spherical Earth globe with no visible distortion, he spent hundreds of hours determining the exact position on a globe from which the most land can be viewed. Like most of his work, that formidable task also became a learning experience as Bucky used it to acquire a true "feel" for the Earth as a solitary spherical Planet rather than a divided amalgamation of nations and regions.[66]

Fuller imagined in 1927 what Le Corbusier experienced in real life in 1929 during his flight over the South American continent. Having experienced the vast geography of the delta of the Parana, the estuary of the Rio de la Plata, the Pampas, Buenos Aires, São Paulo, and Rio de Janeiro from an airplane, Le Corbusier made a connection between his gliding "like a bird over all the bays" and "the ideas of modern planning."[67] If a type of modern planning was inspired by a desire for grand interventions from heights, as Le Corbusier's scheme for Brazil shows, Fuller's drawing foreshadowed "globalization," with many of its present-day ramifications, by broadening the nature of such interventions to a planetary scale. This change of scale in Fuller's drawing was couched in what Emily Rosenberg has called twentieth-century America's "postfrontier" political urge to venture out beyond its own shores.[68] Fuller's futuristic foibles sought to engage the whole of humanity and conceptualize it within the scope of planetary geography, a concern well served by his godlike viewing angle in the drawing.

In another drawing, *4D Lightful Towers and 4D Transport* (circa 1928), the planet became a tabula rasa for the contemplation of housing by means of strategic distribution of 4D towers (Figure 2.14). The monoplane as an enabler of a "trackless" transportation system created an invisible grid covering the whole planet. Fuller's drawing seemed to float in a kind of nongravitational space, accentuated by two icons of modernity: the skyscraper and the airplane.[69] They represented new possibilities of inhabiting the planet and new forms of mobility. Fuller's aerial gaze stripped the earth of its essential geography, turning it into an experiment that paralleled his solipsistic view of his own life as a clean slate. As with his own life, he wanted to redraw the planet's potentials. To this end, Fuller employed an aviator's perspective on the world. He wrote an "epic poem on the history of Industrialization" (1939), employing the figure of the aviator to narrate the modern world in the twentieth century.[70] Much later

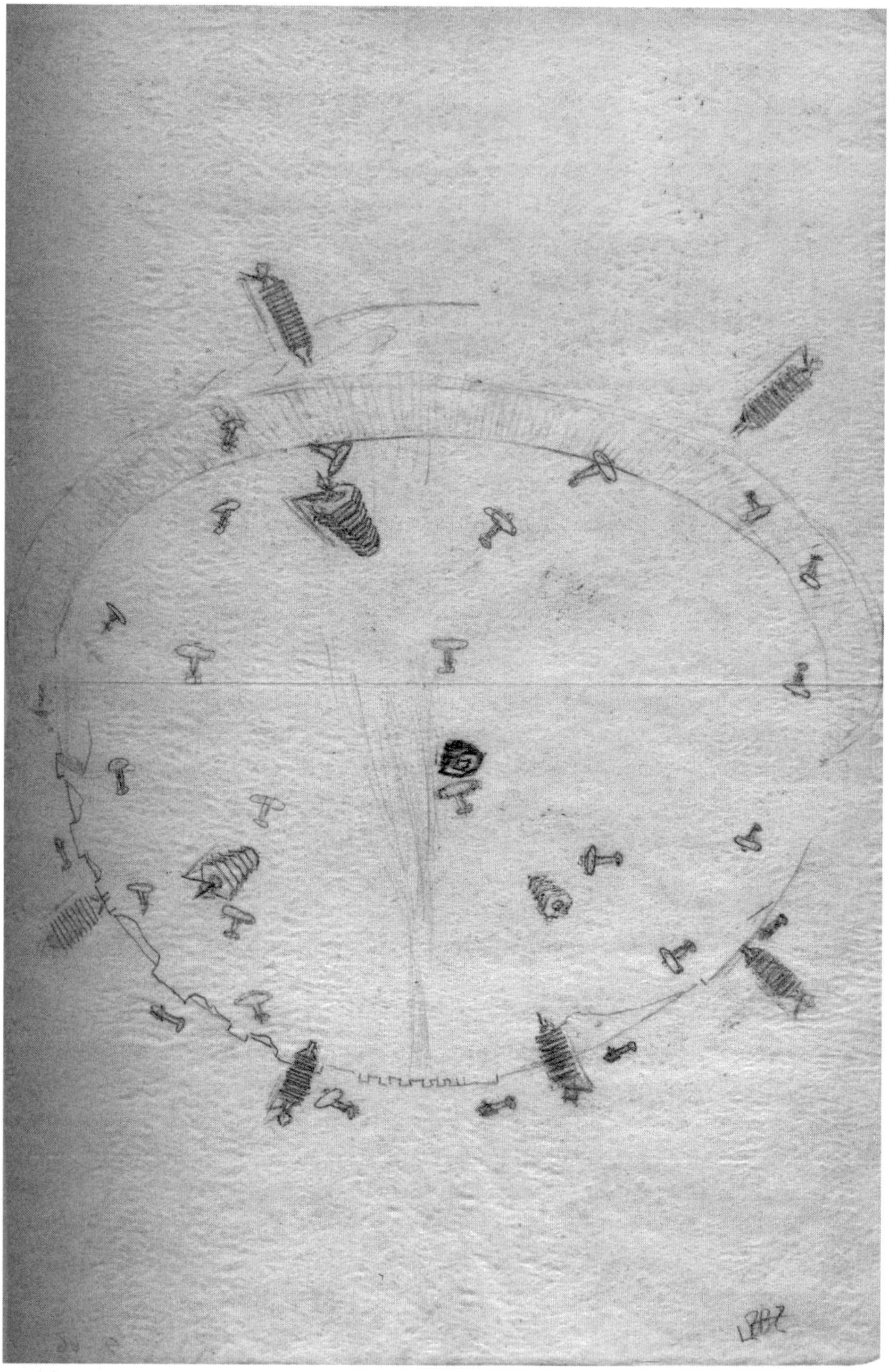

Figure 2.14. Buckminster Fuller's drawing of the Earth (circa 1927). Reprinted in K. Michael Hays and Dana Miller, eds., Buckminster Fuller: Starting with the Universe, *A Whitney Museum of Art Book (New Haven, Conn.: Yale University Press, 2008), 81, plate 5. Courtesy of The Estate of R. Buckminster Fuller.*

in his life—during the heyday of lunar expeditions in the late 1960s—Fuller still viewed the aviator as a distillation of collective human efforts: "Self-sustained directional control of airborne man represents the integrated product of a fabulous history of personal investment by innumerable individuals in daring sequences of single-handed and necessary failures, attendant upon undertakings only warrantable to the individual himself."[71] For Fuller, the airborne man—even though a product of collective human ingenuity—ultimately presented the epitome of the self-driven individualist.

This particular view provided the foundation for Fuller's ideation of the genius; his take on Leonardo da Vinci (about whom Fuller read in the 1920s) is a case in point.[72] Leonardo was a product of the human family, yet his genius was expressed uniquely in *his* ability to dream the impossible. He was, therefore, the summation of the collective and the individual. Fuller applied the same concept to Henry Ford, "the greatest artist of our day."[73] His idolization of Leonardo and Ford was strategic and, to some extent, self-serving, for his portrayals of them were wistful versions of Fuller's portrayal of himself.[74] Not only did Fuller seek out models to emulate, but he also sought to develop a personal theory of evolution. The genius's ability to see things in totality from above was a sign of his supreme cognitive advancement. In the chapter "Genius and Talent" in *Nine Chains to the Moon,* Fuller discussed what could be called an evolutionary triad comprising the genius, his hereditary genealogy, and his eye capable of multiple perspectives:

> The GENIUS, as discovered by genetics, is characterized by a combination of highly divergent physical life cells that are representative of widely cross-bred parent chains. These cells engage in a ceaseless polar tug-of-war for dominance of the specific human offspring and the result is a dual or multiple personality manifestation. . . .
>
> Dual or multiple personality provides, then, two or more viewpoints,—equivalent to the two eyes of a range finder, an instrument which mechanically widens the distance between the two human eyes; or the multiple eyes of the Fairchild aerial camera. . . . Genius's dual or multiple personalities may be said to be representative of a breadth of viewpoint, more-than-average, high worldly, and having an exquisite sense of Timeliness.[75]

Fuller's statement was tinted with autobiographical hues. He found in the genius's purported superhuman perspective a comforting refuge from his own sense of personal failures. Embracing a larger-than-life cause appealed to him as a ruse for both allaying his anxiety and establishing himself as a public intellectual befitting his pedigree. But this argument was much more ambivalent than merely an escape into an autobiographical fictionalization of his grand role. His strategic autoconstruction was certainly aided by the rebellious bones that he already had in his body.

FULLER'S POLITICS OF STANDARDIZATION

Fuller's desire to standardize his design for a single-family house was shaped by various ethical imperatives, ranging from a principled commitment to technological advancement (for which the airplane was a shining example) to the moral responsibility of supporting his family and, more broadly, remedying prevailing social pathos. Even though the project was presented to the public as a cure-all for what Fuller had thought was a pervasive lack of health and hygiene in the American household, it was purportedly triggered by a tragic event in the Fuller household. Fuller claimed that his young daughter's death as a result of living in poor "tenement housing" in New York City inspired him to think about an ideal house of the future: "Alexandra died in 1923, and I couldn't help feeling somehow responsible, that if she had had a proper environment she would have lived."[76] Two variants of the same ideas on home, "Cosmopolitan Home Corporation: Lightful Products" and "Lightful Houses," emerged from three months of disciplined work in the aftermath of his "crisis."[77] These written notes were the basis for what he called the "4D essay" and the 4D House (the precursor to Dymaxion House).

It is worthwhile to examine briefly Fuller's idealistic attempt to create a single-family house against the reformist culture of the 1920s. His prophecies about the liberation of the domestic environment from its fixation on ground, its standardized modular components for efficient construction, the lightness of its structure, and the full automation of all household chores were part of a modernist fascination with the creation of an ideal "tomorrow." As if echoing the excesses of the Jazz Age, in the midst of an "endless parade of predictions about the shape of homes to come," the term *home of tomorrow* was invested with a host of extravagant ideological intentions.[78]

The fertile imagination of the home of tomorrow can be assessed against a number of societal factors.[79] First, the growing urban population in early twentieth-century America caused an acute shortage of housing in cities. During this time, a shift occurred in the diagnosis of degrading urban tenement housing, from putting the blame on the poor to a focus on unhygienic environmental conditions that allegedly resulted in various social pathologies and diseases.[80] Second, there was a general consensus that a predetermined, stylistic approach to home building had compromised the housing industry for too long. Third, the technological revolution, especially scientific management, inspired many modernist designers to seek out new solutions to the housing problem through standardization. And finally, a consumerist psychology ushered in the teleological belief that if the new house could be presented as an icon of technological advancement—somewhat in the vein of Le Corbusier's dictum, "A house is a machine for living in"—it would appeal to the burgeoning mass market. In all this, many social reformers considered a good house design to be *the* solution to a host of urban problems, especially the "decaying character" of urban dwellers. As Gwendolyn Wright notes:

> The abysmal poverty, disease, and discontent of the inner city were attributed to overcrowded tenement dwellings. Those who wanted to uplift the victims of poverty now viewed tenement-house reform as a key to changing the residents' lives. Housing conditions were evidence of every failing of character, the cause of every social problem, and the surest path to improvement. "Home improvement" for the urban poor, like "home improvement" for the middle class itself, was considered the direct route to virtue; bad home environments were the inevitable road to despair.[81]

This sentiment pertaining to the unhygienic environment as the root cause of various urban ills was pervasive among city reformers. Fuller, however, enlarged the scope of the problem by linking poor urban conditions to formalist and historicist architecture's failure to respond to "real" urban crises. The adaptation of advanced technology to solve housing problems would reverse the so-called downward spiral of society. The 4D House exemplified this faith in technology. Fuller wrote to his sister in August 1928, "Ferro-concrete architecture may be likened unto the plastic cocoon of the archaic worm from which will emerge the 4D butterfly."[82]

In May 1928, Fuller presented the 4D House to a small architectural audience at the American Institute of Architects' annual convention in St. Louis.[83] The AIA leadership ridiculed his theme of standardization in the 4D House, favoring autonomy of creative expression over prototyping as the true calling of their profession. Fuller characteristically turned the disfavor into an inevitable consequence of a self-righteous crusade against establishment dogmas. Ironically, St. Louis—etched into his mind as the unforgettable site of his first public rejection—had earlier been the totem in his call for a new *Spirit of St. Louis* in architecture. Fuller's search was reminiscent of Lindbergh's own pursuit of sponsorship for his flight from St. Louis's business community, which sought to advertise the city as the epicenter of American aviation.[84]

After returning from St. Louis to Chicago, Fuller mimeographed two hundred copies of his 4D essay and sent them out, unsolicited, to a large number of recipients, some of whom were familiar to the author and some of whom were not. His avowed goals in this mission were to introduce his vision for an

American house, to garner broad support for his ideas, and to network with influential people and potential financiers.[85] Important recipients of Fuller's essay included Bertrand Russell; Henry Ford; the author Jean Toomer; architects Raymond Hood, Harvey W. Corbett, Claude Bragdon, Hugh Ferriss, and Russell Walcott; Harvard president A. Lawrence Lowell; University of Chicago president Max Mason; former AIA president Thomas R. Kimball; and the translator of Le Corbusier's *Towards a New Architecture,* Frederick Etchells (Figure 2.15). The range of recipients—from architects and medical doctors to authors and philosophers, from industrialists to academics—suggests that Fuller was interested not merely in presenting a business plan or an architectonic solution to the housing crisis but also in engaging a diverse intellectual audience to promote his ideas on human progress.

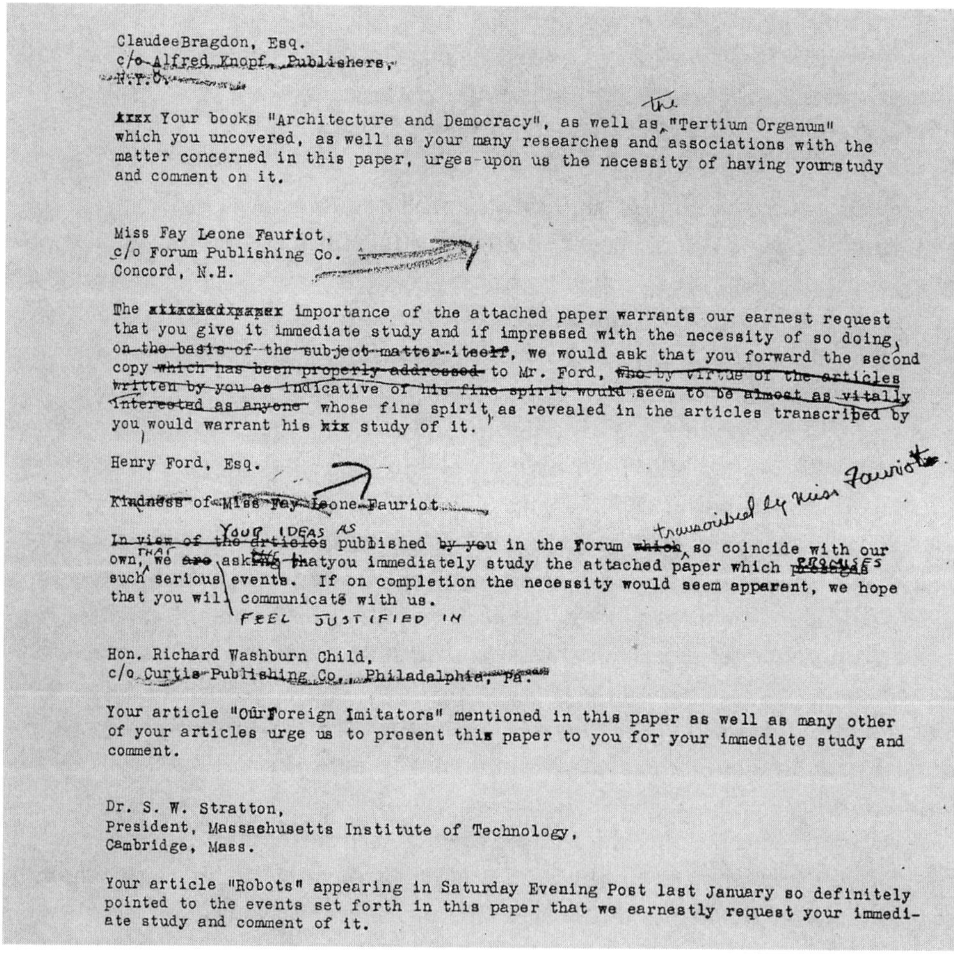

ClaudeeBragdon, Esq.
c/o Alfred Knopf, Publishers,
N.Y.C.

xxxx Your books "Architecture and Democracy", as well as the "Tertium Organum" which you uncovered, as well as your many researches and associations with the matter concerned in this paper, urges upon us the necessity of having your study and comment on it.

Miss Fay Leone Fauriot,
c/o Forum Publishing Co.
Concord, N.H.

The xxxxxxxxxxxxx importance of the attached paper warrants our earnest request that you give it immediate study and if impressed with the necessity of so doing, on the basis of the subject matter itself, we would ask that you forward the second copy which has been properly addressed to Mr. Ford, who by virtue of the articles written by you as indicative of his fine spirit would seem to be almost as vitally interested as anyone whose fine spirit as revealed in the articles transcribed by you would warrant his xxx study of it.

Henry Ford, Esq.

Kindness of Miss Fay Leone Fauriot.

In view of the articles YOUR IDEAS AS published by you in the Forum which transcribed by Miss Fauriot so coincide with our own, THAT we are asking that you immediately study the attached paper which xxxxxxx PRECIPITATES such serious events. If on completion the necessity would seem apparent, we hope that you will FEEL JUSTIFIED IN communicate with us.

Hon. Richard Washburn Child,
c/o Curtis Publishing Co., Philadelphia, Pa.

Your article "OurForeign Imitators" mentioned in this paper as well as many other of your articles urge us to present this paper to you for your immediate study and comment.

Dr. S. W. Stratton,
President, Massachusetts Institute of Technology,
Cambridge, Mass.

Your article "Robots" appearing in Saturday Evening Post last January so definitely pointed to the events set forth in this paper that we earnestly request your immediate study and comment of it.

Figure 2.15. Buckminster Fuller's edited correspondence list for the "4D essay." The Special Collections of Green Library, Stanford University. Courtesy of The Estate of R. Buckminster Fuller.

The 4D essay was part autobiography, part manifesto embedded in a cluster of impassioned ethical aporias, and part advertisement aimed at garnering sponsorship for Fuller's business plan to build an industrially reproducible prototype house. It gave the impression of a young provocateur's tirade against the bourgeois stylism of the American housing industry, as well as the profit-driven corporate–banking nexus that produced it. There existed, Fuller noted, a fateful relationship between the overcrowded and unhealthy condition of the American home and the "herding instinct of the animal man."[86] If civilization was going through a crisis, it was perpetuated by a fundamental misunderstanding of the idea and role of home as a microcosm of modern life. Home was where the problem was, so fixing the problem would reverse the degeneracy of civilization.

At the epicenter of Fuller's contention was an attempt to articulate a panoramic theory of the ideal development of humanity: "The serious matter of creating an industrial house that shall at the same time possibly compass and effect *[sic]* millions of lives in what they have to do and what they choose to do, effect *[sic]* the upbringing of millions of children, demands of us the very broadest thought which we can master."[87] This intervention in the upbringing of humanity was based on an ethical imperative to build "character." Fuller wrote: "Until people dare to stand on their own and have the courage of their own convictions and express these convictions to the world, they will never acquire character until this is done."[88]

Fuller's fundamental contradiction was that he viewed the temporal adjustment of the concept of the house with an evolving technological zeitgeist as a way to achieve an immutable final condition, "character," his own metonym for "truth." Or, more broadly, humanity—its developmental potential resting on the core idea of the house—was constantly adjusting itself within given circumstances to eventually attain perfection, be it character, truth, or a utopia. But does not an irrefutable causality, or telos, nullify the very concept of change through which the final condition has been achieved? This was one of the great ironies of the progressivist evolutionary discourse.[89] The contradiction of telos and change permeated both the opposing camps of the debate. On one side, the Aristotelians believed in a teleological model of change driven by a suprarational purpose that pushes change toward a discernible goal. On the other side, the proponents of empiricist scientific revolution argued that change occurs by happenstance, or without what Bacon called a "final cause." Darwin contradicted himself when he straddled the two camps: on one hand, the endless adaptability of species and, on the other, "all corporeal and mental endowments will tend to progress towards perfection."[90] Hegel's theory of history as a perpetual march of reason toward the finality of modern freedom, in contradistinction to his dialectical logic, faces a similar epistemic riddle.[91]

Fuller confronted this riddle by embracing American industry's fledgling mantra of standardization and filtering it through an ethical basis of reproduction.

Although he acknowledged the popular apprehension of standardization—for it seemed antithetical to the individualist ethos—Fuller saw in the project of standardization a panacea for the problems of civilization. When an object becomes standard, it has followed an arduous path of trial and error, testing out all kinds of adaptive mechanisms, to reach *the* state of perfection. The telos of the object's adaptive history means that it has attained a redemptive status when it could be mass-produced for the greatest good. Analogous to the pursuit of "truth," the standard is the only attainable condition because, for Fuller, there cannot be two simultaneous standards, just as there cannot be two "truths." He wrote:

> There is only one truth, always, [that] is revealed in the multitudinous functions of living just as there is only one straight line which is the shortest distance between two points, the public adopts this one way of doing the thing best, and when enough people have recognized the truth to warrant someone investing capital in the production of a mechanical means of performing that true function, we then have standardized production.[92]

A contributor to the magazine *House Beautiful* wrote in 1929: "As Mr. Fuller would insist, standardization rightly understood is not the imposition of a dully identical pattern on ever-various life. Before a standard can be attained, an ideal must in some part have been captured. Standardization then repeats economically and in quantity as much of the ideal as has been mastered, making it available for large numbers of people, and providing a base for advance."[93] In other words, the process of standardization enables the mass production of an "ideal" condition that would otherwise not be democratically accessible to a vast majority of people.

The notion of an ideal condition, as in standardization, had a peculiar social Darwinist subtext. Just as the human body continually adapts itself for the fittest survival in nature—reflecting the climactic stage of natural selection, or nature's standardization—the best architectural or mechanical design similarly represents the most advanced process of adjustment to a technological world, thereby warranting its standardization. In one way or the other, biological and mechanical truths are just two expressions of the same search for one Truth, one standard. In his case, Fuller advocated human agency as a necessary catalyst in the search for an architectural standard. If nature perfects a species by eliminating all things that stand in the way of its most efficient adaptation to the environment, Fuller argued that designers must proactively remove all physical anomalies, or what he called the "drudgery" of the designed object. In a biological context, drudgery meant the abnormal obstruction and resistance points in the path of the human body's evolutionary mobility.[94] Drudgery was a code for speed-reducing, filth-catching, and undesirable protrusions in the body's exterior. Therefore, the drudgery-free morphology of a species reflected its most advanced evolutionary stage after nature had experimented with all possible forms and shapes for its most efficient survival in a competitive environment. To eliminate the drudgery of a body or an object is to *streamline* it, accelerating its evolutionary motion toward its final, standard form.

The fact that the theoretical roots of streamlining lay in aerodynamic principles associated with aviation technology buttressed Fuller's advocacy of standardization in the American housing industry.[95] Fuller strengthened his discourse of standardization by repositioning it within what many modern design theorists considered a natural analogy between architecture and transportation machines during this period. Sheldon Cheney identified the airplane as the mechanical "standard" in the evolution of technological truth, which was merely the material condensation of irrefutable natural laws.[96] When mechanical truth was achieved through the replication of a bird's principles of flight, an archetypal human dream became possible by means of a heavier-than-air machine. The airplane was, therefore, the consummate result of both cosmic and mechanical truth. Fuller echoed this sentiment: "The aeroplane continually approached perfection, and as it approached perfection by the process of applications of truth so has it approached one final design."[97] In Fuller's rhetoric of evolution, the "final design" could be standardized and reproduced for maximum distribution among the masses.

His views on standardization confronted the binary opposition of evolutionary changeability on one hand and immutable perfection on the other. Fuller imbued the concept of standardization with an ethical program aimed at harnessing perfected technology for the optimal benefit of humanity. He saw the possibility of mitigating the fear of losing autonomy of creative expression due to the advent of industrial mass production through the standardization of a singular "perfection." He held this to be the reflection of the highest form of autonomy, or überautonomy so to speak, of the type or the self. He sought to canonize the "final design" with this überautonomy, couched in a humanitarian mission to reform society. With the industrial mass production of the ideal type—be it a house, an airplane, or the human body—society as a whole would reap the benefit of the best product. If Fuller's ambitious ethical-humanitarian program behind standardization promised lofty solutions to the problems of civilization, it was not surprising that he marketed his own single-family house for standardization by virtue of its technological superiority. His genius lay in conflating his architectural proposal with such uplifting philosophical grandeur (for which he used the airplane as a metaphor) that the boundary between his personal goals and civilization's purported goals dissolved, creating a singular narrative of renewal.

A DYMAXION FLIGHT

Le Corbusier's cri de coeur toward the "eyes which do not see" the great potentials of transportation machines for architectural development set a tone of machine aesthetics in the 1920s. In *Towards a New Architecture,* Le Corbusier rejected the myopic "eye" of historicist architecture in favor of a new perspective that could visualize the functionalist logic of the airplane as the basis of a modern house:

> Let us look at things from the point of view of architecture, but in the state of mind of the inventor of airplanes. The lesson of the airplane is not primarily in the forms it has created, and above all we must learn to see in an airplane not a bird or a dragon-fly, but a machine for flying; the lesson of the airplane lies in the logic which governed the enunciation of the problem and which led to its successful realization. When a problem is properly stated, in our epoch, it inevitably finds its solution. The problem of the house has not yet been stated.[98]

Fuller read *Towards a New Architecture* immediately after its 1927 English translation was published and absorbed its core ideas.[99] In fact, he sent out his 4D essay to the translator of the book, Frederick Etchells, soliciting comments from both Le Corbusier and the translator with a kind of great-men-think-alike hubris: "Your splendid introduction to Monsieur Le Corbusier's 'Towards a New Architecture' leads us to forward the attached confidential paper which presages most serious events. We would appreciate your comments as well as those of Monsieur Le Corbusier."[100] In *4D Time Lock,* Fuller's lament about architecture's failure to create the modern house based on the utilitarian principles of transportation machines reflected a Corbusian rhetoric. Fuller wrote:

> In hospitals, moving picture theaters, the modern drugstore, restaurant, ocean liners, airplanes, etc., where the need for expeditious and healthy handling of masses of people and catering to their needs takes place, we find great improvements can only be applied to the house when complete redesigning of the building takes place, through centralized economic authority. In pursuit of these concepts it is possible to research, analyze, and design a harmonious and efficient home, with mechanical solutions from submarine, airplane, hospital and theater.[101]

Both Le Corbusier and Fuller saw the machines of mobility as instrumental to a new world order, a new system of organizing the world for the maximum optimization of its resources. Their focus on transportation machines was not just about a simple transferal of locomotive technology to housing design, but about a much larger epistemological argument favoring mobility over the sedentary, lightness over the heavy, and functionalism over a bourgeois attachment to historical style.

In this context, Fuller's neologism *lightful* acquired a particular autobiographic flourish, wherein weightlessness was a symbolic act of flying away from what Italo Calvino called the claustrophobia of conformist life patterns.[102] *Celeritas* offered an alternative to *gravitas*. This view in many ways fueled the avant-garde search for "lightness" in modern architecture.[103] Many modernist architects considered lightness of materials and construction as signs of a building's efficiency, fast constructability, elimination of pastiche decoration, and, more conceptually, a welcome release of architecture from stifling historical affiliations. Heaviness, on the other hand, was deemed antithetical to modernism's aesthetic fluidity. The concept of lightness was sometimes incorporated in architecture through the use of glass, columns as opposed to walls, and a philosophical reconsideration of how a building conventionally stood on the ground. Glass facades ensured "dematerialization" of architecture through transparency, while "floating" architecture on stilts simulated a decentered space that released the building from what Schopenhauer called architecture's archetypal modulation of burden and support.

Weight was antithetical to Fuller's worldview. He considered weightiness—his metonym for conformism, historicism, and earth dependence—an impediment to *his* world. He wrote: "There is no virtue in weight for itself. The progression of humanity is from stony darkness of complete and awful weight, to eternal light which has no weight."[104] Lightness meant less dependence on the ground, implying increased mobility toward standardization. Fuller's views on mobility—which the Italian Futurists had already made an unmistakable image of radical progress—were directly proportional to his views on the reduction of weight. Thus, lightness and mobility offered an opportunity to create a new type of foundation for architecture, and by extension civilization, by rejecting the very notion of a foundation. "Just as in the aeroplane and automobile elimination of weight, which can only produce friction, and spell inefficiency in transportation and all goods including housing must be transported, so will the weight of materials become more and more an important factor in civilization. Materials will be used by virtue of their weight and their successful fulfillment of a set function."[105]

Within Fuller's idealistic world, lightness took on a larger cast of human progress. Dymaxion House would weigh only three tons, as opposed to the one hundred tons of the typical American house. In some of his 4D drawings, Fuller criticized architecture's alleged petrification into the "tailor-made archaic contraption" of columns and entablature that had seen "no structural improvement in 5000 years" (Figure 2.16). The satiric lightness of his 4D tower in relation to the

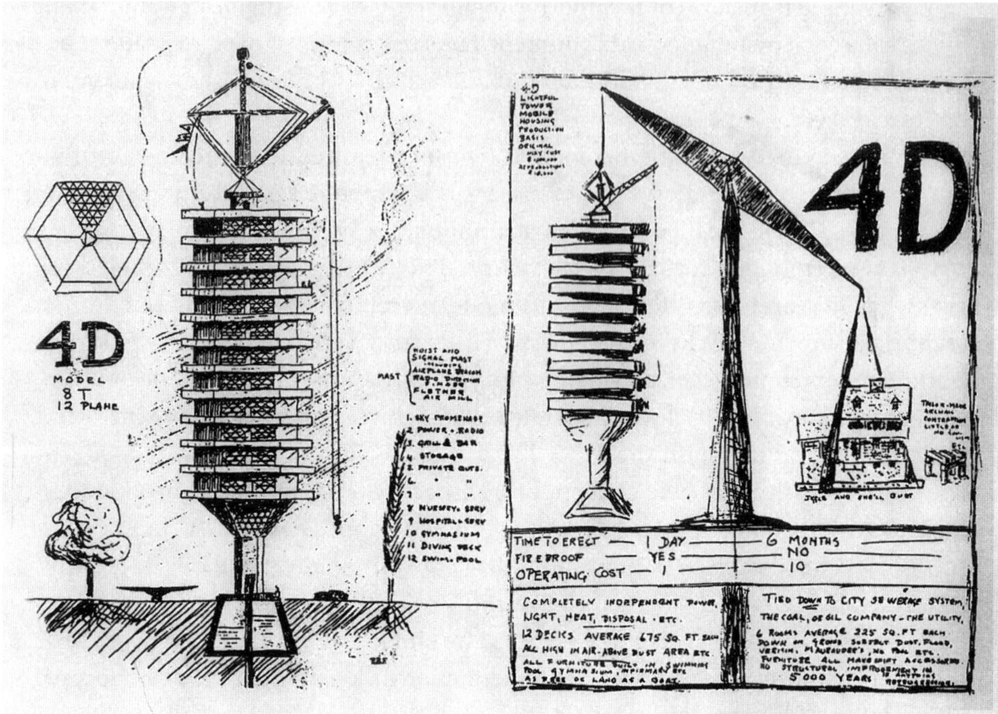

Figure 2.16. Buckminster Fuller's sketch of the 4D tower and a traditional house on a scale (circa 1928). Fuller claimed that it would take only a day to build the 4D tower, as opposed to the much longer construction time of a traditional single-family house. The 4D tower would be "completely independent . . . , light, neat . . . 12 decks average 675 sq. ft. each." In contrast, the traditional house would be "tied down to city sewerage system, the coal, or oil company—the utility, 6 rooms average 225 sq. ft. each. . . . No structural improvement in 5000 years." Reprinted in Joachim Krausse and Claude Lichtenstein, eds., Your Private Sky: R. Buckminster Fuller, the Art of Design Science *(Lars Müller Publishers, 1999), 96. Courtesy of The Estate of R. Buckminster Fuller.*

weightiness of the two-story traditional house on a lopsided scale suggested architecture's potential release from a millennial entrapment while showing the 4D tower's expeditious constructability anywhere in the world. Dymaxion House did not stand up; it hung like a "sky house" from a duralumin mooring mast by means of high-tensile wire and was "assembled in twenty-four hours, furnished to set the most modern equipment to shame, and produced in mass at $3,000 each, or 50 cents a pound."[106] Minimal contact with the ground allowed the house's constructional mobility and rapid deployment without much dependence on the local utility grid.

With the usual audacity of a radical, Fuller claimed that the prototypical automated house would alter the nature of domestic architecture and the quality of life in and around it by rendering redundant all human labor necessary to perform household chores. Such freedom would then enable the occupants of the house to spend their time not in menial cleaning jobs but in cerebral pursuits, further facilitated by a technologically induced healthy and controlled interior environment. Fuller's rejection of weightiness and orthodoxy was reflected in another drawing. An airplane on the upper left released the "lightened" infant of a future race, most likely a twin reference to Fuller's newborn daughter and to generations to come (Figure 2.17).[107] The Lindberghian monoplane presumably played the role of civilization's womb, from which the "superchild" of tomorrow emerged, as if to herald the arrival of a brand-new era.

But no drawing represented Fuller's avian ambition more poignantly than his comical drawing (included in *4D Time Lock*) in which a mammoth dirigible

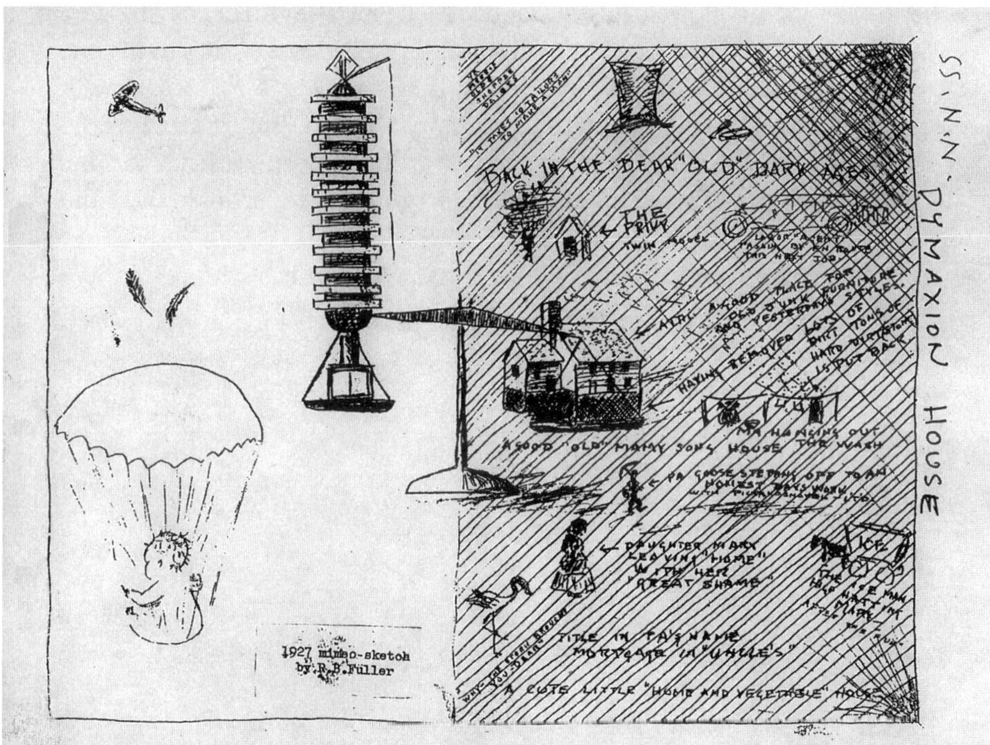

Figure 2.17. Buckminster Fuller's comparison of the 4D tower and a traditional single-family house. In this drawing, Fuller criticized the unchanging patterns of a traditional house and the pastoral household chores that took place around it. The traditional dwelling was "a cute little 'home and vegetable' house" that transported people "back in the dear 'old' dark ages." The 4D tower, in contrast, implied healthy living and a new age, symbolized by a child with a parachute. Reprinted in R. Buckminster Fuller, 4D Time Lock *(Albuquerque: Lama Foundation, Biotechnic Press, 1929/1972), 109. Courtesy of The Estate of R. Buckminster Fuller.*

audaciously planted a ten-deck 4D apartment tower in remote locations as distant as the Arctic (Figure 2.18). The logic for "air delivery" of tall towers, as Fuller pointed out, was to avoid the cumbersome obstructions of natural terrain that would prohibit the movement of large factory-made houses. He wrote:

> In 1927 I thought of the idea of trying to deliver large structures by air to be above all obstacles. I found I could make a ten-deck building so light that it could be carried by the Graf Zeppelin, suspended horizontally under the Zeppelin's belly. The Graf, upon reaching the arctic site, could drop a bomb, make a crater, lower the building into the crater, plant it like a tree and fly back home leaving the building occupied.[108]

Fuller's biographers viewed this outlandish concept as evidence of the eccentricities of a radical designer.[109] Fuller himself provided some clues to his inspirations for the peculiar set of air-delivery drawings. Six serialized drawings narrated a seamless story: the temporal story of the dual but intertwined lives of flying machines and skyscrapers. The drawings began with an incoming dirigible carrying a 4D tower suspended horizontally under its belly. Fuller's

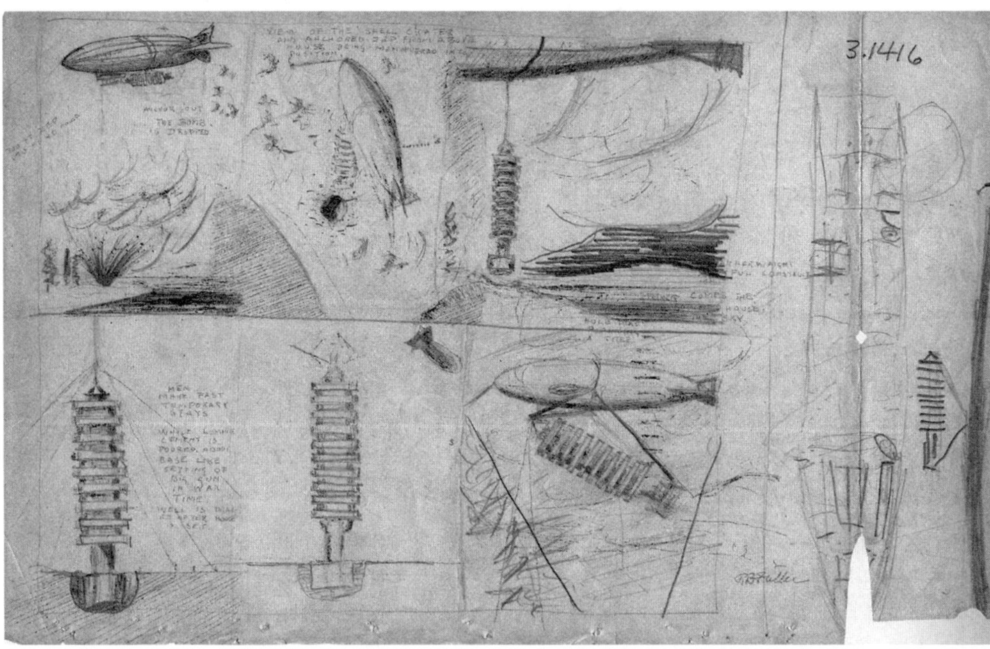

Figure 2.18. Buckminster Fuller's drawing of the air delivery of 4D towers (circa 1928). Reprinted in K. Michael Hays and Dana Miller, eds., Buckminster Fuller: Starting with the Universe, *A Whitney Museum of Art Book (New Haven, Conn.: Yale University Press, 2008), 82, plate 6.*

handwritten caption reads: "Down comes the 4D tower house[s] from the sky, Featherweight 'Lightful Construction.'" The next two drawings showed how the dirigible dropped a bomb to create a crater, and the remaining three drawings detailed the process of stabilizing the tower into the foundation hole as the dirigible returned for its next mission.

It is most likely that Fuller's air-delivery drawings were inspired by the ways aviation permeated the veins of popular culture, especially newspaper comic strips.[110] After World War I, in which the advent of airplanes changed battle dynamics, comic strips with aviation themes swelled to feed the robust public appetite for aviators and their stories of derring-do in the sky (Figure 2.19).[111] Many of the creators of aviation strips were pilots who "flew the real birds," often creating angles of view in their strips that simulated actual flight experience. The first aviation strip in America, *Tailspin Tommy,* appeared in the summer of 1928, in the direct aftermath of Lindbergh's flight. It was drawn by an aviator named Hal Forrest, who had flown planes in World War I and had been a stunt pilot in the movies.[112] Interestingly, the strip's hero, Tommy, engaged not just in the task of flying but also in enforcing a moral code in the aerial world, fighting lunatic scientists, and discovering lost races of humankind. With the subsequent arrival in the comics of such flying heroes as Buck Rogers and Flash Gordon, a heady worldview in which these aviating heroes fought the forces of evil and saved humanity became an integral part of the aviation-strip narrative.

Fuller must have been intrigued by this new art form, which seems to have influenced his drawings. In a lecture in Mexico in 1963, he recounted the social impetus behind what he called "funnies philosophy": "Undeniably the 'funnies' are the most generally inspected portions of our daily newspapers, and may be considered the economic frosting that sells the cake. It is more than significant that these funnies . . . have become serials of handy philosophy."[113] Fuller's funnies chronicling the air delivery of 4D towers, with all their comical pretense of practicality, can be seen as a microcosm of his larger ideas not only

Figure 2.19. The aviation-themed comic strip Tailspin Tommy *(created by Glenn Chaffin and Hal Forrest) was inaugurated in the* Boston Globe *on July 23, 1928.*

on the planetary potentials of aviation but on developing a method of self-analysis. In his air-delivery funnies, as an overseer of the dirigible's daring feat, Fuller himself "flew" around and gazed at the aerial action from unexpected angles. In the second panel, he hovered over the dirigible itself like an omnipotent force and appeared to be bringing the dirigible-tower-crater axis into alignment from a godlike point of view.

Fuller's longing to circumnavigate the dirigible itself was awash in a typical Fullerian desire to create an avian self-historiography, reminiscent of his hero Leonardo da Vinci's reveries of flight.[114] Sigmund Freud, studying Leonardo's childhood dreams of flying, argued that his anatomical studies of the flight of birds and flying machines were a coded desire for his own dream of flying and that the flight itself was, in turn, a commentary on the perch from which the flight takes off.[115] In other words, the flying bird became the "eternal glory of the nest where it was born." Leonardo's childhood desire to take wing, Freud suggested, was no less than an infantile attempt to mimic adulthood and to construct a bridge to sexuality. Freud's proposition—while it appears to be flawed with its excessive emphasis on the "infantile erotic roots" of dreams of flight—illuminated Fuller's dreams of flight in the aerial delivery of 4D towers.

If Fuller alluded to his imagined flight in the air-delivery funnies, they could also be viewed as representations of another dream. Was Fuller's dream of flight a reincarnation of his yearning for self-discovery in the grandiose role of servant of mankind? When he raised himself to the sky in the second panel to offer the "view of the shell crater and anchored zeppelin from above" while the "house [was] being maneuvered in position," he unconsciously elevated us as well to see the grandeur of his own cosmic ambition to be a master solver of planetary problems. In proposing himself as a galactic bird in the drawings, not only did he deliver the 4D towers, but he also proposed a self-historiography, one in which he tapped into the venerable mythology of the aviator.[116] Fuller's dirigible was an emblem of his vision for what he wanted to be. Fuller *was* the dirigible. In the guise of a dirigible, Fuller unveiled a planetary human habitation program (albeit with ecological violence).

Was Dymaxion House an autobiographic gesture? Was there a connection between Dymaxion House's symbolic defiance of gravity and Fuller's rejection of the habitual? Fuller's promotion of Dymaxion House coincided with the stock market crash in October 1929. In those anxious economic times, his radical criticism of establishment ideologies and market-based economies struck a populist chord. If the unorthodox morphology of Dymaxion House reflected a defiant spirit, it was perhaps intended as an intellectual critique of what Harvey Goldberg called the pervasive "weight of conformism" that bogged down the great American promise to create a self-critical and humane society.[117] Charting a history of American "radical" thinkers—from Henry Demarest Lloyd to Heywood Broun, from John Brown and Theodore Dreiser to Thorstein Veblen—Goldberg defined the radical as an iconoclastic thinker (unlike the term's current negative associations with extremism) who took on the dominant ideologies of a conforming society. One of the defining myths of American radicals, according to Goldberg, was that they steadfastly adhered to the ideals of a future-oriented, progress-driven, and just society. This was a daunting task, especially because these radicals also fought against the peculiar paradox of social conformity: "If the great genius of the United States for technological change has yielded an especial receptivity to the scientific innovator, its firmly entrenched economic and social interests have organized strong opposition to critics of the culture or opponents of the power structure."[118] The difficulties of the radical, then, stemmed from a conflict between cultural changeability affiliated with innovation and society's demand that people conform to its norms and conventions. Thus, as Thomas P. Hughes argues, inventors have generally been estranged from society:

> In their withdrawal, the inventors were like avant-garde artists resorting to the atelier or the alternative life-style of a historic Montmartre, Schwabing, or Greenwich Village. Aware of the unorthodoxy of their ideas, inventors and artists intensified their feelings of being outsiders by their physical withdrawal. Working in their retreats, intellectual and physical, they created a new way, even a new world, to displace the existing one. The withdrawal to isolated spaces of their own choosing and design not only removed inventors and artists from the constraining influences of the status quo but also sheltered them from the hostility or ridicule of those whose established views and institutions the inventors' new ideas would undermine.[119]

The irony of this observation is that the reclusive radical eventually tends to be consumed by the very normative structures of capitalistic systems that he opposed in the first place. The alleged oligarchic control of American society by figures entwined with the powerful triad of economy, politics, and military—the "power elite," to use C. Wright Mills's term—demonstrates society's capacity to internalize the radical's criticism to reinforce the very foundation of the hegemonic ideology.[120]

Where does Fuller stand in the history of American radical thinking? If we are to agree with Goldberg's definition of the American radical or Hughes's articulation of the outsider, Fuller fits the mold. He both romanced and went against his time. His manifesto *4D Time Lock* is replete with radical ambitions and subversive pronouncements, questioning collective shortsightedness that hobbled both the housing industry and society. As an embodiment of this "outsiderist" ambition, Dymaxion House sought a dramatic alteration of the

American housing industry, which Fuller criticized as the wobbly fulcrum of an intellectually stunted society.

Yet a closer look reveals how Fuller's radicalism went beyond Goldberg's and Hughes's definition of the American dissenter. Fuller's program was a creative blend of iconoclasm and a strategic shaping of his self-image within which he orchestrated a conscious critique of radicalism itself by undermining the very premise of go-against-the-tide with a humanist viewpoint of middle-class revival. The final paragraph of *4D Time Lock* illustrated this observation: "The 4D book quite evidently has not been designed to flatter any banker, society tin ear, or other material tyrant . . . nor to receive any minor endorsement of radicalism, but rather, to be the epic of the great middle class, which is Humanity."[121] But the great epic of the middle class, as Edward H. Carr argued, was also a manipulative story of creating a mass society, a blob of people bound together by the "rationally calculated use of irrational methods of persuasion."[122] Fuller's rebellion against hegemonic "mob exploiters" got stuck in the same trap of maintaining social cohesion through persuasion. Thus, his radicalism straddled two conflicting goals: on one hand, he revolted against the culture of ideological homogeneity, and on the other, he viewed middle-class conformity (and the consumerist culture it created) as an advantage. He knew that the commercial success of his *radical* house depended on a conforming (and buying) middle class. Fuller was ready to play it both ways.

ASCENDING WITH NEW YORK CITY

Fuller relocated to New York, from Chicago, in July 1929. The Big Apple's skyline was about to be crowned by the Chrysler Building's art deco spire. The stock market crash was only three months away. Early the following year, construction of the world's tallest building would begin. Greenwich Village was a hothouse of bohemian cultures. The time seemed to be propitious for Fuller. His fascination with ascension acquired a mystical dimension as he came in contact with the Gurdjieff circle in Greenwich Village and experienced firsthand the rise of the Empire State Building.

In the "Art News in Brief" section of the *New York Times* on October 27, 1929, an entry read: "Romany Marie announces the showing of the works of Lajos Tihanyi, painter from Paris; Isamu Noguchi, sculptor, and Buckminster Fuller, the inventor of the Dymaxion House."[123] Romany Marie, a self-styled doyenne of Greenwich Village and former acolyte of the anarchist Emma Goldman, ran an exotic tavern that was a magnet during the interwar years for avant-garde artists, cultural and social radicals, and bohemians. Historians of New York City discuss Romany Marie's Tavern (which moved to multiple places within the Village) as an essential fixture of the radical Village vernacular and a "notorious tearoom of the 1920s and 1930s" (Figure 2.20).[124] The tavern, as

Caroline Ware wrote, "became a symbol of the repudiation of traditional values. Here congregated those for whom the traditional pattern in which they grew up had become so empty or distorted that they could no longer continue a part of it and submit to the social controls which it imposed. Many who were drawn to the Village came to seek escape from their community, their families, or themselves."[125] With its mystical interior decor, further dramatized by the popular hostess's Gypsy garb, foreign accent, and motherly demeanor, Romany Marie's Tavern was the hub where the socially agitated could tease out their ideas and exhibit their iconoclastic artwork.

For Fuller, Romany Marie's Tavern was a natural destination where he could carve out an intellectual niche for himself during the tempestuous social climate of the late 1920s.[126] The tavern was therapeutic, for it provided Fuller not just with a sympathetic audience full of "long-haired men and short-haired ladies" but also with an opportunity to tease out many new ideas.[127] As an invitee roster for a 1939 Romany Marie's Tavern reunion demonstrates, among the many so-called old habitués of Romany Marie's were Diego Riviera, Marcel Duchamp, Theodore Dreiser, Sinclair Lewis, Ernest Hemingway, Hugh Ferriss, Joseph Stella, John Sloan, Berenice Abbott, and Muriel Draper (Figure 2.21).

At the tavern, Fuller's philosophical exchanges with mystical guru Gurdjieff and the members of his circle allowed him to revisit his earlier interest in the esoteric properties of the fourth dimension, especially higher space philosophies.[128] One of the key Gurdjieffites who took particular interest in Fuller's ideas, especially Dymaxion House, was the American author Jean Toomer, who

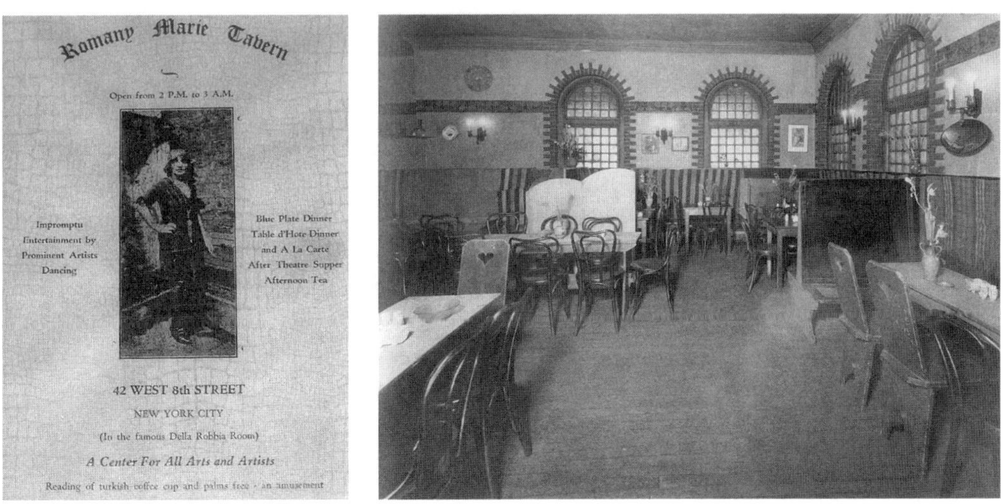

Figure 2.20. An advertisement for Romany Marie's Tavern and the interior of the tearoom. Reprinted in Robert Schulman, Romany Marie: The Queen of Greenwich Village *(Louisville, Ky.: Butler Books, 2006), center insert. Courtesy of the Estate of Robert Schulman.*

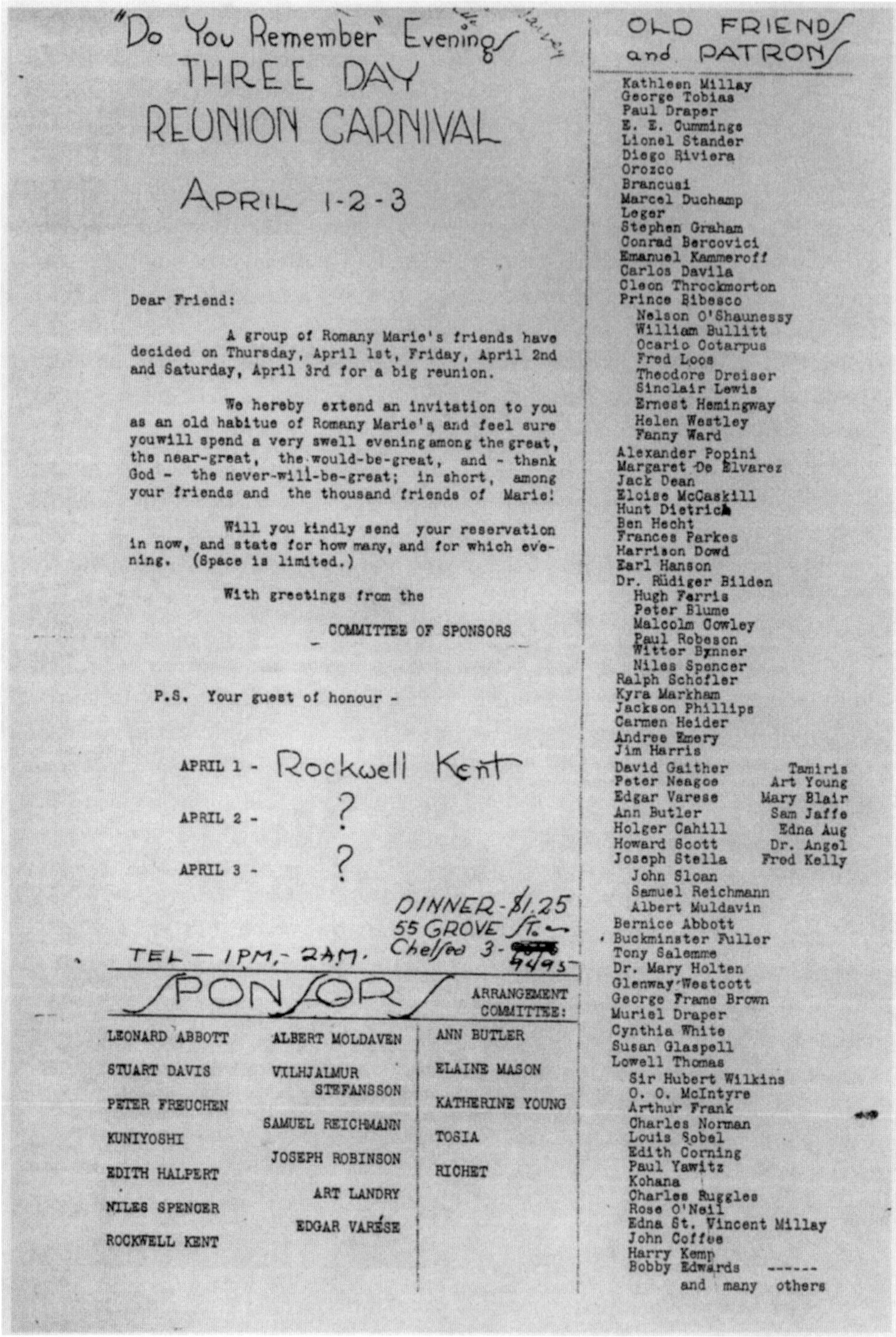

"Do You Remember" Evenings
THREE DAY
REUNION CARNIVAL

APRIL 1-2-3

Dear Friend:

A group of Romany Marie's friends have decided on Thursday, April 1st, Friday, April 2nd and Saturday, April 3rd for a big reunion.

We hereby extend an invitation to you as an old habitue of Romany Marie's and feel sure youwill spend a very swell evening among the great, the near-great, the would-be-great, and - thank God - the never-will-be-great; in short, among your friends and the thousand friends of Marie!

Will you kindly send your reservation in now, and state for how many, and for which evening. (Space is limited.)

With greetings from the

COMMITTEE OF SPONSORS

P.S. Your guest of honour -

APRIL 1 - Rockwell Kent

APRIL 2 - ?

APRIL 3 - ?

DINNER - $1.25
55 GROVE ST.
TEL — 1PM,- 2AM. Chelsea 3-4495

SPONSORS

LEONARD ABBOTT
STUART DAVIS
PETER FREUCHEN
KUNIYOSHI
EDITH HALPERT
NILES SPENCER
ROCKWELL KENT

ALBERT MOLDAVEN
VILHJALMUR STEFANSSON
SAMUEL REICHMANN
JOSEPH ROBINSON
ART LANDRY
EDGAR VARÉSE

ARRANGEMENT COMMITTEE:

ANN BUTLER
ELAINE MASON
KATHERINE YOUNG
TOSIA
RICHET

OLD FRIENDS and PATRONS

Kathleen Millay
George Tobias
Paul Draper
E. E. Cummings
Lionel Stander
Diego Riviera
Orozco
Brancusi
Marcel Duchamp
Leger
Stephen Graham
Conrad Bercovici
Emanuel Kammeroff
Carlos Davila
Cleon Throckmorton
Prince Bibesco
Nelson O'Shaunessy
William Bullitt
Ocario Cotarpus
Fred Loos
Theodore Dreiser
Sinclair Lewis
Ernest Hemingway
Helen Westley
Fanny Ward
Alexander Popini
Margaret De Elvarez
Jack Dean
Eloise McCaskill
Hunt Dietrich
Ben Hecht
Frances Parkes
Harrison Dowd
Earl Hanson
Dr. Rüdiger Bilden
Hugh Ferris
Peter Blume
Malcolm Cowley
Paul Robeson
Witter Bynner
Niles Spencer
Ralph Schofler
Kyra Markham
Jackson Phillips
Carmen Heider
Andree Emery
Jim Harris
David Gaither
Peter Neagoe
Edgar Varese
Ann Butler
Holger Cahill
Howard Scott
Joseph Stella
Tamiris
Art Young
Mary Blair
Sam Jaffe
Edna Aug
Dr. Angel
Fred Kelly
John Sloan
Samuel Reichmann
Albert Muldavin
Bernice Abbott
Buckminster Fuller
Tony Salemme
Dr. Mary Holten
Glenway Westcott
George Frame Brown
Muriel Draper
Cynthia White
Susan Glaspell
Lowell Thomas
Sir Hubert Wilkins
O. O. McIntyre
Arthur Frank
Charles Norman
Louis Sobel
Edith Corning
Paul Yawitz
Kohana
Charles Ruggles
Rose O'Neil
Edna St. Vincent Millay
John Coffee
Harry Kemp
Bobby Edwards ------
and many others

Figure 2.21. A list of invitees to the 1939 Romany Marie's Tavern reunion. Reprinted in Robert Schulman, Romany Marie: The Queen of Greenwich Village *(Louisville, Ky.: Butler Books, 2006), center insert. Courtesy of the Estate of Robert Schulman.*

crossed paths with Fuller first in Chicago and then again in New York City.[129] Toomer spent a transformative period at Gurdjieff's institute in Fontainebleau in France, returned to New York City in the fall of 1924, and promoted the mystic's ideas in America through lectures and writing. In a striking parallel to Fuller's desire for self-discovery, Toomer also believed that "an individual must master the private failings that hinder his personal development before undertaking any public responsibilities . . . [and that] it was every individual's duty, but particularly the artist's, to cast off the false features of the self in order to attain . . . 'self-purity, self-unification, and self-development.'"[130] Furthermore, after absorbing Gurdjieff's ideas of self-development through a "hyperconscious" control of the mind, Toomer remained committed to the lofty social mission of the artist: "Because the artist has gifted sight, because he is the vessel through which the multitudes find their voices . . . it was incumbent upon the artist to become an example for the multitudes and to assume the burden of development. It was the duty of the artist to purge himself of all psychological and spiritual impurities before he could address the problems of his age."[131] Fuller's *4D Time Lock* is fraught with similar ideas of the artist's responsibility toward society. In fact, Toomer sought to bring together Fuller and another New York figure, Hugh Ferriss, for he saw a Gurdjieffite thread passing through their common interest in self-development as a precondition for serving humanity.[132] Evidence of Fuller's interest in Toomer's theorizations on self-discovery can be seen in his cataloging of the prepublication leaflet for Toomer's limited-edition book on life's essences, titled *Essentials* (1931), in Dymaxion Chronofile.[133] Described in the leaflet as "as intense as Nietzsche, as pointed as Bernard Shaw," Toomer's program of self-control for developing clairvoyant faculties might have provided Fuller with a theoretical framework for defining himself as a visionary artist.

Fuller read the esoteric and theosophical writings of Claude Bragdon, who moved to New York City in 1923 from Rochester.[134] The fact that Fuller included Bragdon in the first batch of recipients of his 4D essay suggests that he valued the mystical architect's opinion. Fuller wrote this message to Bragdon as an introduction for the 4D essay: "Your books 'Architecture and Democracy,' as well as 'Tertium Organum' which you uncovered, as well as your many researches and associations with the matter concerned in this paper, urges upon us the necessity of having your study and comment on it."[135] Whether or not Fuller judiciously read Bragdon's 1922 English translation of Russian philosopher and Gurdjieff acolyte Peter Ouspensky's *Tertium Organum,* it is telling that Fuller drew some parallels between his own work and that of Bragdon and Ouspensky.[136]

The conceptual similarities between some of the core ideas in *4D Time Lock* and *Tertium Organum* (meaning a higher form of logic) merit a critical look at Ouspensky's book. Suspicious of positivism's alleged claim to absolute knowledge, Ouspensky argued for a "cosmic consciousness" that would help humankind rediscover the intuitive principles "to organize life on earth." He believed the principal aim of human existence is to develop an inward eye, a higher consciousness that propels the ascension of the human mind above animalistic and fragmentary individual consciousnesses that are tantamount to blindness, even when one has the physical properties of sight. Borrowing on the ideas of cosmic consciousness from the nineteenth-century British Canadian psychiatrist Richard Maurice Bucke, Ouspensky traced startling affiliations between the redemptive powers of cosmic consciousness and contemporaneous technological innovations in aviation.[137] Ouspensky quoted Burke:

> The immediate future of our race is indescribably hopeful. . . . Before aerial navigation national boundaries, tariffs and perhaps distinctions of language will fade out. Great cities will no longer have reason for being and will melt away. . . . In contact with the flux of cosmic consciousness all religions known and named today will be melted down [in a similar vein]. The human soul will be revolutionized . . . all intermediaries between the individual man and God will be permanently replaced by direct unmistakable intercourse. . . . The world peopled by men possessing cosmic consciousness will be as far removed from the world of today as this is from the world as it was before the advent of self-consciousness.[138]

Burke's analogy was simple: just as human flight would collapse artificial boundaries of nationalities and isolated cities, cosmic consciousness would create a planetary brotherhood of minds. But the essence of the argument lay elsewhere. Both Ouspensky and Burke argued that only a handful of men could eventually ascend to claim the power of cosmic vision in the evolutionary journey from animal consciousness to self-consciousness to cosmic consciousness. According to Ouspensky, the acquisition of cosmic consciousness was premised on a fundamental idea of humanity's unequal cognitive abilities: "*The new conception of humanity disposes of the idea of equality,* which after all does not exist, and it [the new conception] tries to establish the signs and facts of the differences between men, because humanity will need soon to divide the 'progressing' from the 'incapable of progress'—the wheat from the tares, for the tares are growing too fast, and choke the growth of the wheat."[139] The cosmic consciousness was thus the sacred province of an exclusive group of thinkers, the "progressing" type whose cognitive operations were radically superior to those of the fast-growing "incapable of progress" type. The eugenic undercurrent in Ouspensky's ominous division of humanity into "progressing" and "incapable of progress" could be found within early twentieth-century mystical philosophies.

Interestingly, such divisions of humanity along a "thinker" and "nonthinker" vertical axis were also common in various ideologies of progress articulated by aviation enthusiasts. As Joseph Corn notes, "According to some prophets [aviation enthusiasts], airplane flight would even alter human nature, eventually producing a new breed of human being, far superior to earth-bound types."[140] During the interwar period, various aviation, geographical, and city-themed magazines zealously popularized the theme of the aviator's holistic perspective as an emblem of some kind of superconsciousness.[141] Even though skeptical and apprehensive of technology's alleged complicity with the excesses of materialism, many mystical philosophers and modernist artists alike shared a fascination with aviation's purported ability to usher in a spiritual consciousness. Against this background, the analogy between the aviator and the cosmically aware mystic hardly seemed audacious. Fuller, too, was swayed by this analogy. But, for him, the agency of the person, seeking a higher consciousness or furthest evolution, was crucial:

> So important an evolutionary event (as successful flight by man) may only be won in principle from the potentials of universe through a complex interaction of a plurality of initiatives daringly taken by individuals. Self-sustained directional control of airborne man represents the integrated product of a fabulous history of personal investment by innumerable individuals in daring sequences of single-handed and necessary failures, attendant upon undertakings only warrantable to the individual himself.[142]

Fuller's notion of "self-sustained directional control of airborne man" had roots in Ouspensky's esoteric belief that man's intellect was capable of autoguiding the course of its own evolution toward a higher plane of cognition, unlike the mind of an animal. But this quality was not seen as an automatic condition of all men, for there was an inherent disparity between who could evolve proactively and who could not. Ouspensky observed: "The mind of a man has far more power over itself; it can assist in its own evolution, and can also *impede* it . . . And the individual who is not evolving does not remain in a static condition, but goes down, *degenerates*."[143] Apprehensive of the state in which mankind existed at the mercy of nature's laissez-faire design, Ouspensky believed that it was imperative for the capable to propel their self-development consciously. From this select group of self-guided men would come a "new master," "who may not at all like everything that we have built." For Ouspensky, "the future belongs not to *man*, but to *superman*."[144]

In early twentieth-century mystical literature, it was common for the idea of a superman to serve as a supreme form of intellection situated outside humanity's epistemological framework. If the future was uncertain, then this uncertainty and the attendant anxiety could be intellectually counterbalanced by the supposedly infinite or unknowable mental aptitude of a superman. Fuller gravitated toward the notion of a higher being's "unknowability" as a ruse to break free from the confinement of social norms. In the chapter "Land to Sky, the Outward Progression" in *4D Time Lock,* Fuller echoed Ouspensky's dismissal of all types of consciousness prior to achieving a superconsciousness. The "selfconsciousness of bestial man," Fuller wrote indignantly, could allow only a narrow anthropomorphic view of God, tragically mired in all human limitations. But "the completely developed individual, freed of all sect dogma and credo, and mindful of the temporal mask, which faith alone may pierce, perceives God as the great common spirit of love that compasses the universal sphere, the infinity beyond the stars, towards which we ever expand."[145]

Fuller buttressed this observation by reading Einstein's 1931 essay "Religion and Science."[146] Discussing the common antagonism between religion and science, Einstein identified three stages in the evolution of a religio-spiritual consciousness: first, the fear of adverse forces that compels primitive man to conceive an anthropological image of a savior who redeems if worshipped; second, the morality of good and sin that provokes the idea of a God who either rewards or punishes; and finally, a cosmic religious consciousness that transcends a God-centric worldview in which man's action is determined by neither fear nor the desire for reward, but rather by the will of the self to grasp the holistic methodologies of nature's or a supreme force's working. Einstein wrote:

> The religious geniuses of all ages have been distinguished by this kind of religious feeling, which knows no dogma and no God conceived in man's image; so that there can be no church whose central teachings are based on it. Hence it is precisely among the heretics of every age that we find men who were filled with this highest kind of religious feelings and were in many cases regarded by their contemporaries as atheists, sometimes also as saints. Looked at in this light, men like Democritus, Francis of Assisi, and Spinoza are closely akin to one another.[147]

It is this "type" of men with the "highest kind of religious feelings" that Fuller had in mind as he imagined the task of building a new world, a master builder of a propitious future. In the early years of the Depression, he came in contact with theosophists, Gurdjieffites, bohemians, and social rebels in Greenwich Village—encounters that emboldened him to cast the master builder in the mold of a superconscious being. His reading of both Ouspensky's ruminations on superman and Einstein's thoughts on the "heretics of every age" served his quest for a master builder and provided a template for his self-definition.

During his frequent visits to Romany Marie's Tavern, Fuller led a "nomadic" urban life around Greenwich Village. In late 1931, he found an unusual lodging opportunity that seemed to fuel his fascination for seeing cityscapes from above. It was an affordable but gratifying refuge more than two hundred feet above the ground, on the roof of a modern industrial warehouse—the Starrett-Lehigh Building—completed in 1931 and designed in the then-unorthodox International Style by architects Russell G. and Walter M. Cory (Figure 2.22).[148] Constructed to serve as a railroad freight terminal above an existing open-air freight rail yard, the Starrett-Lehigh Building covered a whole city block between 11th and 12th Avenues and 26th and 27th Streets in the West Chelsea area of Manhattan. An innovation in warehousing technology, the iconic building included the provision of railcars running directly into the first floor, while a large 30,000-pound elevator system lifted delivery trucks directly to various levels for efficient and speedy unloading of goods.[149] The building was included in "Section Two: The Extent of Modern Architecture" in the Museum of Modern Art's 1932 landmark exhibition *The International Style,* curated by Henry-Russell Hitchcock and Philip Johnson.[150]

Completed at the outset of the Depression, the Starrett-Lehigh Building remained mostly unoccupied. The building's two-square-block flat roof was punctuated by only two structures, apart from the bulky tanks that supplied water

Figure 2.22. R. G. & W. M. Cory, Starrett-Lehigh Building, New York City (1931). Reprinted in Terence Riley, The International Style Exhibition 15 and the Museum of Modern Art *(New York: Rizzoli, Columbia Books of Architecture, 1992), 175.*

to the building. One of those two rooftop structures contained the elevator machinery, and the other was an empty storeroom, which Fuller immediately wanted to make his future home. With his usual persistence, Fuller convinced the property manager to rent it to him for about thirty dollars per month. The commanding view of the city from atop the twenty-story building and the relative quiet of the roof in comparison with the raucous streets below seem to have appealed to him. Fuller biographer Sieden describes his discovery thus: "Upon reaching the warehouse roof, he was astounded by the panoramic view, which stretched from New York harbor in the south to the Hudson River in the west, and he decided that the roof of the building would be an exciting place to live."[151]

Fuller studied the blueprint of the Starrett-Lehigh roof and "redesigned" the austere room he had rented.[152] Before long, he cut a window into the storeroom, transforming the grim space into a livable room with a view. The "Pent House," as he called it, became his bohemian playground with loud nightly soirees (without any nagging neighbors to complain about noise and the outrageous behavior of party attendees), as the roof provided an urban oasis from the gloomy streetscape of Depression-stricken New York City while unfolding an empowering aerial vista of the city. The habitat was a kind of poor man's version of what Carol Willis has called the "garden in the machine," a patch of tranquil nature aloft on a tall building from which to rediscover the modernist city below.[153] Within the vertiginous morphology of 1920s New York, elevated gardens, terraces, and lofts became tropes of the growing ambitions of the city. Height provided mystique and an alternative to the crowded city on the street. As Willis notes:

> During the 1920s . . . some New York artists began to perceive these exclusive settings not as interior stages, but as platforms from which to behold the theater of the city beyond and below. In the nineteenth century, the intrusion of the machine into the metaphorical American garden produced, as Leo Marx and others have argued, a sense of conflict and tension. In the modern period, for those who celebrated the city, the inclusion of nature in pastoral patches like the skyscraper gardens tranquilized urban anxieties and offered a new American ideal, the machine-age metropolis.[154]

Fuller's Pent House on top of the Starrett-Lehigh Building hardly qualified as a "pastoral patch," yet it provided him with respite and space for escapades with friends.[155] Like other artists and architects of the time—Ferriss, Bragdon, Sheeler, Stieglitz, and O'Keeffe—Fuller was influenced by the experience of seeing the vast panorama of New York City from the concrete roof of the Starrett-Lehigh

Building. The *roof* became another crucial element in Fuller's self-definition. If the airplane views that he enjoyed during his 1922 flight over the Northeast littoral occasioned the development of an aesthetic of ascension, the Starrett-Lehigh roof presented a handy location for the continued formation of that aesthetic.

But this formation was not trouble-free. If Fuller's aerial gaze from the Starrett-Lehigh scanned the city and bolstered his self-image, the gaze also alleviated the pain of an intense extramarital affair he was having at the time with a much younger woman named Evelyn Schwartz, an aspiring actress who worked at New York City's Macy's department store.[156] The relationship unfolded against, and engaged with, the vertiginous drama of New York City itself—especially its latest wonder and the pinnacle of its spectacular skyline, the Empire State Building.[157] Around the time the then-tallest building in the world was formally inaugurated on May 1, 1931, Fuller and his young companion exchanged passionate letters. The Empire State Building and its observatory became their rendezvous. By the time Fuller moved into the Pent House in early November, his relationship with Schwartz had become so intense that it threatened Fuller's marriage. As one of his letters to Schwartz—juxtaposed with a sketch of New York City's skyline as viewed from the Pent House—reveals, Fuller's elevated perch provided a peculiar calming moment through an uplifting view of the Empire State Building, as well as the Macy's department store, nestled in the Empire State's shadow (Figure 2.23).[158]

From the quiet of the Pent House, Fuller imagined a virtual urban triangle that connected the masculine symbolism of the Empire State Building, the "feminine mystique" of Macy's, and Fuller himself on the roof of the Starrett-Lehigh Building. Fuller wanted his gaze reversed so that he could be seen and reimagined by his lover from her work at Macy's, while the Empire State Building acted as an urban negotiator in this visual dialogue. Fuller's desire to be seen through the glass of his "windows" invoked Lacan's mirror metaphor, in which "the subject anticipates in a mirage the maturation of his power."[159] The truth of the self lies not in the authentic reproduction of the self by the mirror but rather, according to Lacan, in the mirror's temporal dialectic in which the future self already exists. Thus, in the dualism of the Empire State Building's height and his admirer's adulating gaze, Fuller imagined a pre-presence of the self in a befitting heroic mold. When their affair waned during the final days of 1931, as Schwartz became increasingly frustrated by Fuller's noncommittal attitude, an anguished Fuller found refuge in the loneliness of his Pent House.[160] Sometime around Christmas, he wrote a rambling and incoherent letter to Schwartz in which he tried to convince her that their shared commitment to humanity was much greater and more urgent than the mundane necessities of institutional marriage.[161]

In the feverish cascade of ideas in his letter, Fuller sought to reassure himself of the validity and magnanimity of his life's larger mission to be useful to

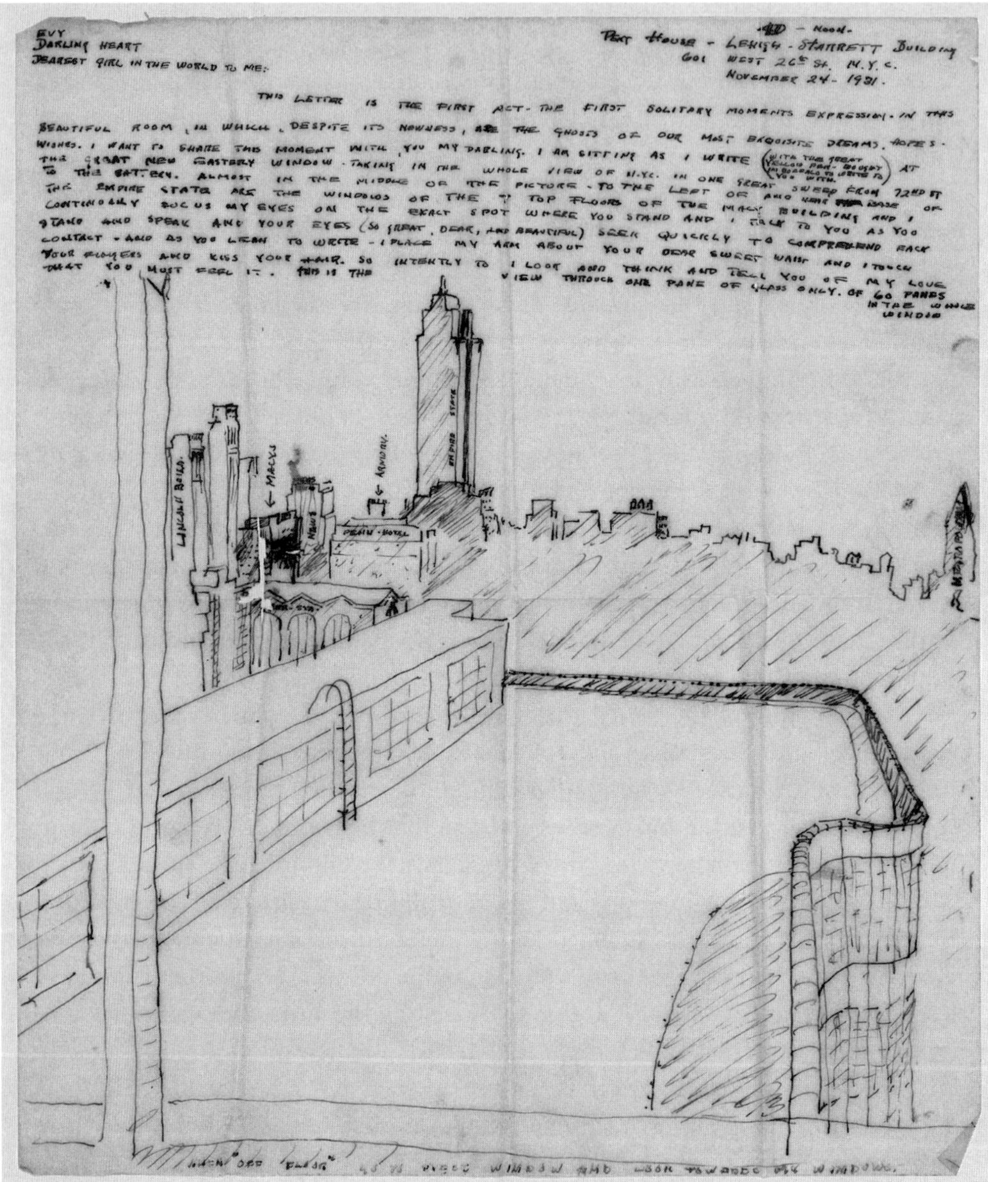

EVY
DARLING HEART
DEAREST GIRL IN THE WORLD TO ME:

– NOON –
PENT HOUSE – LEHIGH – STARRETT BUILDING
601 WEST 26th ST. N.Y.C.
NOVEMBER 24 – 1931.

THIS LETTER IS THE FIRST ACT – THE FIRST SOLITARY MOMENTS EXPRESSION – IN THIS BEAUTIFUL ROOM, IN WHICH, DESPITE ITS NEWNESS, ARE THE GHOSTS OF OUR MOST EXQUISITE DREAMS, HOPES – WISHES. I WANT TO SHARE THIS MOMENT WITH YOU MY DARLING. I AM SITTING AS I WRITE (WITH THE GREAT YELLOW PEN – BOUGHT IN BUFFALO TO WRITE TO YOU WITH.) AT THE GREAT NEW EASTERLY WINDOW – TAKING IN THE WHOLE VIEW OF N.Y.C. IN ONE GREAT SWEEP FROM 72ND ST TO THE BATTERY. ALMOST IN THE MIDDLE OF THE PICTURE – TO THE LEFT OF AND [illegible] BASE OF THE EMPIRE STATE ARE THE WINDOWS OF THE 7 TOP FLOORS OF THE MACY BUILDING AND I CONTINUOUSLY FOCUS MY EYES ON THE EXACT SPOT WHERE YOU STAND AND I TALK TO YOU AS YOU STAND AND SPEAK AND YOUR EYES (SO GREAT, DEAR, AND BEAUTIFUL) SEEK QUICKLY TO COMPREHEND EACH CONTACT – AND AS YOU LEAN TO WRITE – I PLACE MY ARM ABOUT YOUR DEAR SWEET WAIST AND I TOUCH YOUR FINGERS AND KISS YOUR HAIR. SO INTENTLY DO I LOOK AND THINK AND TELL YOU OF MY LOVE THAT YOU MUST FEEL IT. THIS IS THE VIEW THROUGH ONE PANE OF GLASS ONLY. OF 60 PANES IN THE WHOLE WINDOW

Figure 2.23. Letter from Buckminster Fuller to Evelyn Schwartz, November 24, 1931. The Special Collections of Green Library, Stanford University. Courtesy of The Estate of R. Buckminster Fuller.

humanity. As a *New York Times* reporter wrote of Fuller's letters in 2008, these "seemingly crazy writings were important because they showed that in recurrent dark periods Fuller was not trying only to persuade others his ideas were important, but to persuade himself that he mattered. The letters . . . were a form of self-encouragement as Fuller struggled to find a reason for going on."[162] Fuller's "great window" on the city and the power of his gaze that could zoom in to single out his lover from a vast urban visual pool both allayed his sense of abandonment and facilitated his attempt to figure out who he was and what his purpose in life would be.

A photograph of Fuller standing on the roof of the Starrett-Lehigh Building with the Empire State Building in the background can be seen as a conscious attempt by Fuller to solidify his self-image as a towering figure, which suited his fancy of serving humanity (Figure 2.24). In this photo, most likely taken in November or December 1931, Fuller—upright, neatly dressed, and kind of

Figure 2.24. Buckminster Fuller standing on the roof of the Starrett-Lehigh Building (1932). The Special Collections of Green Library, Stanford University. Courtesy of The Estate of R. Buckminster Fuller.

"Gulliverized" against the silhouette of the Empire State Building—presented a paradox. He wanted the totemic symbol of New York (the skyscraper) included within the picture frame to legitimate the essential modernity of the self, yet he consciously sought his own aggrandizement in the photograph, as the camera focused on him dwarfing the world's tallest building. Fuller looked not at the skyscraper but rather steadfastly at the photographer. The absence from the photo of the platform on which he stood was most likely intended to represent Fuller as a soaring figure, while the skyscraper's role was primarily prosthetic, to prop up Fuller's ethereality and ascension. The skyscraper was the signifier and Fuller the signified, to borrow from Roland Barthes's semiological model.[163]

In his effort to define himself, Fuller co-opted the Empire State Building's success in punctuating anew New York City's skyline and offering the city a "stairway to heaven."[164] On May 1, 1931, the day the Empire State Building was inaugurated, ESB Corporation president Al Smith gasped from the eighty-sixth-floor observatory: "Being this high up sort of pulls the island together—constricts it—lays it out like a kid's map."[165] Smith's utterance was part of an urban mythology invested in the putative virtuosity of verticality. Fuller's photograph tapped into this mythology as a way to forge a lofty biographic narrative that would later serve his projected role as a grand patriarch of the planet.

VERTICALITY TO GALACTIC: *NINE CHAINS TO THE MOON*

After a three-year period of writing and rewriting, Fuller published his first book, *Nine Chains to the Moon,* in September 1938 (Figure 2.25).[166] It covered a wide array of topics, ranging from "What is a house?" to the economics of "dollarability" to the evolution of "genius and talent." Underneath the polymath's intellectual fireworks remained the theme of the aesthetics of ascension. This aesthetic attitude was shaped by an inquiry into different variants of the master builder, the genius, or "phantom captain," to use Fuller's term, whose definition drew on such seemingly unrelated topics as space exploration and eugenics, both popular in New Deal America.

The book elicited a wide range of reviews, from appreciative to lukewarm to dismissive. Sinclair Lewis, America's first Nobel Prize winner in literature in 1930, encapsulated the book's appeal thus:

> Imagine a three-ring circus—high-wire acts, blaring bands, clowning, and all one man as the whole show. That will give you a feeble idea of "Nine Chains to the Moon" by R. Buckminster Fuller, engineer extraordinary, inventor of the Dymaxion car and house, seer, and scientifico—a philosophical whirling dervish spinning with a high-tension hum on the edge of the Deep End. It is at once a guidebook and dream book of the future, a purge

Figure 2.25. R. Buckminster Fuller, Nine Chains to the Moon *(Philadelphia: J. B. Lippincott Co., 1938).*

> of the past, a debunker of architecture, economics, political science, and your pet preconceptions.[167]

The *New Yorker*'s reviewer called the book an übermanifesto of the future: "Mr. Fuller has the Leonardoesque mind (but much more eccentric), a profound faith that society will shortly be rationalized without any messy revolutions into a completely efficient super-service state. This extraordinary book invites at least the cursory attention of all future-minded citizens."[168] Sharing Fuller's impassioned iconoclasm, Frank Lloyd Wright wrote about the book with a personal touch: "Buckminster Fuller—you are the most sensible man in New York, truly sensitive. Nature gave you antennae, long-range finders you have learned to use. I find almost all your prognosticating nearly right—much of it dead right, and I love you for the way you prognosticate."[169]

Whether embraced or criticized, *Nine Chains to the Moon* ensured Fuller's wider visibility in the public domain. By the time he began writing the book in 1935, he had already fashioned a reputation as a maverick and was a frequent invitee in both academic and professional lecture circuits. He lectured not just on the promise of his industrially reproducible house but also on a much wider spectrum of ideas related to human progress through optimal technological adaptability. He had come to terms with the promise of Dymaxion House as an *idea* for nurturing humanity rather than as a practicable and profit-making business enterprise. Fuller's brief editorial stint at *Shelter* in 1932—despite the magazine's failure due to his financial mismanagement—helped him propagate his views from a formal platform. In 1933, collaborating with ship designer and friend Starling W. Burgess, Fuller built the three-wheeled streamlined vehicle called the Dymaxion Car, which was exhibited at the Chicago World's Fair (1933–35). The streamlined automobile demonstrated his futuristic commitments, even though an accident during a test drive killed its driver and hurt two foreign passengers (who had arrived in the United States on the *Graf Zeppelin*), causing the project's demise.[170] Fuller also played a key role in building the research and development division of Phelps Dodge Corporation, then the third-largest copper company in the world.

Nine Chains, therefore, presented Fuller at a crossroads: the anxiety of fashioning himself as a social reformer receded into the past while new opportunities unfolded before him. He was now regarded (as well as ridiculed) as a "prophet" of the future, or, as Sinclair Lewis dubbed him, "a philosophical whirling dervish." When Fuller's publisher told him, rather unceremoniously, that he was not "on the list of people who understand Einstein, so we can't publish [the book]," Fuller went to Princeton University to meet Einstein and, in a legend-building conversation with him, convinced the physicist of the urgency of his prognostications (or so he later claimed).[171] Einstein was one of the protagonists of *Nine Chains*, and the physicist's analysis of human nature as two basic emotion types—fear/passive and longing/active—shaped the tenor of Fuller's argument in the book: the onus of humanity's development lay on the shoulders of the "longing" type, a select brotherhood of master builders who act on their own volition, rather than on society's normative expectation of them.[172]

The book appeared to be the manifesto of an oddball genius: forty-three short chapters on a bewildering spectrum of topics. The book's core argument, distilled from the metaphors, fables, and Fuller's linguistic follies, is that an intellectual suspension from orthodoxy is necessary to create a new civilization, and that new thoughts will be possible when humanity ascends to look both holistically inward and outward to the universe for new resources. According to Fuller, the architecture of both physical shelter and thought must transcend the parochialism of narrow Earth-centric views of the universe. The book's title explained this thought:

> The title, *Nine Chains to the Moon,* was chosen to encourage and stimulate the broadest attitude toward thought. Simultaneously, it emphasizes the littleness of our universe from the mind viewpoint. A statistical cartoon would show that if, in imagination, all of the people of the world were to stand upon one another's shoulders, they would make nine complete chains between the earth and the moon. If it is not so far to the moon, then it is not so far to the limits,—whatever, whenever or wherever they may be.[173]

By emphasizing the "littleness of our universe," Fuller sought to magnify the power of the mind. Nine complete human chains between the earth and the moon was then a coded advocacy to think vertically and outwardly, an essential element in cerebral prowess that enables "personal contact with all astronomical bodies." This cosmic connectivity, in turn, would inform building practices. Fuller argued, "Scientific shelter design . . . is linked to the stars far more directly than to the earth. STAR-GAZING? Admittedly. But it is essential to accentuate the real source of energy and change in contrast to the emphasis that has always been placed on keeping man 'down to earth.'"[174] One consistent thread in *Nine Chains* was Fuller's belief that humanity's vertical movement was a defining element in the evolutionary annals of the most advanced human race. In a section titled "We Call It 'Earth,'" he wrote:

> By means of his harnessed inanimate servant, power, and his extended mechanisms, man has now explored, measured, and "set" under control much of his earth's crust and his once-"outside" universe, entirely despite the inertia of vanities,

> superstitions, exploitation, humpty dumpty moralities, laws and destructive selfishness. He has flown in his imagination-conceived, intelligence-wrought, de-selfed mechanisms at 72,000 feet above the earth's surface, almost three times the height of the earth's highest mountain, and sixty times higher than the Empire State Building. Yet this is an insignificant feat compared with flights and heights to be attained in the NOT FAR AHEAD "NOW," in new intelligence-to-be-wrought mechanisms of flight.[175]

What were the origins of Fuller's musings on extreme altitude or human chains to the moon? Although they were fraught with subjective excesses, Fuller's verbose depiction of a stratospheric airplane in the guise of an "imagination-conceived, intelligence-wrought, de-selfed mechanism" capable of flying at 72,000 feet and his fictional human chains to the moon were couched in the American romance of space as a new extraterrestrial frontier.[176] As the natural satellite of Earth, the moon had already been the object of much fantasy, especially among science fiction writers (Figure 2.26).[177] This fascination created a burgeoning industry of pulp science fiction. Speculative stories of lunar voyages captivated an American readership eager to escape from the anxious times of the Depression. Popular books and magazines regularly published the most up-to-date research in astronomical phenomena and commentary on the future of space exploration.

H. G. Wells, whose work was a major influence on Fuller's thoughts, wrote about the clash of galactic civilizations in *War of the Worlds* (1898) and *The First Men in the Moon* (1901), which was reprinted in the January 1927 issue of America's first science fiction magazine, *Amazing Stories* (Figure 2.27).[178] In having his protagonists embark on a lunar journey by means of a gravity-defying material, cavorite, Wells broadened the horizon of man's galactic thinking while offering a "reverse" gaze on humanity from the vantage of the Selenites, the so-called lunar population. In *The Shape of Things to Come* (1933), Wells probed a startling affiliation of aerial transportation with the advent of a utopian civilization ruled by a dubiously benevolent regime called the "Air Dictatorship."[179] In the summer of 1934, during a visit to America, Wells posed before Fuller's recently completed streamlined vehicle, the Dymaxion Car, and the caption on the resulting photo in the June 2 issue of the *Saturday Review of Literature* read: "The Shape of Things to Come Confronts Mr. Wells." Wells and Fuller's futurist discourses converged at the point where their prognostications on humanity's progress revealed their common interest in both the utilitarian potential and the political symbolism of transportation machines. Wells's fantasy of lunar voyage and Fuller's human chains to the moon stemmed from their core interest in the cognitive evolution of humanity.[180] That human beings were thinking about reaching the moon was an indicator of the evolutionary mobility of humankind.

Figure 2.26. The cover of Amazing Stories Quarterly, *Spring–Summer 1933.*

Figure 2.27. Galactic travels by rocket were a leitmotif in pulp magazines from the late 1920s. The original cover of The First Men in the Moon, *by H. G. Wells; reprinted in* Amazing Stories, *January 1927.*

But the moon was not a just a staple of the utopian genre during the interwar era; the goal of reaching the moon by means of gravity-defying projectiles had already begun to inspire scientists.[181] A group of science fiction writers founded the American Interplanetary Society in 1930 in New York City to promote space expeditions. The American physicist Robert H. Goddard, an admirer of Jules Verne and Wells, was obsessively but secretly seeking to discover a method of reaching extreme altitudes.[182] With the support of the Smithsonian Institution, Goddard administered the "first successful launch of a liquid fuel rocket" in 1926, from a field near Auburn, Massachusetts.[183] The media generally interpreted Goddard's effort to create a gravity-defying rocket capable of reaching the moon as far-fetched, since the feasibility of a lunar expedition still existed between fiction and reality. Goddard's research failed to garner adequate federal grants or corporate patronage. Yet Charles Lindbergh, convinced that the "future of rocket flight" would produce human benefits, came to Goddard's aid in 1929 and secured funding from the Carnegie Institute and Guggenheim Foundation to support America's fledgling science of rocketry.[184] Even when the American economy took a perilous downturn during the Depression, Lindbergh managed to keep the Guggenheim funding flowing for Goddard's rocketry laboratory in Roswell, New Mexico.

Whether Fuller ever met Lindbergh or Goddard personally is unclear from archival sources, but Fuller's reference to the "nine chains to the moon" reflected his eagerness to validate his self-image as a visionary by delving into the contentious and futuristic discourses of the era. As Howard McCurdy has observed: "The leaders of the rocket societies proclaimed a gospel of remarkable power. Humans, they said, would carry out expeditions of discovery in space as ambitious as those of earlier explorers on Earth, maintaining the spirit of adventure and discovery they had inspired."[185] Despite the allusion to the moon as a new frontier in *Nine Chains to the Moon,* Fuller's book was concerned less with the rocket that would carry humanity to the moon than with the visionary who conceived rocketry itself. In other words, Fuller engaged the discourse of reaching the moon by human chains or rocketry as a prelude to his research proper that sought to provide an understanding of, say, Goddard's intellect as a forum for a broader discussion on the development of an advanced human race.

In this sense, the machine, for Fuller, was a technological extension of the inventor or the genius.[186] The rocket was a tangible expression of Goddard's superior thinking. The engineering excellence of the *Spirit of St. Louis* was nascent in the genius of Lindbergh. The tool was a mechanical analogue of the inventor, even if the tool began to acquire social agency to eventually extend the latitude of the human mind. Fuller's view of technology suggested that man ultimately had the power to steer his own development by consciously maneuvering the world of tools and his environment. In *The Post-industrial Prophets,* William Kuhns called this Fuller's "ecological pattern transformation," a sort of post-Darwinian mode of self-evolution in which man adapts to his environment under his own volition.[187] When man creates the radio or the airplane, he ushers in new ways to adapt to, and transform, his environment. Kuhns wrote:

> Fuller sees the most recent major ecological pattern transformation in the revolution achieved by the automobile and the airplane. Fuller's ecological concept of evolution resembles Julian Huxley's "social revolution" and Teilhard de Chardin's "noosphere." All three suggest that man can consciously pattern his own evolution and that technology represents the new nerve fibers of an emerging super-organism. This latter emphasis is especially strong in Fuller.[188]

In a somewhat tautological argument, within what Kuhns called a "super-organism," technology is produced in such a way that it is eventually *internalized* within man's mind/body consciousness. In *Nine Chains,* Fuller called this superorganism the "phantom captain," which maintained or drove the techno-biological body, and "with the phantom captain's departure, the mechanism becomes inoperative and very quickly disintegrates into basic chemical elements."[189] In other words, the phantom captain breathes life into the seemingly vegetative world of physical bodies. The body and other technologies, including architecture and means of transportation—two key examples in the Fullerian narrative—are machinations of the phantom captain in its bid to harvest the environment for humanity's maximum mental development. If architecture provides a kinesthetic metaphor for the phantom captain's ship, "the goal is not 'housing,' but the universal extension of the phantom captain's ship into new areas of environment control, possibly to continuity of survival without the necessity of intermittent 'abandoning ship.'"[190]

Fuller's esoteric argument boiled down to this: it was only with the attainment of superconsciousness—the phantom captain—that man could ensure an expedited self-development. The evolutionary underpinnings in Fuller's narrative seemed unmistakable: "The goal is the emergence of humanity. The means is industrial. Not re-form, but to form. Evolution tends toward the accelerated development of new form, embodying one or many of the basic elements, but in ever new streamlined alignment."[191] In the final chapter of *Nine Chains,* "Anthem," Fuller's statement on what he called "controlled evolution" echoed the eugenicist mind-set of the time: "I may factually prove for man's experience 'conceived' knowledge, my ability to control the course and rate of the time expansion veins of realization (not its checking or warping) by the intuitively accredited SELECTION BY INTELLECT instead of procrastinatingly submitting to inevitable non-rationalized descriptive natural erosion expansion-entropy."[192] Fuller's confidence in his "ability to control the course" of evolution with the "selection by intellect" resonated with eugenic proclivities that called for the application of scientific techniques to breed superior human genes. Although evolutionary themes underpinned his earlier manifesto *4D Time Lock,* those themes became more pronounced in *Nine Chains.*

In 1935, the year Fuller began writing *Nine Chains,* Alexis Carrel (1873–1944), the French-born Nobel Prize–winning scientist and controversial member of Manhattan's Rockefeller Institute for Medical Research, published his best-selling book *Man, the Unknown.*[193] Given his allusion to Carrel's book in *Nine Chains,* as well as the parallels between the two books with regard to the dystopic condition of the modern man, Fuller most likely read the book with interest and absorbed its core ideas.[194] In *Man, the Unknown,* Carrel embarked on a populist campaign in support of eugenic policies as the basis for creating the oligarchy of an "enlightened elite." Carrel predicted an apocalyptic future unless a powerful elite assumed the charge of not only leading humankind but also remaking it by purging it of its sociobiological impurities. *Man, the Unknown* was a smash hit in the United States and Europe and was translated into twenty languages in more than fourteen countries. In America alone, the book sold nearly a million copies and rose to the top of the *New York Times* nonfiction best-seller list. A week after the publication of *Man, the Unknown,* on September 16, 1935, *Time* magazine featured Carrel on the cover with an oblique reference to his political philosophy: "Human beings are equal. But individuals are not."[195] Despite its racial politics and the derision it attracted, Carrel's controversial theory was just one thread in a wide-ranging discussion in interwar America—a populist and nativist discourse on racial hierarchy that was championed in the public domain and by government policies. The book's urgent call for fixing human civilization resonated with many people troubled by the financial crisis of the 1930s. In Carrel's manifesto, Fuller found a theoretical explication of his own drive to remedy civilization.[196] The book's alarmist theme of the decline of Western civilization due to modern society's failure to bring about man's spiritual and moral development in corresponding proportion to his material and scientific advancement, and how only the concerted propagation of a master class would save the civilization, appealed to many radical reformers of the period.[197]

In the chapter "What Is a House?" in *Nine Chains,* Fuller drew parallels between Carrel's mystic-scientist hero able to lead Western civilization to a utopia and his own "architect-engineer" who could shoulder the sacrosanct duty of evolving "an adequate shelter design that will make possible the rational and spiritual self-realization toward which man has ever so longingly striven."[198] Carrel and Fuller shared, with a somewhat similar mystical fervor, a belief in the power of a lofty brigade of geniuses to do miracles for humanity. Fuller employed the notion of the iconoclastic architect-engineer capable of renovating civilization in Carrel's sense of the hallowed genius, a political philosophy based on the fundamental precept of differential human categories. For civilization to rise from the morass of degeneration, Carrel argued, Western societies must understand the inherent defects of the tenets of equality embedded in the philosophy of democracy: "Human beings are equal. But individuals are not. The equality of their rights is an illusion. The feebleminded and the man of genius should not be equal before the law."[199] Although Fuller moderated Carrel's reactionary politics, he too believed in the superiority of certain racial categories, as well as in the technologically induced development of the genius, albeit through the creation of an ideal shelter.[200] For him, the idea of a decaying civilization—plagued by the profit-driven industrial–capitalist network (his villain in *Nine Chains* is Mr. Fincap, whose name is a satirical acronym for financial capitalism) and mediocrity in leadership—urgently called for a lofty figure like the architect-engineer. In his daring imagination, Fuller himself fulfilled this calling.

Fuller was not in any way an ideologue of racial hierarchy, and he did not seek a eugenic revolution through a propagandistic campaign like those of Carrel and Lindbergh. It is hard to pinpoint Fuller's political ideologies, if any, beyond his self-righteous fury born of a host of personal grievances and failures; yet he unwittingly absorbed many well-circulated ideas of the time without necessarily understanding their long-term social implications. These ideas acquired the appearance of normalcy by rapidly permeating the cultural veins of American society. Although Fuller was speaking within the context of house design, it is not hard to see how the following statement reflected the social contour of the period's eugenic landscape: "It is a law of evolution and design that designs, whether by man or by 'nature,' are reproduced in direct proportion to their mechanical adequacy of satisfaction of universal requirements, whether it be a book, a rose, a pencil or a baby."[201]

Like Lindbergh and Carrel, Fuller endorsed the idea that renovation of humanity would be possible through the growth of a "northwest civilization" inspired by the spirit of New England Puritans. Carrel's preface to the 1939 edition of *Man, the Unknown* was instructive:

> In the United States, the upper third of the population reproduces much less rapidly than the lower third. Europe and the United States are thus undergoing a qualitative as well as quantitative deterioration. On the contrary the Asiatics and Africans, such as the Russians, the Arabs, the Hindus, are increasing with marked rapidity. Never have the European races been in such great peril as today.[202]

In this scenario of gloom and doom, Carrel argued, it was imperative for a proactive social movement to enforce eugenic principles in a civilized world. In a somewhat parallel vein, Fuller praised New England Puritans—or the "longing type," using Einstein's category for motivated people—as pioneers in building the New World in the image of a mentally evolved and racially pure utopia, a project in which imported African "slaves" became a mere prosthetic hand for the Puritans' body. In a chapter titled "Longing Crosses the Sea," Fuller argued that the building of the New World took place not because of the Old World's necessities but because of the fortitude of "mind-over-matterist" pilgrims and seafarers:

> Colonization of the new continent was accomplished . . . by an extraction of a relatively "pure" longing type from out of the confused, "Middle-Ages" world. The colonizers longed for liberty; longed for freedom of philosophic thought, or physical exploration, and for economic freedom, either by deed or invention. They longed, too, for freedom in which to breed a new race, no longer enchained by confused religions, traditional esthetic art, and superficial, material, intrinsic values. . . . America was colonized, on the one hand, by communes of thinkers, and, on the other, by buccaneers who "blackbirded" slaves from Africa to till the soil of Virginia. It is significant that the mind-over-matterists, the Pilgrims, and others who in due course developed the inanimate slave, landed in the north, whereas materialists who perpetuated the animate slave settled in the south. This was in accord with the northwest progression of the mind-over-matterist and industry. . . . Shiploads of immigrants, eschewing all raiment and gadgets reminiscent of the decadent artistic glamour of the civilization they had left behind, landed *en masse,* and, with equanimity, faced the hardships and difficulties of communal survival. Being of the longing type and embodying in large measure the scientist-artist, they succeeded rapidly and efficiently industrializing their survival problems. . . . the Pilgrims did not come to this country as conscious pioneers of the industrial principle, but by segregation through longing happened to be its most able progenitors.[203]

In developing a theory of the "longing type" or the "scientist-artist," Fuller accepted the prevailing theory of racial categories in which the northwestern European stock represented the most advanced human condition and the New World's colonization and beneficial development were the natural consequence of that condition. To buttress this argument, Fuller introduced a model of civilization's progress: the "northwest spiral of civilization."[204] That is, material advancement of each age perpetually "crops out to the west and north of the last development," while the rest of the world becomes the passive recipient of this advancement or awaits colonization for its enlightenment. Advanced technologies, such as radio and aviation, were the material reflection of the collective "longing" mind that enabled the northwest spiral of civilization.

In Fuller's model of progress, America was cast with a Turnerian nostalgia; that is, the country's westward expansion began beneficently from the northern Puritan enclave in New England, and the frontier reached all the way to the West Coast. In the "democratic" expansion of a bountiful land was the manifestation of the mental prowess of a "longing" type. Fuller, however, was not a historian; his interests were neither exclusively in civilization per se nor in the historian's critical appraisal of the western frontier. His theorization on the growth of civilization was concerned about the strategic prophecies of a futurist who operated with the working assumption that there had always been a malleable future in every step of America's formation. If that useful future were the handiwork of a few evolved men rather than of the masses, then the mental alacrity of those few must be perpetuated in order to legitimate the very theories of a workable future.[205] This is where Fuller's theorization on civilization intersected with those of other eugenic-minded reformers of the interwar era. Carrel ominously asked, "Are we capable of renovating ourselves, of avoiding the cataclysms which are imminent, and of continuing our ascension?"[206] And, even though their goals were different, Carrel's question seemed similar to Fuller's plea to create a society ruled by the dynamic "longing" type, the pure product of the northwest spiral of civilization. In explaining the northwest spiral, Fuller wove his vision of the future by drawing, if inadvertently, on the period's populist eugenic discourse.

Following in the footsteps of social Darwinists, Fuller subscribed to the idea that evolutionary advancement could be scientifically measured by the mind always moving in the "direction of least resistance." Fuller proposed a sweeping history of mankind set on a self-propelled course of least resistance: "The history of man's CREATIVE effort is the story of his struggle to control 'direction' by the ELIMINATION of known RESISTANCES."[207] Unfortunately, in Fuller's somewhat glorified narrative of "least resistance" in the arc of human progress, ideologically inconvenient "aberrations" such as the violence wrought in the project of colonialism and the forced displacement of African natives and indigenous populations were left out of historical consideration. Thus, in the operational history of the creative mind that propels the northwest spiral of civilization, the "elimination of known resistances" acquired a disturbing eugenic hue.

But is it fair to implicate Fuller in the complicated histories of the American eugenics movement and American expansionism, with the benefit of our enlightened hindsight? We can comprehend the full scope of Fuller's humanist program, envisioning a mentally advanced society, only by critically appraising Fuller against the social currents of his time, rather than against the cloistered mythology he and his biographers created. The political meaning of his espousal of the aviator as a model for human evolution, or the architectural determinism embedded in Dymaxion House as an enabler of evolutionary consciousness, cannot be separated from the eugenic impulses of technological utopianism during the interwar years. Fuller's romance of the aerial frontier, captured in his pithy mantra "Vertical is to live," was enmeshed with his meritocratic views and his endorsement of elitist stratification of society. Only when we trace those connections and intersections do we have a useful history of Fuller or what Edward Said called the "tragic limitation" of historical figures' inability to see beyond the dominant ideologies of their own eras.[208] The ideological weight of the northwest spiral of civilization—placing Fuller's New England ancestry at the eye of that spiral—was so heavy that Fuller failed to see the consequences of colonial and ecological violence that such a model of human growth entailed. Yet this failure does not hide his technohumanist view that the world's resources could be optimized through "creative control" enforced by the "scientific-minded."[209] In fact, only when we examine Fuller engaging with various ideological currents of his time do we begin to see a flesh-and-blood Fuller, not a cloistered genius but an "idea entrepreneur" with a complex history of contradictions, conflicts, ambitions, manipulations, and good intentions.

But, ultimately, the utility of Fuller's own early autobiography lies not in his celebration of the airplane and verticality as essential tropes of American modernity, or in his Dymaxion House and his intellectual flirtation with evolutionary ideologies, or even in his exuberant nonsense in *4D Time Lock* or *Nine Chains to the Moon*. It lies in the *folie de grandeur* of his storytelling. Fuller's autobiographical narrative culled a range of interrelated objects and ideas to construct a self-portrait of a visionary builder, as evolved as an aviator with holistic views of the world and omnipotent to the point of being able to rebuild civilization. The fact that Fuller created an aggrandized self-myth of a master builder to overcome personal failures is less appealing than *how* he created that myth by selecting certain images and ideas and then arranging them in a way that neatly constructed his tragic-heroic image. Virginia Woolf called this biographical technique "creative fact": "Almost any biographer, if he respects fact, can give us the creative fact; the fertile fact; the fact that suggests and engenders."[210] Fuller knew how to transform the factuality of architecture and his life into fertile thoughts on human progress. His placement of a toy airplane in the miniature Dymaxion House was a conscious biographic gesture, one with which he narrated his life in ascensional terms. The mathematical empiricism of nine human chains to the moon was no less a totemic symbol of his ambition to renovate humanity. To transform the fact into the fertile fact, Fuller needed an uplifting vertical frontier.

THREE

The Master Builder as Superman

Norman Bel Geddes's Futurama

"I HAVE SEEN THE FUTURE"

On April 30, 1939, President Franklin Delano Roosevelt inaugurated the New York World's Fair, designed to be "the biggest world's fair in history."[1] While it celebrated a historic moment—the 150th anniversary of George Washington's inauguration as the first U.S. president in New York City—the fair was planned to show that, as Roosevelt said during his opening speech, "the eyes of the United States are fixed on the future."[2] Conceived by the New York elite and their corporate sponsors to dispel the economic woes of the Great Depression by revitalizing consumerist beliefs, the 1939 fair, with all of its futuristic follies, has acquired an iconic status in the history of world's fairs.[3] The organizers of the fair appropriated the task of transforming the fairground at Flushing Meadows in Queens from what was known in popular parlance as Corona Dumps into a didactic instrument of rational and progressive planning.[4]

The fair presented seven thematic zones: Transportation, Production and Distribution, Government, Food, Community Interests, Communications and Business Systems, and Amusement. The zones vied to outdo each other by showcasing fantastic exhibits and simulated journeys into what the event's organizers called the World of Tomorrow. According to a special survey conducted among World's Fair visitors by the American Institute of Public Opinion under the direction of George Gallup in May 1939, Futurama—presented as part of the automobile giant General Motors' Highways and Horizons exhibit in the Transportation Zone—"far outranked" all other exhibits in popularity.[5] As superlatives abounded, the media seemed unanimous in their designation of Futurama as the "number one hit show" of the fair.[6] Among the forty-five million visitors during the fair's two seasons, in 1939 and 1940, nearly twenty-five million reportedly saw Futurama. In E. L. Doctorow's best-selling novel *World's Fair,* Depression-era boy-protagonist Edgar sums up Futurama's allure: "That was everyone's first stop. . . . I didn't mind the long wait we had, practically an hour."[7]

Futurama's spectacular reception was not a surprise, however. The exhibit's creator, American industrial designer Norman Bel Geddes, was well-known for his flamboyant design style. His clients, colleagues, and friends alike considered him to be extravagant, visionary, and a "master salesman."[8] His obituary in *Industrial Design* in 1958 summed up his life thus:

> Norman Bel Geddes charged through his 65 years powered by what friends and former associates have unanimously called "genius." He was guilty of all the excesses which that label implies—massive fantasies, a cavalier attitude toward money, and an absolute conviction of the brilliance of his endeavors. But beneath these eccentricities lay a passion to create, a passion whose giant scope aroused public awareness to the possibilities of industrial design. The geography of Geddes' ideas roamed from functional ashtrays to twenty-engine airplanes.[9]

Henry Dreyfuss, a fellow industrial designer, noted wryly that as a trailblazer, Bel Geddes lived "in an ivory tower . . . showering ideas on those of us who stood below."[10]

Bel Geddes's professional career began in the period from 1913 to 1916, when he held positions at multiple Detroit-area advertising agencies. After stints in stage design in Los Angeles—where he met Frank Lloyd Wright in 1916—Bel Geddes eventually gravitated to New York City, where he spearheaded a new direction in stagecraft. In 1927, in the midst of a thriving career in stage design, he suddenly changed course to embrace the fledgling profession of industrial design, wherein he pioneered the aesthetic development of streamlining.[11] Presenting alluring impressions of velocity, efficiency, and progress, streamlining became a ubiquitous visual hallmark of the 1930s. For many, Futurama represented the culmination of Bel Geddes's streamlined projection of a hyperefficient America.[12]

Designed to be a panorama of the future, Futurama was spectacular in all its details and experiences. Under General Motors' corporate sponsorship, Bel Geddes hypothesized the future as a one-acre animated model of an American utopia as it might appear in the year 1960 to citizens traveling in an airplane, a transportation technology that, in the late 1930s, still fascinated a whole generation of Americans (Figure 3.1). Touted in a General Motors leaflet as "the largest and most lifelike model ever constructed," Futurama was a 35,738-square-foot visual extravaganza containing approximately "500,000 individually designed houses; more than a million trees of 18 species; and 50,000 scale-model automobiles, of which 10,000 are in actual operation over super-highways, speed lanes and multidecked bridges."[13] Reportedly built in eight months by nearly eight hundred people, Bel Geddes's exhibit translated the future into a

Figure 3.1. Spectators at the 1939 New York World's Fair gazing down at the Futurama model. Harry Ransom Center, The University of Texas at Austin. Courtesy of the Edith Lutyens and Norman Bel Geddes Foundation.

captivating spectacle in which a wholesome ecological package of "abundant sunshine, fresh air, and green parkways" seamlessly blended with a continental highway network, power infrastructures, streamlined skyscrapers, and futuristic airports.

Convinced that total and rational organization was the path to progress, Bel Geddes cast a continental planning grid over the entire country. In developing Futurama, he drew on an eclectic range of works by architects, planners, and experts, some of whom he knew personally. At a conceptual level, he was influenced by the period's predictable roster of planning prophecies: the center-oriented zoning of Le Corbusier's Ville Contemporaine de trois millions d'habitants (1922) and its reincarnations in the 1930s, the decentralist agro-urbanity of both Frank Lloyd Wright's Broadacre City (1935) and Lewis Mumford's regional planning (Mumford scripted the popular documentary *The City,* which debuted at the fair), and the science fiction fantasy of H. G. Wells's things to come.[14] "In the manner of a super-salesman," to quote the *New Yorker,* Bel Geddes, however, balanced out the impracticalities typically affiliated with prophecies by meshing them with a pragmatic investigation of landmark works in highway engineering. To this end, he tapped into the traffic research of Miller McClintock, chairman of the Harvard Traffic Bureau and a consultant for an earlier planning project by Bel Geddes, and the autonomous freeway concept of Benton MacKaye's "townless highway" (1930).[15] Thus, fusing an array of ideological contents and practical projections, Futurama aspired to be the culmination of early twentieth-century planning visions.

Scholarship on Futurama generally overlooks this fusion by overstating the role of the automobile in the exhibit's planning.[16] A national highway system crisscrossed Bel Geddes's America, presenting a trope of corporate capitalism's economic mantra of unfettered movement of goods and people. Yet Futurama was much more than highway engineering. It was strategically designed to be seen as a comprehensive "investment in the future," to quote Alfred P. Sloan Jr., GM chairman at the time.[17] This investment neatly blended technology's public benefits and corporate interest into one convincing civilizational ensemble. Bel Geddes's exhibit prophesied a synthetic American utopia built on an assortment of cutting-edge technologies: remote-controlled multilane highways, power plants, farms for artificially produced crops, rooftop platforms for individual flying machines and autogyros, and various gadgets, all of which were intended to create an ideal built environment and, ultimately, to reform society. Bel Geddes presented his ideas of the future with captivating realism and a sense of immediacy, striking a popular chord with an American audience eager to see beyond the economic convulsions of Depression-era America.

Yet it was not the spectacle of the future itself but the technique of seeing the future that made Futurama the "smash hit" of the 1939 New York World's Fair, attracting on average more than two thousand spectators every hour, or twenty-eight thousand a day. Carried above the gargantuan model by means of a suspended, winding conveyor belt in a manner that simulated the experience of flight, spectators attained an airplane view of the so-called World of Tomorrow (Figure 3.2). The eighteen-minute ride—along with a masterful manipulation of light, sound, and scale—created the illusion of a day-to-night aerial journey over the varied and meticulously crafted terrain of an American utopia. Futurama's simulated flight offered an effective visualization technique, as Bel Geddes noted in his book *Magic Motorways* (1940): "The visitor to a great American city in 1960 approaches it by air, in order to see the layout of the new design more readily."[18] Given that Bel Geddes belonged to the early twentieth-century airplane generation and was an amateur designer of futuristic aircrafts, his fascination with human flight was predictable. Yet, while the airplane was a ubiquitous symbol in the 1939 World's Fair's futurist agenda, Futurama's theatrical replication of the experience of flight on a mass scale provided novelty.

Bel Geddes self-consciously exploited the strategy of aerial viewing as the primary means to prove the superiority of his own vision of utopia over the existing paradigms, which apparently suffered from defective planning: "As

Figure 3.2. Spectators on the conveyor belt hovering over the Futurama model. Harry Ransom Center, The University of Texas at Austin. Courtesy of the Edith Lutyens and Norman Bel Geddes Foundation.

the spectator circles high above the city, he is able to compare the congested, badly planned areas of the 1930's with the well-organized districts of the newer city."[19] By virtue of their elevated position, Futurama's spectators grasped the coherence of Bel Geddes's World of Tomorrow, which disclosed multiple scales simultaneously, with city blocks in proportional relation to a highway system, as well as minutely observed, artificially controlled trees in glass domes. Was Bel Geddes's utopia constructed as a grand aerial epic, legible only to a flying spectator? Were Futurama's hovering spectators conceived as an integral part of the show's ambitious futurist projections?

When spectators exited after viewing the show, each was presented with a souvenir pin that read, "I have seen the future" (Figure 3.3). Although intended

Figure 3.3. Each Futurama spectator received this souvenir pin after visiting the show. Author's collection.

as a memento in keeping with the fair's official theme, "Building the World of Tomorrow with the Tools of Today," the pin highlighted two crucial and mutually reinforcing concepts underlying this much-celebrated exhibit. The first was the idea of the future as spectacle, and the second was the process of *seeing* that spectacle, suggesting an active, participatory viewership, implicitly manifested in the triumphant utterance "I have seen the future." The phrase underscores the process of seeing while drawing our attention to the spectacle of the future. In other words, *how* the future was seen had perhaps become more appealing than *what* was seen in it. There were other captivating futurist spectacles at the 1939 New York World's Fair—among them Democracity, created by industrial designer and former Bel Geddes acolyte Henry Dreyfuss, which was housed in the Perisphere (the iconic gypsum-clad white globe of the fair's Theme Center); and industrial designer Walter Dorwin Teague's City of Light inside the Consolidated Edison Building. But no other exhibit celebrated the act of seeing the future and, more important, the spectators themselves as vigorously as Futurama. The souvenir pin's self-assured pronouncement made an implicit argument that seeing the future was beyond the specular and intellectual capacity of ordinary folks. Therefore, by visiting the abstract and elusive space of the future, the spectator was in a privileged position to claim superior knowledge of things to come. The pin's populist slogan was a conscious strategy on the part of General Motors and Bel Geddes to convince the spectator that he (as we shall see, it was mostly a masculinist discourse) had just returned from a heroic odyssey to the future. As much as the exhibit itself, the spectator was "designed" to be an element—or even the protagonist—of Futurama's utopian narrative.

By placing the Futurama spectator in a virtual cockpit from which he surveyed the vast model, Bel Geddes consciously summoned the omnipotent ocular experience of the aviator, able to inspect the world from the sky. Like millions of American boys in the early twentieth century, he grew up admiring the Wright brothers. He spoke of his "Icarus syndrome" in his autobiography, *Miracle in the Evening*: "The Wright brothers had flown off a big cliff at Kitty Hawk, North Carolina, in their big kite. The latter intelligence was particularly interesting and exciting for me, particularly since Aunt Hattie had read to us the myth of Icarus who attempted flight with wings glued to his arms. It made me happy to think that someone had finally done the job."[20] During the heyday of aviation, Bel Geddes experienced firsthand the American fascination with the aviator, whose veneration reached absurd heights when the popular media began to describe the man in the cockpit as a new type of human being, who could both fulfill the eugenic dream of the perfect man of tomorrow and bring about a new civilization with the power of his godlike gaze on the world. In the aftermath of Charles Lindbergh's solo Atlantic crossing in 1927, Bel Geddes was swayed by the deification of the pilot of the *Spirit of St. Louis*; as his personal collection shows, he even bought a signed copy of Lindbergh's autobiography.[21]

Bel Geddes drew on Le Corbusier's portrayal of the aviator as the alter ego of an all-seeing builder. In *Aircraft* (1935)—a book Bel Geddes owned—the Franco-Swiss architect portrayed the airplane as an anthropological representation of the master builder's superior eye: "The airplane, in the sky, carries our hearts above mediocre things. The airplane has given us the bird's-eye view. When the eye sees clearly, the mind makes a clear decision."[22] Le Corbusier's reflection during a flight in 1929 over Rio de Janeiro encapsulated how modernist planners sought to cast themselves as aviators: once he was on the airplane, "the conception of a vast programme of organic town-planning came like a revelation."[23] Nowhere was this "avian" psychology more poignantly expressed than in Futurama, where the aesthetics of ascension exemplified a new kind of aviator hero who could be seen to resemble the early twentieth-century master builder, seeking to rebuild the world from his high perch of authority. In many ways, Futurama's visual technique proposed a quintessentially modern observer who exercised the same idealistic and authoritarian gaze that fueled the master builder's reformist dreams. The aviator's aesthetic experience of altitude appealed to the master builder's encyclopedic ambition, particularly in light of modernism's prescription of rational and geometric reordering of the existing city as a panacea for urban and social pathologies.

The 1930s popular superhero theme—as represented especially by the American icon Superman—offers a useful historical and theoretical vantage from which to explore Futurama's politics of aerial viewing.[24] As a populist personification of America's loftiest ideals embedded in the pursuit of truth, happiness, and justice for all—the need for which was strongly felt during the Great Depression—Superman's flying eye on beleaguered American municipalities had a peculiar resonance with that of the master builder, seeking to fix urban problems from godlike heights. Cultural historians view the superhero's rise (from the drudgery and immorality of a lower realm) as an allegory of escape from America's economic and social plight during the Depression; however, the notion of flight also produced heroic visions that conjured many urban utopias in the fields of city planning, literature, film, and science fiction during the 1920s and 1930s.

The ability to see things from above (and the purported moral authority that accompanied it) not only gave rise to a modernist logic of looking at the world but was also often seen as the necessary attribute of an all-knowing builder, representing a superior human race. The American historian Robert Rydell has observed the complicity of the eugenic mind-set with the interwar world's fairs' rhetorical production of sanitized "worlds of tomorrow."[25] As much as they were corporate propagandas to advance a consumerist economy, the world's fairs also provided ideal forums for popularizing the concept of "designing" highly evolved citizenry. Was Futurama's viewing platform—with a streamlined row of seated spectators, disciplined and smoothly gliding over a vast model—a modern allegory of the assembly line, delivering not consumer products but rather ideal producers? An analysis of the exhibit, especially its viewing technique in keeping with the cult of the aviator, can reposition our understanding of interwar modernist visuality as one in which the aesthetics of ascension and its associated revisions of culture, morality, and even evolution played a central role.

FUTURAMA'S VOYAGE EXTRAORDINAIRE

The long queue of Futurama visitors, inching along the serpentine ramp that penetrated the streamlined pylon facade of the General Motors Building, became a part of the iconography of the 1939 New York World's Fair (Figure 3.4).[26] As various press releases authored by Bel Geddes show, he took great pride in the public and media brouhaha that swirled around Futurama. He relished the fact that upon popular demand, the GM Building's hours of operation were extended to eleven a day to accommodate a growing number of visitors to Futurama.[27] He also noted the clandestine activities that reportedly ensued. People who did not wish to stand in line for extended periods paid as much as five dollars to get a spot near the entrance. Sightings of such illustrious personages as Robert Moses, J. P. Morgan Jr., and John D. Rockefeller Jr. in the Futurama visitors' line lent credence to the ideological program of Bel Geddes's exhibit.

Figure 3.4. Fair visitors waiting in line to enter the General Motors Building to view Futurama. Harry Ransom Center, The University of Texas at Austin. Courtesy of the Edith Lutyens and Norman Bel Geddes Foundation.

From beginning to end, the experience of Futurama was one of sheer theatricality thanks to Bel Geddes's pioneering efforts in the field of three-dimensional theater design (as opposed to conventional frontal picture-frame proscenium theater design) during the 1920s.[28] In Futurama, his creation of what one author has called "synthetic space," a streamlined spatial arrangement that led to "a blurring of distinctions between the visitors and their environment," had its root in the "New Movement" in theater design.[29] The proponents of the New Movement advocated noncommercial stage design dictated by the specific artistic and emotional content of the play itself. At the center of this reformist movement was the "New Stagecraft," which sought to create suggestive and synthetic stages by architectural means and to include "audience and performance within a single Modernist volume, with equally good sightlines and seating for all."[30] Bel Geddes was introduced to the tenets of the New Movement by a colleague, the American art critic Sheldon Cheney, who wrote *The New Movement in the Theatre* (1914) and founded the influential magazine *Theatre Arts* in 1916.[31] As a "theater architect" prior to becoming an industrial designer in 1927, Bel Geddes became active in the reform movement and created a number of stage designs that helped popularize the New Stagecraft in America. He believed that the "single rigid formula" of the proscenium–audience "clips the inspirational wings of writers."[32] He created a sensation in the American theatrical landscape with his synthetic design for the Austro-German theater director Max Reinhardt's production of *The Miracle* in 1924.[33] With his customary aesthetic flamboyance, Bel Geddes transformed the entire stage and auditorium of New York's Century Theatre into a unified Gothic cathedral, embedding the audience within the play's action.

Bel Geddes's conceptualization of Futurama harked back to his theatrical experience. The whole Futurama ensemble in the General Motors Building was created as a stage set in which spectators were also actors. A walk-through of the exhibit demonstrates this phenomenon. The snaking ramp deposited spectators at the mouth of another zigzagging ramp inside the building that descended into a chiaroscuro auditorium or "map lobby" (Figure 3.5). As they slowly came down, they gawked at a vast cutout map of the United States that seemed to be suspended in a void, perhaps a two-dimensional precursor to Bel Geddes's continental project that spectators were about to witness. The existing roadways lit up "in a spidery green filigree on the map," in contrast with Bel Geddes's bold, straight highways of tomorrow, glowing in red electric bands.[34] A soft voice prepared spectators for an impending flight to the future, ironically, with nostalgia for the western frontier: "The History of American roads is the history of our civilization as it marched westward from the Atlantic to the Pacific—roadways forging ever onward through mountain, desert, and forest barriers, leaving in their wake great thriving cities, industrial centers, and prosperous farms."[35] Just when spectators were inspired by the patriotic

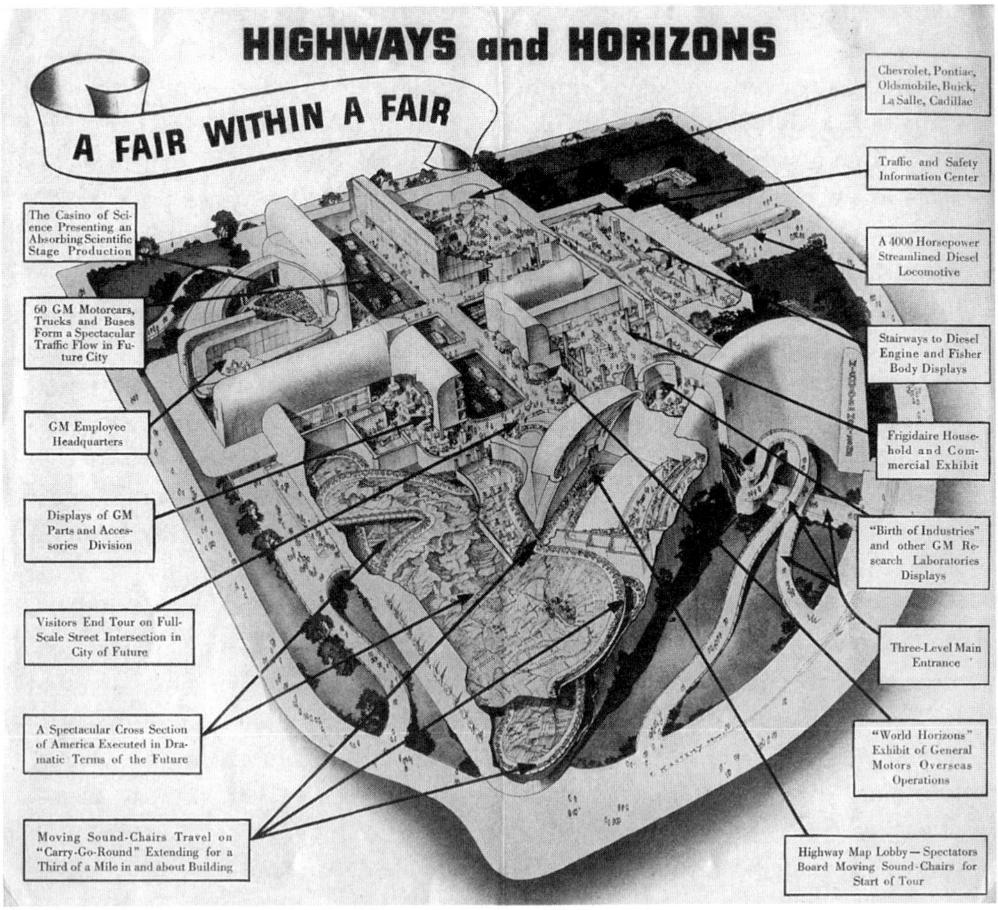

Figure 3.5. An axonometric cutout of Futurama and the General Motors Building. Your Guide to General Motors Highways and Horizons Exhibit, New York World's Fair 1939, *General Motors publicity brochure (1939).*

register of nineteenth-century frontiersmen, a kinetic platform with upholstered chairs positioned in pairs glided by, below the giant map, to whisk the Futurama visitors to the future. Since the conveyor and the chairs were in constant motion, passengers were loaded into them by means of "flat moving belts flush with the floor which travel[ed] at the same rate of speed as the chairs."[36]

Dubbed the "carry-go-round" or "mobilounge," the 140-ton conveyor system was a futuristic behemoth. Constructed by Westinghouse Elevator, it carried 552 seated spectators at a time, covering a winding path a third of a mile long above the model at different heights. Consisting of a continuous chain of platforms, it snaked along a 1,600-foot double-rail track system with the flexible mobility of a conveyor-elevator-escalator combination.[37] As if it were a theatrical and technological analogue of what scholars of American utopian novels called "ahistorical devices of time travel," Futurama's conveyor belt performed the peculiar role of an imaginary bridge that transported spectators from the real world of 1939 to the utopian space of 1960.[38] Once spectators took their seats, the conveyor belt passed through a semidark vestibule—that ambiguous transitional space connecting the real with the fictional—while an avuncular voice from a sound system concealed at shoulder level in each pair of chairs acted as a private guide to spectators as they traveled along an aerial route. The sound system itself was a novelty.[39] Developed by Electrical Research Product, the main unit of the system weighed 20 tons and was capable of delivering 150 different simultaneous narrations synchronized with the particular vistas spectators were viewing along the track. Continuous auditory comments ensured a multisensory experience of the mammoth display, for the eye and ear worked in concert, buttressing the spectator's psychology of participation in the exhibit's utopian development.

It all began with a cheerful invitation to a simulated airplane journey across America: "Come tour the future with General Motors! A Transcontinental flight over America in 1960. What will we see? What changes will transpire? This magic Aladdin-like flight through time and space is Norman Bel Geddes' conception of the many wonders that may develop in the not-too-distant future. Now we have arrived in this wonder world of 1960!"[40] The wonder world revealed itself as a miniaturized utopia to the eyes of admiring spectators, aloft in a masquerading airplane that Bel Geddes set on a meticulously calculated twenty-four-hour aerial trajectory. Achieving a realistic experience of flight by means of a conveyor belt posed a staggering technical challenge, especially when the trajectory changed altitude frequently to provide distant and close-up perspectives. Furthermore, not only did the moving belt have to trek a serpentine and up-and-down route, but it also needed to travel forward and ascend and descend considerably, as entailed by the nature of specific sights.

There was also the challenge of simulating the sun's path as accurately as possible, a challenge that Bel Geddes met by putting together a comprehensive illumination plan that achieved a consistent direction of the sun's rays.[41] By careful manipulation of nearly five hundred concealed floodlights, he created the impression that the journey began on a sunny afternoon, perhaps to ensure maximum visibility at the outset. Filters over floodlights varied in accordance with the change in the quality of light, from afternoon to dusk to dawn and afternoon again. While the aerial journey itself followed Futurama's continental development of the so-called magic motorways, spectators received a comprehensive view of an all-engulfing utopia that meshed the hygiene of country living with the fruits of technology-enhanced urban life.[42]

Futurama's journey began with a sharp swing of the conveyor belt above an idyllic countryside marked by "sunshine, trees, farms, hills and valleys, flowers, and flowing streams" and inhabited by a farming community (Figure 3.6). As the mobile aerie reached different altitudes, spectators were allowed aerial

Figure 3.6. A partial view of the Futurama model. Futurama, *General Motors publicity brochure (1940).*

perspectives on a "vast cross section" of 1960s America: an experimental farm and greenhouse; an aeration plant, purifying and distributing water for irrigation; multilevel suspension bridges; a university center for higher learning; and a cloverleaf street intersection—all taking their places seamlessly in a geographic grid set by safe, radio-controlled seven-lane highways. "Night falls on the countryside," and spectators could now zoom in on housewives "serving supper to hungry families and farm hands." There was an "amusement park in full swing," with "boys and girls shrieking with glee on a pretzel-like sky-ride." Spectators "flew" over a "prosperous and thriving steel town, with efficient and safe access to all advantages within driving distance." At daybreak, "traveling high above the mountains and valleys below," they relished a "bird's-eye view of a paradise for vacationers." A succession of views of tranquil pastoral settings—a religious retreat, "seemingly growing from the rocks"; "a picturesque resort town"; "a mountain lake dam"; and a "winter hotel lodge"—gradually built up a lofty expectation for a grander and befitting finale to an aerial odyssey.

Creating the illusion of airplane views of an entire country was a daunting undertaking. The conveyor belt provided this experience. Sliding at a rate of approximately 120 feet per minute, spectators looked down through a continuous curved pane of glass toward the model, which comprised 408 topographical sections, each measuring 15 feet by 5 feet (Figure 3.7).[43] These sections were developed on the basis of aerial photographs of different regions of the United States provided by the pioneering company Fairchild Aerial Surveys of New York. Bel Geddes commissioned the company in early August 1938 to shoot 119 aerial photographs, for a fee of four dollars each, in different parts of the United States.[44] The many areas that were photographed included Central Park and mid-Manhattan (New York), Jersey City (New Jersey), Mount Wilson (Illinois), Baltimore (Maryland), Yosemite Valley (California), New Orleans (Louisiana), and St. Louis (Missouri). Additionally, stadiums, swimming pools, bridges, airports, cloverleaf highway intersections, and recreational parks were shot to inform the realistic aerial representation of these civic amenities in Futurama. Notes such as the following provided guidelines for making the model as a series of bird's-eye scenes. "When you see roads at certain angles in the far distance, no matter what the roads are made of, they appear to be grayish-white." "Examples of activity which we saw at 500 feet: Man throwing food to chickens which clustered around him, cows drinking, cows lashing and swinging tails, and a horse and wagon moving slowly along with a man walking beside it." "Red is the predominant color in both city and country." "There is an appalling uniformity of houses and appalling density and quantity of buildings in the city—they stretch out as far as you can see without thinning out at all." And, "Once the city stops, it ends quite suddenly, and real country starts—the city doesn't gradually thin out."[45]

Bel Geddes and his crew studied the photographs thoroughly to establish Futurama's scalar variation and environmental effects, ingeniously transforming the conveyor ride into the impression of a continental flight over various types of terrain and urban regions. Bel Geddes described the conveyor system as a surrogate "airplane eye" through which Futurama appeared as "a spectacle that unfolds a new kind of civilization in which industry, finance, and labor will all find greater employment—a vision of new frontiers of progress waiting to be conquered by those who will pioneer around the corner of tomorrow."[46] The designer also struck a deal with Eddie Rickenbacker, a veteran World War I ace, to fly members of his office over many areas on the East Coast so that they could observe the sprawling geography below and take notes on its various aspects.[47]

In developing a mobile eye suspended in the sky and surveying America from coast to coast, Bel Geddes was conscious of the significance of the western narrative in the nation's development. This consciousness, however, posed a conceptual problem that required a peculiar mediation between the past and future. The official "end" of the westward movement in the last decade of the

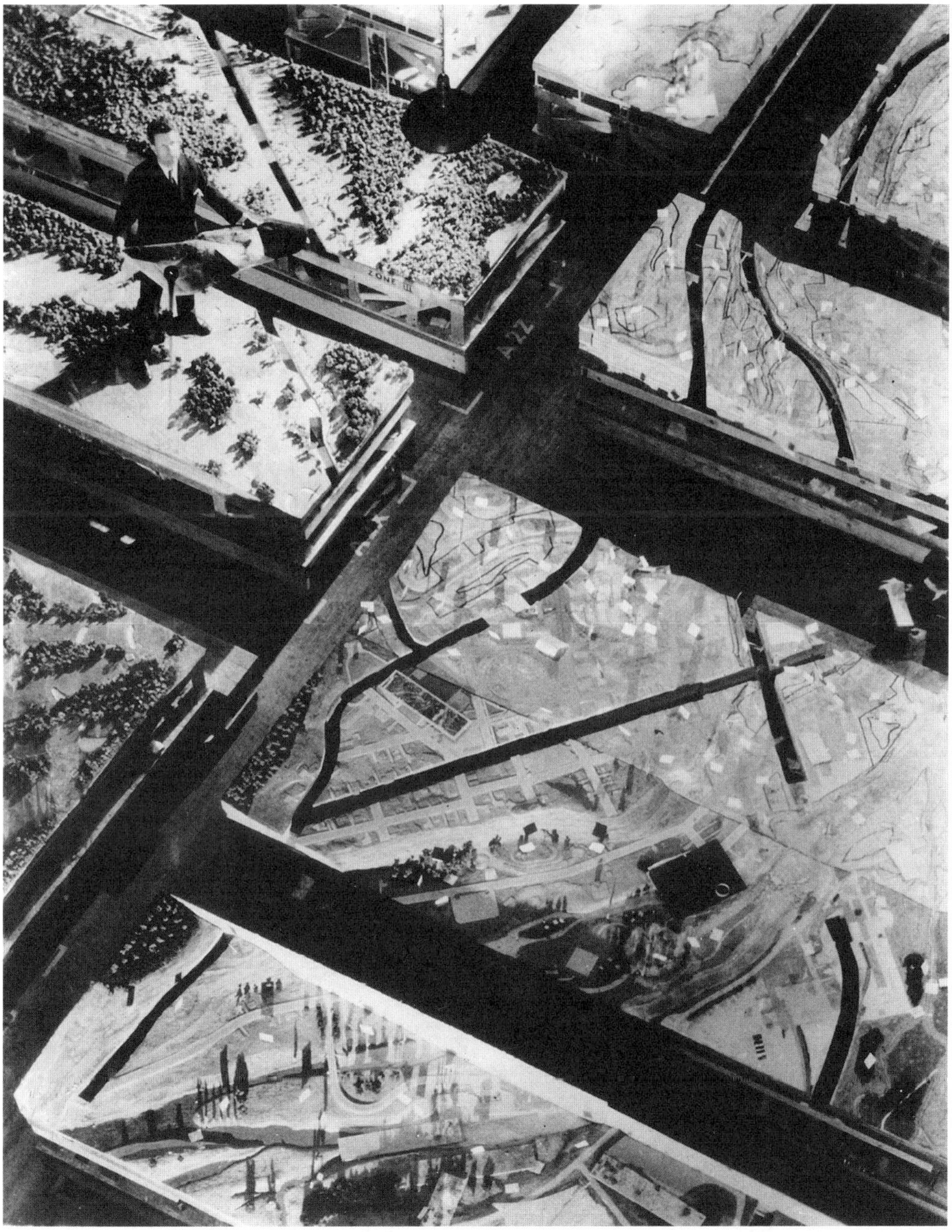

Figure 3.7. Topographical sections of the Futurama model. Harry Ransom Center, The University of Texas at Austin. Courtesy of the Edith Lutyens and Norman Bel Geddes Foundation.

nineteenth century caused a cultural anxiety, as the frontier was, as Frederick Jackson Turner argued, a shaper of the American character.[48] Therefore, the frontier's closure would mean its atrophy. Many urban observers negotiated this anxiety with the assumption that the American city would replace the western frontier.[49] For Bel Geddes, the great metropolis of 1960 was Futurama's logical final act, a new frontier in which the collective experience of taming the Wild West could be set to work more vigorously and efficiently for a brighter future. Instead of rejecting the past, Bel Geddes put it to work to build the future. It was a strategy that convinced fairgoers to gaze into the future through the patriotic filter of their own history. That the climactic destination of Futurama's trajectory would be the great metropolis seemed inevitable. Furthermore, by offering the metropolis as a new frontier to be surveyed and mapped from an aerial vantage point, Bel Geddes's exhibit combined what Albert Boime has called the frontiersman's "magisterial gaze" on the vast reaches of the wilderness with the modernist planner's emblematic eye of aerial inspection and reform (Figure 3.8).[50] With its nationalistic subtext in the form of

Figure 3.8. "Metropolis of the Future." Futurama, *General Motors publicity brochure (1940).*

manifest destiny, the frontier history promised, as Futurama's trajectory showed, new utility for the future.

As the journey neared the metropolis, glowing on the horizon with its sleek skyscrapers, the exhibit's concealed sound system whispered: "We will bank high over the city for a spectacular view of its many wonders. Today—in 1960—it is much larger, divided into three units, residential, commercial and industrial. The city of 1960 has abundant sunshine, fresh air, fine green parkways, recreational and civic centers—all the result of thoughtful planning and design."[51] Presented with a sentimental faith in the unending benefaction of rational planning, the great metropolis offered a harmonious blending of ecology with industry, the past with the future. As the telos in the narrative of progress, Futurama's city of 1960 served Bel Geddes's ambition to ground the World of Tomorrow with the mythos of the frontier.

Futurama's teleology also complemented the interests of General Motors. Bel Geddes's theatrical method provided spectators with the illusion of cobuilding the future and brought to the fore a new paradigm of participation in the visual politics of display. Such a participatory method signaled a break with world's fairs' traditional focus on finished exhibits that generally produced admiring, but ultimately disengaged, spectators. As the cultural historian Warren Susman has observed, the 1939 New York World's Fair endeavored to redefine the exhibit–spectator relationship by celebrating itself as the "people's fair."[52] This redefinition, which was one of the guiding principles of the 1939 fair, was tied to a new emphasis on the display process rather than on the exhibits themselves. Susman notes:

> The real genius of the exhibitors at the Fair . . . was their understanding that the machine itself was not to be central, as it traditionally had been in all world's fairs since 1851 and the Crystal Palace. Rather, they realized that in a consumer-oriented society people ended up more fascinated with process than with machines. This Fair *showed* its visitors the processes. In this respect the intellectuals who planned the Fair and the people who attended it may have found some measure of interaction.[53]

Susman further argues that the 1939 fair could be credited for "a new element of singular innovation." This innovation was best articulated in an editorial published in the August 1940 issue of *Architectural Record*: "The greatest discovery in New York was the discovery of the crowd as actor and as decoration of great power."[54] Susman identifies Futurama as one of the most successful exhibits "where the crowd took on a decorative pattern or where it was effectively used to fill space to excite the interest of other visitors." Employing people as actors rather than as passive consumers in the exhibit's broader scheme of things certainly leveraged the corporation's goal of creating an economic utopia by doing away with the distinction between the consumer and the producer. During the 1930s, many corporations realized, as Roland Marchand has observed, that in the age of consumerism, the public would take most interest not in the isolated finished product but rather in the ways the product would both contribute to and neatly fit into the American Dream.[55]

Futurama reflected this broader shift in display engineering, serving as a shrewd corporate advertisement that strategically blurred the boundaries between education and promotion. When presenting his ideas for the exhibit to the executives of General Motors in early 1938, Bel Geddes reiterated that his project was conceived as a "scientific and educational" panorama viewed through the eyes of a spectator "as though he were in an airplane."[56] Audacious as he was, he recommended a game plan to not present Futurama as a General Motors display at all, but rather to set it up as the result of an independent research project (conducted by him) that the auto giant had the honor and privilege to use as its feature exhibit.[57] Such an emphasis would not only help remove the suspicion with which the public viewed corporate advertisements but also encourage fairgoers to see Futurama as an objective solution to problems that affected their present and future. This approach initiated an immediate and sustained convergence of mutual interests, prompting the GM executives to increase the initial construction budget for Futurama from two million dollars to more than seven million dollars. The top executives—William S. Knudsen, president of General Motors; Richard H. Grant, vice president for sales; and Charles F. Kettering, vice president for research—were convinced of the long-term benefits of a display technique that would yield a comprehensive understanding of a highway empire (dominated by their new line of sleek GM cars).[58] The aerial perspective offered by the conveyor belt would, they assumed, allow spectators a synoptic view of the role of a national highway system in realizing the World of Tomorrow, ensuring a rapidly climbing sales curve for GM autos.

Furthermore, Futurama's engineering of a dramatic spectacle and bold aviation theme would, on one hand, lend credence to GM's unwavering commitment to the future; on the other, it would empower ordinary fairgoers with the feeling of being cobuilders of an ideal future.[59] Knudsen argued: "It will allow the man in the street to project himself into the future, to be part of that future, and to see the motor traffic of the future in action, together with its corresponding social and economic implications. The exhibit will offer a dramatic visual demonstration of how progress in transportation is related inseparably to progress in civilization."[60] For his part, Bel Geddes saw in Futurama a golden opportunity to distill his prior experiences in theater design, industrial design, architecture, urban planning, and transportation and highway engineering into a seminal project. The theatrical production of an idealized future and, more important, the creation of a protagonist spectator-builder of this future would be the hallmarks of Futurama.

In the climactic narration of the multimedia exhibit—"We will bank high over the city for a spectacular view of its many wonders"—Bel Geddes essentially re-created the aviator's visual regime and the modus operandi of the master builder. In this sense, the Futurama spectator's experience of aerial viewing was enmeshed in broader conceptualizations of early twentieth-century modernist visuality, revealing the crucial presence of the aesthetics of ascension in the imagination of the future world. If the protagonist of the aesthetics of ascension was the aviator, Bel Geddes consciously deployed this figure—masquerading as Futurama's spectator—as capable of reaching both the literal and the conceptual vantage from which he could envision his utopia.

But how does one transform utopian dreams—dreams that historically remained in the realm of literary forms—into a three-dimensional exhibit, realizable in the near future, without losing the sense of its visionary character? What kind of visual condition would not just reveal the utopia in its totality but also reconstruct the spectator as a surrogate of the utopia's creator, poised above his creation? In a bid to address these conceptual challenges, Bel Geddes had Futurama's spectator literally fly to an American utopia. On a practical and visual level, the simulated flight over America of 1960 was intended to resolve the optical limitation of earthbound views or the pedestrian's fragmentary experience of space on the ground. In 1936 Walter Benjamin memorably described the modernist angst at being caught up in what he called the "prison-world": "Our taverns and our metropolitan streets, our offices and furnished rooms, our railroad stations and our factories appeared to have us locked up hopelessly."[61] These were the kinds of spatial concerns that guided Bel Geddes's ascension in Futurama; to rise above the object of inquiry was to see it in its entirety. Futurama's continental motorway may have been a key planning element, yet a road journey itself as a narrative device would have not liberated spectators from the "hopelessness" of the pedestrian's prison-world.

On a conceptual level, however, Futurama confronted a crucial philosophical problem: how to experience the utopia that—as a hypothetical condition with ambivalent assumptions about the philosophical, political, and theological nature of being—eludes a physical articulation. Since the late nineteenth century, American writers of speculative fiction had sought to resolve this problem by employing various literary ruses (such as the voyage extraordinaire, time travel, long sleep, hallucination, and mesmerism) to transport protagonists inside what Frank and Fritzie Manuel have called the geographically and historically problematic "shadowy boundaries of utopia."[62] As they explain:

> In the course of time, "proper" utopias, discussions of utopian thought, and portrayals of utopian states of consciousness have so

> interpenetrated that perimeters of the concept of utopia have to be left hazy. From the time of its first discovery, the island of King Utopus has been shrouded in ambiguity, and no latter-day scholars should presume to dispel the fog, polluting utopia's natural environment with an excess of clarity and definition. Thomas More himself could not get straight the exact length of the bridge that spanned the River Anydrus at Amaurotum in Utopia.[63]

To describe Utopia, More needed to employ a character named Hythloday, a philosopher-traveler who visits the ideal kingdom and describes its social and urban morphology. In *Looking Backward* (1888), Edward Bellamy's idealistic hero Julian West, after a hypothetical centennial sleep, wakes up in a future urban frontier (driven by such radical technologies as aircars, telephones, and television).[64] Bert Smallways, the central character in Wells's novel *The War in the Air* (1908), explores a futuristic city from the window of his "aircraft." Shangri-La, the antimodern utopia depicted in Frank Capra's film *Lost Horizon* (1937), is reached, ironically, when an airplane crashes in an imagined vacuum of history and geography.[65] Experiencing the utopia—which perpetually evades our epistemological boundaries—requires, as in Futurama, an equally fictional mode of spectatorship. A make-believe aerial journey and the gaze from a simulated sky provided Futurama with a symbolic bridge to the utopia. Furthermore, Futurama was premised on the idea that the constructability of a perfect social and spatial system also required the articulation of an ideal observer—in this case, the aviator.

The aviator's view of the world was, however, hardly a utopian dream by the end of the 1930s. Exploratory and commercial aviation, airmail service, aerial photography, and aerial surveys had already made elevated perspectives on geography, landscapes, cities, and buildings part of the popular consciousness. Aware of this mass appeal during the interwar period—especially within the futuristic ambition of the 1939 New York World's Fair itself—Bel Geddes must have viewed Futurama's aviation motif as a crowning moment in the teleological narrative of technology. The fascination with flight may have lost its electric moment by 1939, but Futurama's novelty remained in the theatrical way it mass-produced the experience of the aviator, along with all of its cultural valorization, as a prism through which spectators could gaze into the future.

THE ASCENSION THEME AT THE 1939 NEW YORK WORLD'S FAIR

Bel Geddes's Futurama was by no means the only project to embrace the theme of ascension at the 1939 New York World's Fair. The fair's official dream of "Building the World of Tomorrow" was symbolized at the fairground by ubiquitous imageries of ascension, wings, aerial gazes into the future, and airplanes.[66] An aesthetic of ascension permeated the fair's architecture, sculpture and mural programs, films, official posters, guidebooks, and paraphernalia. The fairground itself became the object of a highly idealized perspective from what an official brochure called "the airplane view."[67] The illustrator Harry M. Petit, well-known for his futuristic, airship-filled "Dream of New York," published by Moses King in 1908, used his aerial drawing of the fairground in the fair's official guidebook to glorify the principles of modern master planning, as evident in the title of the drawing, "From Dump to Glory" (Figure 3.9).[68] A photograph in which New York City mayor Fiorello LaGuardia happily inspects a model of the planned New York fair at Marshall Field department store in Chicago in October 1938, during a national tour promoting the fair, offers a testament to the cavalier ways that the fair organizers envisioned the transformation of the marshlands and dumping grounds of Flushing into a planned park that itself would thematically represent America as a progressive utopia (Figure 3.10).[69] If the world's fairs are walled-off showcases of lofty ideals, as Burton Benedict argues, then LaGuardia's complacent gaze endorsed those ideals while mirroring them by the very visual authority he could exercise over the master plan of the fair.[70]

One of the three films that debuted at the fair, the American Institute of Planners' *The City* (1939), scripted by Lewis Mumford and directed by Ralph Steiner and Willard Van Dyke, is a prime example of the fair's buoyant view of the future.[71] The documentary was idealistic, although framed with a Ruskinian tragic view of technological modernity in which the industrial city became

Figure 3.9. Harry M. Pettit, "From Dump to Glory." Official Guide Book, New York World's Fair: The World of Tomorrow, 1939 *(New York: Exposition Publications, Inc. 1939).*

Figure 3.10. Fiorello LaGuardia inspecting the 1939 New York World's Fair scale model. Photograph courtesy of Frank Cronican.

a wasteland of despair, environmental pollution, and cog-like urban masses. Critics generally interpreted the film as a propagandistic proposal to remedy the unhygienic spaces of the modern industrial city, which Mumford and the small but influential Regional Planning Association of America believed could be saved by a pastoralist greenbelt conception.[72] The good life could be ensured not by wholesale mechanization and automobiles, but by a restoration of the sense of healthy living and social well-being associated with Ebenezer Howard–style community-based garden cities.

Despite its tacit critique of industrial modernism, the documentary revealed a peculiar affinity for technology. The Mumfordian city's nostalgia for the countryside and its purported goodness due to its congruity with nature could be understood from the heights that only an airplane could provide. Toward the middle of *The City,* in a section called "Science Takes Flight," a shimmering DC-3 airplane takes the audience along on a 2.54-minute aerial ride to see the bounties of living harmoniously with nature, away from the chaos of the existing machine city. But the camera is not always inside the airplane, peeking through the window; rather, it is often above it, granting heroic views of both the airplane itself and the fleeting greenbelt urban model below. The view is complemented by the narrator Morris Carnovsky's assured voice (and Aaron Copland's background music): "This new age builds a better kind of city, close to the soil once more. As molded to human wants as planes are shaped for speed." The camera's doubly functional perspective establishes an unmistakable relationship between the evolution of technology and that of the city itself.

The penultimate scene of the documentary includes a close-up shot of two "typical" American boys framed against the sky, both happily looking at a shiny miniature airplane clasped in the hands of one of them. The scene is loaded with dual but mutually inclusive symbolism. First, aerial transportation is the unmistakable representation of the future. And second, the protagonist builders of that future are the boys of today. Informed by the period's eugenicist fascination with the aviator, the scene suggests that the boys will pilot humanity toward a resplendent future. This romance of the airplane as a tool enabling a gaze into an ideal tomorrow was a leitmotif of the 1939 New York World's Fair.

The Austrian-born graphic artist Joseph Binder's official poster for the 1939 fair encapsulates the fair organizers' penchant for representing human progress in ascensional terms (see Plate 7).[73] Binder won first prize in the New York World's Fair poster competition in 1938. With simple graphics, he reinforced the vertical thrust of the seven-hundred-foot Trylon—a three-sided obelisk, part of the fair's Theme Center—with a synchronized squadron of nine airplanes speeding to the new stratospheric frontier in which a single star shines.[74] The pixilated skyline of New York City, composed at the foot of the Trylon, appears to be the urban alter ego of the airplanes' upward movement. The airplanes are also placed at the pinnacle of an evolution in transportation technologies. The message appears to be that in the World of Tomorrow, airplanes will be much faster and more convenient than earlier earthbound modes of transportation—the railway and ocean liner. The fact that Binder's poster was used on the cover of the first edition of the fair's official guidebook alludes to the extent to which the fair organizers subscribed to an ascending World of Tomorrow.

The fair's skyward sentiment was expressed globally in July 1938, when Howard Hughes, the acclaimed American aviator and entrepreneur, flew around the world in record time in a Lockheed 14-N Super Electra named *New York World's Fair 1939*.[75] This flight demonstrated how much the fair's promoters were eager to exploit the American romance with the aviator. A description of Hughes's flight by Grover A. Whalen, president of 1939 New York World's Fair, Inc., reveals how much aviation appealed to the fair organizers as a surefire symbol of the World of Tomorrow:

> Word reached me about this time [1938] that Howard Hughes was contemplating a round-the-world flight. We were well under way with our plans for the New York World's Fair, and quite naturally the thought of tying in Hughes' flight as part of the promotion struck me as a good one.
>
> To our great delight, [Hughes] announced that he had decided to name his plane the New York World's Fair and that

> he would carry the message of the Fair with him to thirty countries. . . .
>
> The incredible accuracy of Hughes' predicted arrival at each of his stops was featured by press and radio throughout the world. His flight played a most important part in heralding the "World of Tomorrow" and helped the Fair greatly.[76]

Whalen was no stranger to extravagant public receptions of celebrity aviators and other dignitaries. As New York City's "official greeter" under Mayor Jimmy Walker, Whalen organized the ticker-tape parade for Charles Lindbergh in New York on June 13, 1927.[77] He accorded one to Hughes, too, in 1938, fully knowing the potential of its effectiveness in publicizing the 1939 fair. Cognizant of the American veneration of the aviator, Whalen shrewdly blended a popular myth with corporate marketing propaganda. Reminiscent of how the St. Louis business establishment sponsored and named Lindbergh's airplane *Spirit of St. Louis,* Hughes's airplane could not be a more apt global advertisement for the fair. After all, the New York World's Fair was the international exposition for 1939—recognized in May 1937 by the Bureau International des Expositions, an international body that governed the participation of nations in world's fairs—and included the participation of sixty foreign governments and international bodies, the highest number until then in the history of world expositions.[78] The figure of the aviator, Whalen reasoned, would appeal to an international audience as a universal icon cohering a global fraternity.

That appeal was fully exploited within the fair's futurist ambitions and evolutionary propaganda, which were often fraught with eugenicist allegories. Numerous ascending (male) figures populated the World of Tomorrow, as attested, for instance, by the cover of *1001 Facts about the World's Fair and New York* (Figure 3.11). With wings attached to their muscular physiques, men of tomorrow glided over the vertiginous morphology of New York City or took off from the Theme Center's ramp, called the Helicline, to reach the higher plateau of a perfect future. The fair's organizers, architects, industrial designers, sculptors, and mural artists cleverly exploited the wing theme as a utopian vessel—harking back to the mythos of Daedalus and Icarus, as well as to Leonardo da Vinci—that would transport humanity to an idealized tomorrow. Examples abound: Robert Foster's free-floating stainless-steel human figure with a cape, which adorned the crown of the Ford Motor Company's pavilion; the architect James Gambrel Rogers's winglike pylons that flanked the entrance rotunda of the Chrysler Motors Building; Joseph E. Reiner's sculpture *Speed,* depicting a woman on a winged horse bursting forth from a fountain pool in the fair's Court of Communications; and the mural portraying a winged figure afloat in the sky on the entrance facade of the Hall of Music, designed by Reinhard and Hofmeister. These representative works evince how powerfully the wing metaphor

Figure 3.11. 1001 Facts about the World's Fair and New York, *2nd ed. (New York: The Dreier Hotels, 1939).*

perpetuated the fair organizers' collective gospel of progress and evolutionary aesthetics.

Gertrude Vanderbilt Whitney, the American sculptor and founder of the Whitney Museum of American Art in New York City, contributed to the fair a sculpture called *To the Morrow,* popularly known as *Wings* (Figure 3.12). The sculpture, spanning twenty-four feet, consists of three wings without a body proper that merge into an inclined base on which two figures—one male and one female, probably meant to be the procreators of a new civilization—stand perfectly poised to ascend to the World of Tomorrow. During the work's dedication, while Whitney dubbed it an homage to the youth, Mayor LaGuardia called it a tribute to aviation.[79] These two perspectives were hardly mutually exclusive. From the vantage of the fair's conscious social engineering, the youth and aviation appeared to be two sides of one evolutionary coin. The attainment of the World of Tomorrow was predicated on a harmonious blending of human and technological evolution.

The wing was hardly a revolutionary symbol of the 1939 New York World's Fair; rather, the fair's organizers and its designers drew on a well-circulating visual emblem that epitomized the aesthetics of ascension during interwar America. The fluid lines of art deco and the unobstructed mobility of streamline

Figure 3.12. Gertrude Vanderbilt Whitney, To the Morrow, *1939. Reprinted in Larry Zim, Mel Lerner, and Herbert Rolfes,* The World of Tomorrow: The 1939 New York World's Fair *(New York: Harper & Row Publishers, 1988), 182.*

aesthetics could not be more convincingly represented by anything other than the wing, steeped in mythologies and religious veneration. For the medal that Bel Geddes designed in 1933 on the occasion of General Motors' twenty-fifth anniversary, he visualized the advancement of motor transportation as the kinesthetic power of a wing placed above a streamlined car (Figure 3.13).[80] In another poignant demonstration of this legacy, the cover of the December 1928 issue of *Popular Aviation,* which commemorated the twenty-fifth anniversary of motored flight, presented a symbolic juxtaposition of the visual history of aviation and a towering winged man of tomorrow, worshipped by inventors and scientists (see Plate 8). The romantic message of the image was, unequivocally, this: the telos of technological evolution was ultimately the modern era's own Icarus, who was both the symbol and the harbinger of a utopian society. The wing was a new halo.

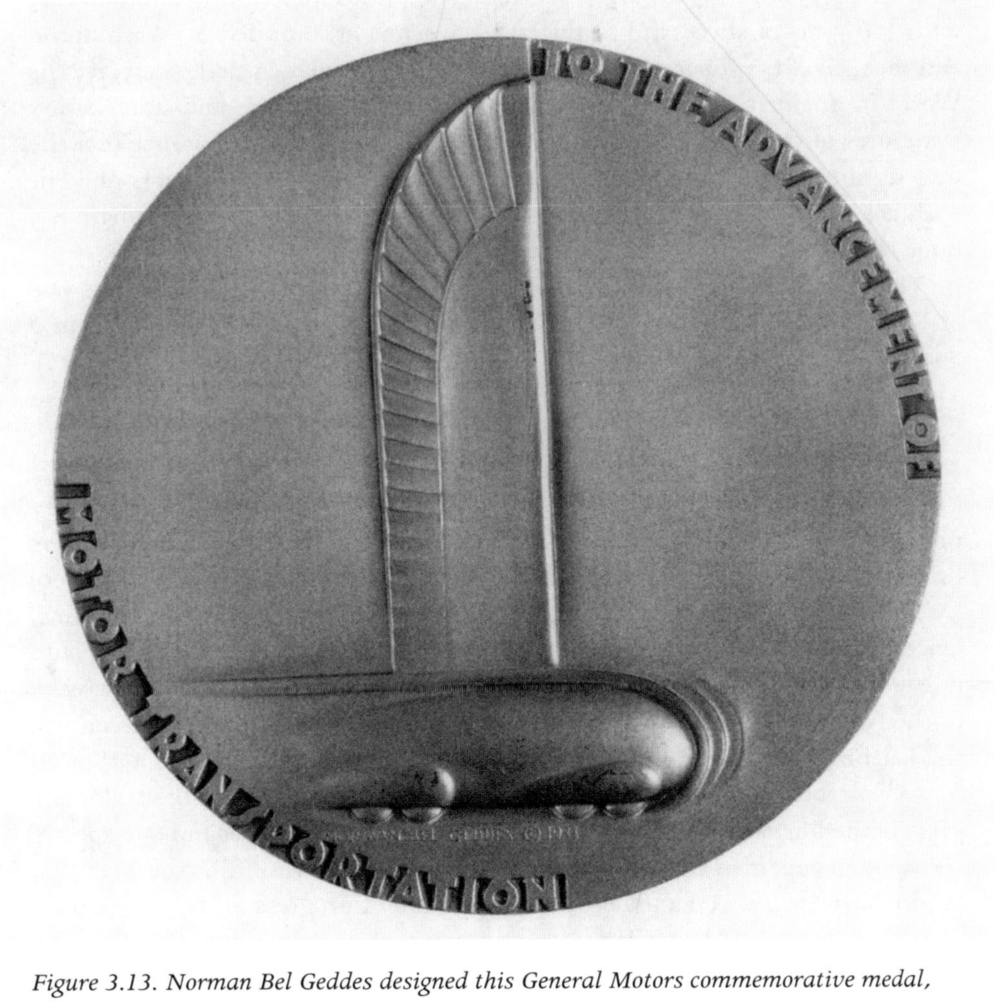

Figure 3.13. Norman Bel Geddes designed this General Motors commemorative medal, sculpted by Rene Chambellan and minted by Medallic Art Company, USA, 1933.

The 1939 fair staged its winged drama with spectacular exhibitionism, both outside and inside the fair's Theme Center, designed by the architectural firm of Wallace K. Harrison and J. Andre Fouilhoux. The Theme Center consisted of three discrete forms, namely, the Trylon, a 700-foot three-sided obelisk; the globular Perisphere, 200 feet in diameter; and the spiral Helicline, a 950-foot ramp.[81] Proffering majestic counterpoints to the dual threats of the Depression and the impending cataclysm of war in Europe through orchestrated optimism, the Theme Center was the navel of the fair's future-gazing campaign. On the central axis of the fair, which ran along the Constitution Mall between the Theme Center and the U.S. government's Federal Building, stood James Earle Fraser's mammoth statue of George Washington clad in his original inaugural robe. The positioning of Washington's statue on the axis was a self-conscious tactic on the part of the fair's planners to produce the illusion that Washington was gazing at the Perisphere, as Robert W. Rydell observes, "his back on years of progress, his eyes on the future. The philosophical suggestion is that with 150 years of successful democratic government, founded by Washington and the men of his generation, behind the nation of today, America can face the World of Tomorrow, represented by the huge, modernistic, and unorthodox structures of the Perisphere and Trylon, with the same cool assurance that the first president exhibits in his massive sculpture."[82] Rydell also notes one fair enthusiast's wry remarks: "Perhaps to George Washington, the Perisphere is a huge crystal ball, telling of the 'shape of things to come.'"[83]

The Perisphere was not, however, merely Washington's crystal ball; it was also like a mythical womb inside which the fairgoers found themselves in a godlike position to foresee the birth of the fair's "official" utopia (Figure 3.14). That utopia took shape in the form of the industrial designer Henry Dreyfuss's keynote exhibit Democracity, described in the lofty language of the fair's *Official Guide Book* as the "symbol of a perfectly integrated, futuristic metropolis pulsing with life and rhythm and music."[84] A highly choreographed journey preceded the drama of beholding Democracity inside the Perisphere. Fairgoers first entered the Trylon at the ground level and then ascended to the Perisphere by means of two escalators—considered the highest in the world at the time. Once they reached the entrance to the iconic white globe, they stepped onto one of two revolving balconies that ran around the entire inner circumference of the Perisphere, like two parallel latitudes. Revolving in opposite directions, the balconies took six minutes to make a complete revolution, the time allotted for the fairgoers to witness, from a simulated sky, the large and realistic diorama of a planned urban and exurban ensemble of 2039. The propagandistic spectacle was accompanied by a musical score by William Grant Still, conducted by Andre Kostelanetz, and a touchy-feely narration of progress in the megalopolis of the future, spoken by the popular newscaster H. V. Kaltenborn.[85] The *Official Guide Book* described Dreyfuss's blueprint of a harmonious civilization from the vantage point of a future aviating citizenry:

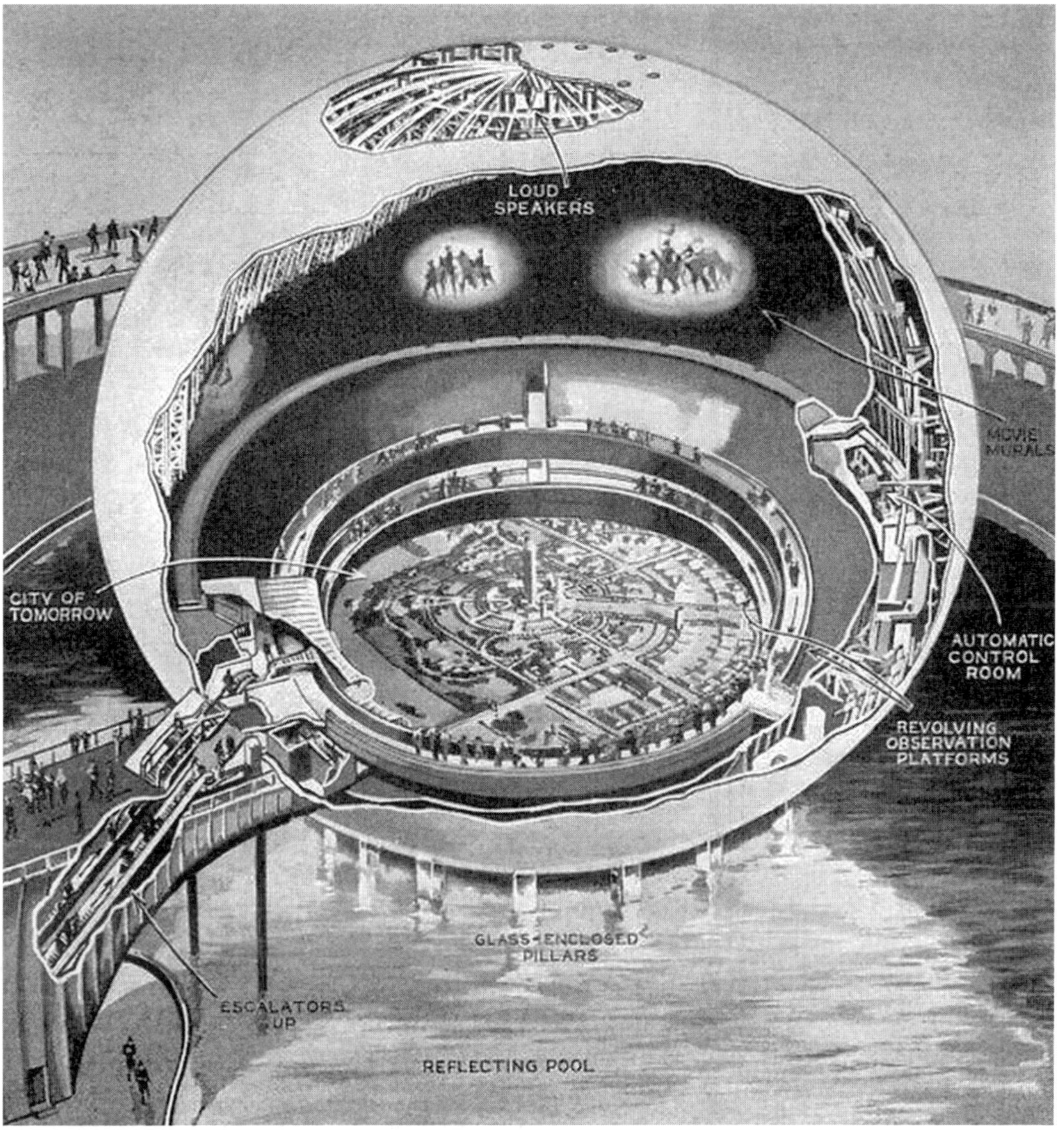

Figure 3.14. An axonometric drawing of the Perisphere showing Democracity in the interior.

> Here is a city of a million people with a working population of 250,000, whose homes are located beyond the city-proper, in five satellite towns. Like great arteries, broad highways traverse expansive areas of vivid green countryside, connecting outlying industrial towns with the city's heart. After you have gazed at the model for two minutes, dusk slowly shadows the scene. The light fails, and the celestial concave gleams with myriad stars. To the accompaniment of a symphonic poem, a chorus of a thousand

> voices reaches out of the heavens, and there at ten equi-distant points in the purple dome loom marching men—farmers, stamped by their garb; mechanics, with their tools of trade. As the marchers approach they are seen to represent the various groups in modern society—all the elements which must work together to make possible the better life which flourishes in such a city as lies below. The symphony rises to diapasonal volume, the figures assume mammoth size; the music subsides, the groups vanish behind slowly drifting clouds, and suddenly a blaze of polaroid light climaxes the show.[86]

Including a visual compilation of the period's popular urban planning ideologies, Democracity extolled the glories of democracy, market economy, and a technologically streamlined life available to all. The exhibit's mantra of progress and marching forward toward an ideal republic was further dramatized by a chorus of voices singing the fair's theme song (rhymed by Ira Gershwin): "We're the rising tide come from far and wide / Marching side by side on our way, / For a brave new world, / Tomorrow's world, / That we shall build today." The theme song's reference to Aldous Huxley's best-selling novel *Brave New World* (1932) invoked the image of progress as a free-for-all solution while avoiding Huxley's dystopic interpretation of corporate social engineering.

Democracity was, above all, a theatrical performance in which the spectators on the revolving balconies were also designed to be seen as part of the exhibit. If the theme exhibit offered an architectural manifestation of the brave new world, then the fairgoers perched on the revolving balconies gazing down evoked, as did Futurama, the image of its master builder. Like Futurama's conveyor belt, Democracity's balconies played out an aviation theme. Such visual technology exemplified the fair's persistent display methods, which engaged spectators, as Warren Susman and Roland Marchand have noted, in ways no previous world's fairs had attempted.[87]

Democracity highlighted how the fair's corporate sponsors and designers endeavored to simulate the experience of the World of Tomorrow by incorporating aviation themes. During the halcyon days of air travel and aerial photography, infusing key exhibits with the futuristic promise of human flight galvanized the fair's visitors. The most popular thematic area of the 1939 fair—the Transportation Zone, which included the General Motors Building, the Ford Building, the Marine Transportation Building, and the Chrysler Motors Building, among others—abandoned the conventional nonparticipatory display of transportation machines in favor of presenting these machines as performative experiences. It is not surprising that, according to one survey, the fair's two most utopian extravaganzas—General Motors' Futurama and the Theme Center's Democracity—topped the list of fairgoers' favorite exhibits.[88]

Donald Deskey's Focal Exhibit of the Transportation Zone, housed in the Chrysler Building, featured a mixed-media history of transportation that climaxed in the departure of a commuter rocket for a suborbital trip to London. The *New York Times* reported: "Here with the startling effects of light, speed, sound, and distance, a rocket ship is loaded into a gigantic gun and launched into the night, to go winging into the vast reaches of the sky toward London."[89] While stratospheric traveling, space colonization, and intergalactic voyages were the staple ingredients of interwar science fiction, they were presented at the fair not just as tenable technologies in the immediate future but also as events that would soon revolutionize space and time. Deskey's stratospheric trip was more about a reconceptualization of geographic distance than about faster transatlantic crossing. Howard Hughes's round-the-world flight had already set an optimistic backdrop for the Transportation Zone against which the fairgoers could inspect a fantastic exhibit like Deskey's, sharing in the process the exhibit's phantasmagoric meditations on a new planetary consciousness.

NORMAN BEL GEDDES'S AESTHETICS OF ASCENSION

The question of evolution, be it technology's or humanity's, has intrigued modernist designers since the early twentieth century. In deliberating this question, Bel Geddes's first book, *Horizons* (1932), showed a remarkable similarity to Le Corbusier's *Towards a New Architecture,* a manifesto of modern architecture that Bel Geddes had studied upon receiving it from his wife as a birthday gift.[90] Nowhere was the convergence of these two men's futurist missions more apparent than in their emphasis on the role of transportation machines, especially the airplane, in shaping the ideal environment of the future as well as humankind's experience of that environment. For both men, automobiles, ocean liners, and airplanes were not simply emblems of modernity but also conceptual sites where a multitude of modernist arguments concerning aesthetics and human progress were debated.[91]

In *Horizons,* Bel Geddes viewed the phenomenon of flight in at least two divergent ways. First, he embraced the concept of the aerial vehicle presenting "the same organic problems in terms of design as do architecture, sculpture, and literature."[92] Second, the flying machine was at the apex of the modern industrial society, with its ability to widen the "horizon that [would] inspire the next phase in the evolution of the age."[93] The first concept implied that the airplane was the consummate expression of the modern era because "the latest generation has been born to the air, as others of us have been born to the railroad, steamship and automobile."[94] Like many of his contemporaries, Bel Geddes viewed the airplane's aerodynamic design as the fulfillment of the modernist dreams of functionality and an economy of aesthetics:

> How out of place the moldings and gadgets that we see on our automobile would appear if we saw them on an airplane! When the airplane was developed, it was an all new problem. Its requirements were such that it never occurred to any one to base its design principles on, for instance, a carriage with wings. One may say that when the design of an object is in keeping with the purpose it serves, it appeals to us as having a distinctive kind of beauty. That is why we are impressed by the stirring beauty of airplanes. The underlying principle of the emotional response that the airplane stirs in us would seem to be the same as that which accounts for the emotional effect of the finest architecture—the form, proportion, and color best suited to that object's purpose.[95]

Bel Geddes's articulation of the airplane as a crucial site of a functionalist epistemology was expected, as was the case with many other modernist artists and architects. His juxtaposition of the rose window of Reims Cathedral and a Lycoming airplane motor sought to demonstrate that a suprahistorical leitmotif propels all monumental human endeavors toward functional and aesthetic perfection (Figure 3.15).[96] The striking visual similarity between a Gothic rose window and an airplane motor stemmed, Bel Geddes posited, from conscious adherence to certain irrefutable laws of natural composition and beauty. According to Le Corbusier, Bel Geddes, and others, the airplane's aerodynamic form—which met its specific functional needs with such economy that it took flight—proffered an occasion for modernist attempts to correlate aesthetic development with human evolution. Aesthetic phenomena, they reasoned, evolved parallel to

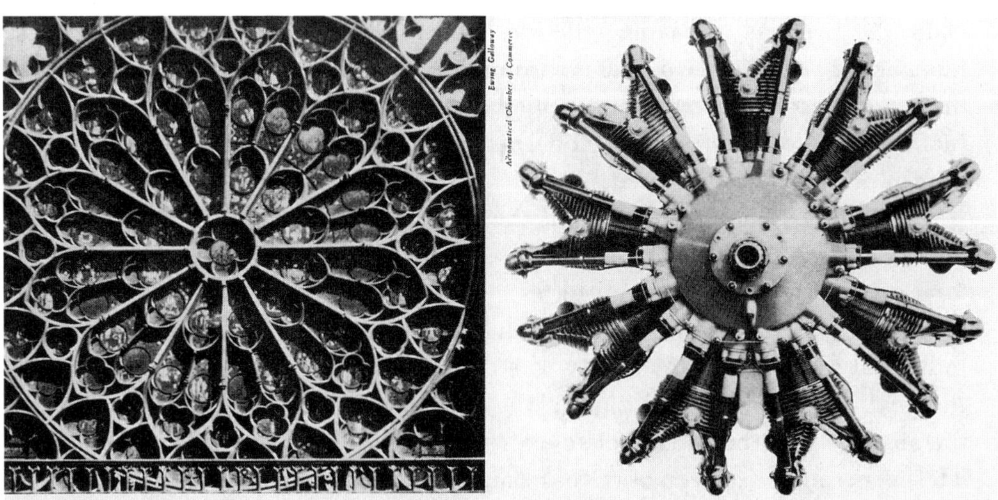

Figure 3.15. The rose window of Reims Cathedral and a Lycoming airplane motor, in Norman Bell Geddes, Horizons *(Boston: Little, Brown, and Co., 1932), 276–77.*

human development, and therefore the finest aesthetic expression would be the one that reflected the most advanced condition of the human species.

In this sense, the airplane, with its progress-defining streamlined form and functional supremacy, symbolized the perfect man, the epitome of its species and climactic result of natural selection. Darwin lurked behind Le Corbusier's anatomical allegory of the airplane: "The aluminium framework of an airplane—search for economy of material, for lightness, always the fundamental, the essential law of nature. Similarly in the marrow of our bones, the same fibres 'of equal resistance' exist."[97] For him, the merging of Icarus and the flying machine was the necessary precondition for an aesthetic revolution: "New machines, new men. They are filled with enthusiasm, the pleasures of daring, of breaking with current stupidities. Once in the air, carried along by the wind, they exult in the daring of their departure."[98] Bel Geddes's colleague and fellow industrial designer Raymond Loewy presented an "evolution chart of design" in 1930 that put humans and transportation machines (including the airplane) on the same quasi-Darwinian evolutionary ladder, arguing that the most functional and aesthetic machine, like its human counterpart, reflected nothing but its inexorable evolutionary climax.[99] As the culmination of functional and aesthetic perfection, the airplane epitomized the most evolved man: the genius artist or a sort of superman, who blazed the evolutionary trail so far ahead of the rest that he resembled a flying machine in human form, a phantasmagoric hybrid aptly evoked by the familiar DC Comics exclamation, "It's a bird, it's a plane . . . it's Superman."

The airplane, for Bel Geddes, was more than a mechanical analogue to the Darwinian body, which shared its functional and aesthetic superiority. His superimposition of a cutout of the most popular commercial airplane of the 1930s, the Douglas DC-3, on a photograph of Futurama, the camera gazing down on an airport of the future—a collage later used in *Magic Motorways* (1940)—was intended as a surrogate for Futurama's spectator and his commanding aerial gaze (Figure 3.16).[100] Was the airplane presented as a totem of the spectator's expanded visual field, his new "horizons," to which the title of Bel Geddes's book alludes? Such a visual field, Bel Geddes added, enabled the visionary to "[batter] down the limitations of the new materials and ideas of their time."[101] When the horizon limits an individual's vision, he claimed, that person "is too likely to be influenced in a transaction by the immediate consequences than to see it in perspective as a part of his life as a whole."[102] Against this claustrophobic experience of "immediate consequences," the designer explained the aesthetics of ascension: "Standing on the shore of the ocean and looking out to sea, his [i.e., the artist's] horizon is two and one half miles away. Leaning on the rail of the promenade deck of an ocean liner and looking out to sea, the horizon is eight miles away. If he climbs to the crow's nest, his horizon has increased more than six times what it was when he stood on shore."[103]

Figure 3.16. Norman Bel Geddes's collage of a DC-3 airplane on top of a photograph of the Futurama "airport of the future." Harry Ransom Center, The University of Texas at Austin. Image courtesy of the Edith Lutyens and Norman Bel Geddes Foundation.

The discourse of the airplane as an enabler of a widening, or disappearing, horizon was common in aviation magazines of the early twentieth century; as *Flying* magazine noted in 1915: "Human flight has opened the sky to man, giving . . . unlimited freedom." *Flying*'s rumination and its Bel Geddesian echo suggested some kind of existential freedom from what the writer and urbanist Paul Virilio has described as an oppressive history of "perspectivalism" in which the horizon separated the infinite frontier of the sky from its finite earthen counterpart. The horizon imposed an epistemic boundary on what was knowable, blocking out the promise of what lay beyond. The visual conquest of the horizon, thus, entailed what Virilio calls a "zenithal perspective," alluding to the transcendental positioning of an omniscient spectator along the zenith.[104] The subjugation of the horizon, as Bel Geddes posited, was to be the primary attribute of the visionary artist—or the "Man of Tomorrow."[105]

It seemed natural that Bel Geddes would view his design work in the idealized image of the Man of Tomorrow. In this context, three designs by Bel Geddes prior to Futurama are particularly instructive: a transoceanic airplane called Air Liner Number 4 (the fourth installment in his aircraft design), an eatery called Aerial Restaurant, and the City of Tomorrow, an advertising campaign for the Shell Oil Company. To a great extent, the theoretical lessons Bel Geddes learned from these projects informed his conceptualization of Futurama.

Bel Geddes's rendezvous with aircraft design was not surprising.[106] Within two years after Lindbergh's maiden transatlantic flight, he tackled the problem of transatlantic flight on a grand scale from the viewpoint of luxury travel. In November 1929, upon the recommendation of the National Advisory Committee for Aeronautics, Bel Geddes employed the Brooklyn-based German aeronautical engineer Otto A. Koller, who had served as chief engineer and designer for the aeronautics division of the German government during World War I.[107] With Koller's assistance in "design, calculation, and construction of the airliner," Bel Geddes developed the design for Air Liner Number 4, an ambitious undertaking that dealt with a host of issues, from basic aerodynamic principles and interior arrangement to broader questions of spatial design in "avian" terms.[108] Bel Geddes himself studied the nuts and bolts of aeronautical design, which is demonstrated by the presence of a sizable collection of books on the subject in his personal library.[109] The craft's shape, dimensions, and capacity were impressive: it was a V-winged leviathan aerial vessel with a wingspan of 528 feet and sleeping accommodations for 451 passengers and 115 crew members. The professional correspondence between Bel Geddes and Koller reveals that Bel Geddes, despite his lack of professional training in aeronautics, was serious about building Air Liner Number 4 (a wind-tunnel model based on MIT specifications was in the works) and procuring the necessary corporate sponsorship.[110] Of course, in the wake of the 1929 stock market crash, no corporate promise loomed on the horizon, but Bel Geddes steadfastly held on to his fantasy project. His interest in aeronautics was, however, fired by something more than an aspiring industrial designer's longing to manufacture an airplane. Following in the footsteps of some of the contemporary modernist designers he venerated, such as Le Corbusier, Erich Mendelsohn, and Bruno Taut, Bel Geddes viewed airplane design as a protracted inquiry into the fundamentals of aesthetic philosophy. Curiously, his design of the transatlantic behemoth bore a resemblance to Le Corbusier's depiction of "the Airplane of To-morrow" in *Towards a New Architecture* (Figures 3.17 and 3.18).

The architecture of Air Liner Number 4 addressed the issues of compact planning, efficient spatial distribution, and avian formal expression. Inside, the craft featured a nine-story architectural honeycomb providing all of the amenities of a modern hotel; a three-story atrium was located at the front and center of the plane, flanked by a promenade at the level of deck 7, with large shatterproof glass windows running the entire length of the main wing. The layout of the multiple dining rooms, dance floors, cocktail lounges, a recreation deck with tennis courts, a gymnasium, a solarium, a library, a veranda café, and rows of private suites revealed its spectacular spatial program. The massive wing rested on two teardrop-shaped pontoons that served as crew quarters and storage space. Mounted above the main wing, a smaller secondary wing accommodated twenty propellers that would lift the amphibian into the sky. Bel Geddes

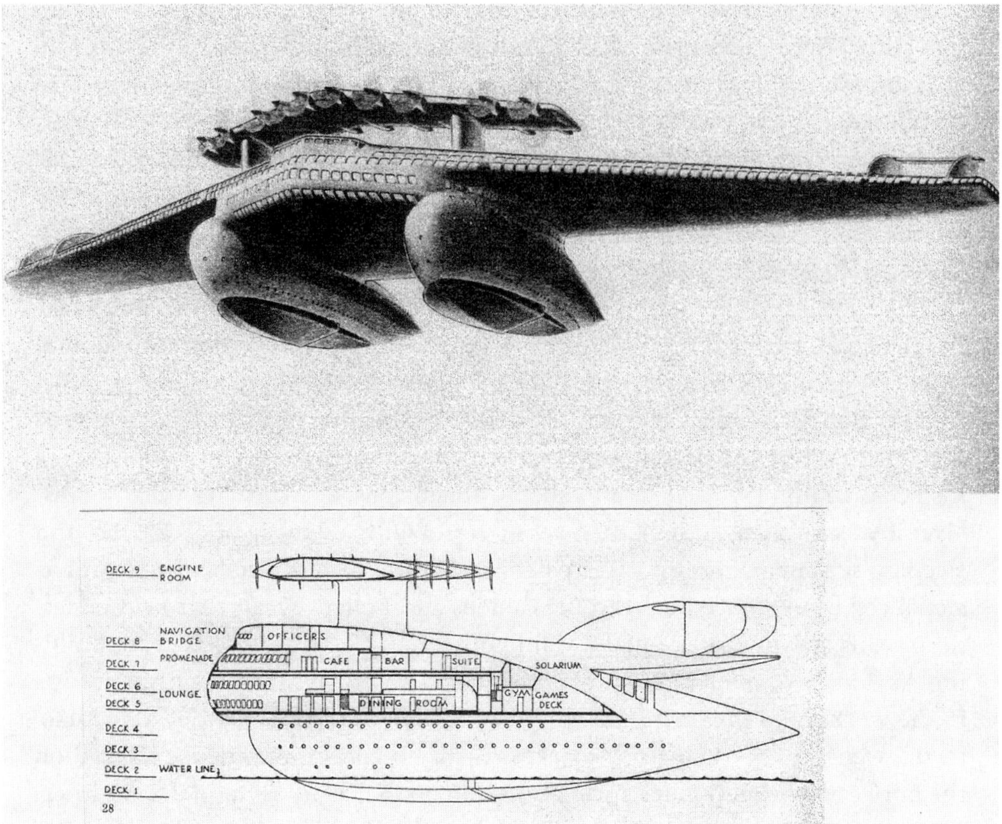

Figure 3.17. Air Liner No. 4 in Norman Bel Geddes, Horizons *(Boston: Little, Brown, and Co., 1932), 110, 115. Harry Ransom Center, The University of Texas at Austin. Courtesy of the Edith Lutyens and Norman Bel Geddes Foundation.*

anticipated the construction of Air Liner Number 4 by 1940, not an outlandish prediction, given that amphibious flying boats such as the Sikorsky S-40 were already in service by the time Bel Geddes and Koller came up with their proposal.

In terms of its architecture, Air Liner Number 4 was like a floating city. The gargantuan craft represented Bel Geddes's conception of an energy-efficient, smoothly functioning, and self-sufficient city—a utopia that heightened the drama of its autonomy by literally taking flight. By aerializing architecture, Air Liner Number 4 instantiated a conceptual shift that, as Virilio puts it, tilted the concept of architecture out of its age-old gravitational axis.[111] Bel Geddes's aeronautics produced the illusion of architecture freed from its archetypal dependence on the ground. If the earth was the foundation of architectonic knowledge—symbolized by the post-and-lintel spatial coordination of Marc-Antoine Laugier's primitive hut—then Bel Geddes's Air Liner Number 4 moved into a category of architecture for which the earth was now literally a receding memory.[112]

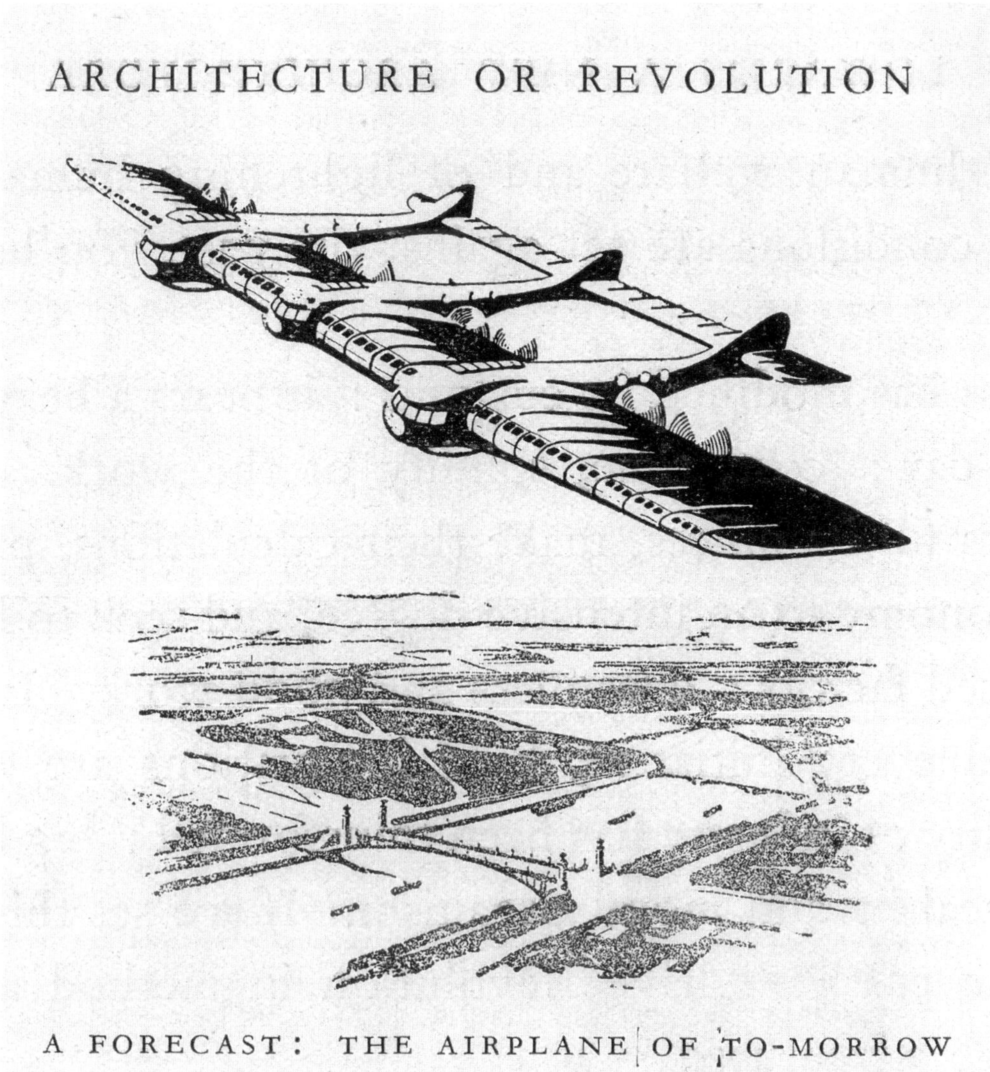

Figure 3.18. Le Corbusier, Transoceanic Airliner, circa 1923, in Towards a New Architecture *(New York: Payson & Clarke, 1927), 283. Copyright 2012 Artists Rights Society (ARS), New York / ADAGP, Paris / F.L.C.*

Many early twentieth-century avant-garde designers viewed architecture and, by implication, humanity as having been perpetually condemned to heaviness and being rooted to the ground, all of which implied immobility and servile attachment to hegemonic norms. As Italo Calvino suggested, the discourse of lightness that permeated modernist sensibilities offered an alternative logic of spatial modulation and "fresh methods of cognition and verification" in reviewing the nature of the universe.[113] Lightness did not simply mean making architectural structures less heavy; rather, it signified finding new ways of conceiving the form/ground relationship through a visual representation of antigravity. Bel Geddes read Nietzsche, the provocateur who suggested that if gravity is a euphemism for habitual servitude, tradition, or even history, then conquering it would indicate a triumph of radical individualism and an unhinged future. Observing the Perisphere and the Trylon—two daring realizations of the concept of "hovering" architecture in the 1939 New York World's Fair's Theme Center—the Russian avant-garde artist El Lissitzky commented: "The liberating of foundations from being tied to the earth goes even further and demands the conquest over gravity itself. It demands floating bodies, a physico-dynamic architecture."[114] Embodying the tenets of hovering architecture while dispensing with architecture's customary dependence on the ground, Bel Geddes's Air Liner Number 4 equated antigravity with a notion of heroic ascension.

Bel Geddes's 1929 proposal for an aerial restaurant was similar in this respect. Designed for the 1933–1934 Chicago World's Fair (but unrealized for functional and economic reasons), the project presented what could be called an avian reinterpretation of architectural form (see Plate 9).[115] Suspended, cantilevered, and floating atop a 278-foot vertical shaft, the revolving Aerial Restaurant—which the *Chicago Daily News* called a "unique structure designed for the Century of Progress"—revealed architecture's readjusted relationship with the earth. It was a mega-entertainment in the sky with a three-tiered restaurant. The lowest tier was 166 feet in diameter, seated six hundred patrons, and contained a sprawling dance floor for two hundred couples. At full capacity, the three diminishing decks could house twelve hundred people while serving clienteles with varying culinary needs. Housing nine service and three visitor elevators, the shaft was 32 feet in diameter and stood on a concentric podium that also housed an open foyer and a parking garage in the basement. A concealed circular kitchen on the ground floor served the hovering restaurant. Bel Geddes presented the sightseeing terraces at the top as the main attractions of the ensemble. A continuous glass window, from floor to ceiling, in all three tiers of the restaurant offered a panoramic view of Chicago. It was a typical Bel Geddesian architectural extravaganza that made quite a media splash.[116]

Aerial Restaurant's architecture appeared birdlike, simulating flight and separating itself from the "native ground"—to use Martin Heidegger's term—or from an originary ground "on which . . . man bases his dwelling."[117] If Heidegger's earth signifies a stable, immovable, and primal ground that anchors and orients architecture within the coordinate visual field of our upright posture, then Aerial Restaurant's architectural typology redefines the earth's significance. Despite being supported by a shaft, Aerial Restaurant appeared to take flight, providing another allusion to the master builder's ascension.

Bel Geddes's avant-garde aspiration for "flying" architecture should also be examined against the fantasy world of the science fiction magazines that began to saturate the American mass market in the 1920s. Hugo Gernsback, the first and most influential publisher of science fiction magazines in America, and Frank R. Paul, an architect better known as a science fiction illustrator, together created some of the most enduring images of floating cities that represented a phantasmagoric parallel to Bel Geddes's avian architecture.[118] The same year that Bel Geddes designed Aerial Restaurant, Paul illustrated the cover of the magazine *Air Wonder Stories* with a "City in the Air" (see Plate 10).[119] Based on a two-part story by a science fiction writer named Edmond Hamilton, Paul's fantastic illustration of a highly mechanized floating city was, interestingly, modeled on New York City. The description inside *Air Wonder Stories* stated: "Here we see the future air city of New York suspended high in the air kept aloft by the cosmic rays and made mobile by the mysterious propeller tubes. The city can rise above storms and, if necessary, above clouds to escape rains and snow. In the center we see the electrostatic tower which gathers the energy for the city's operation."[120] Paul's City in the Air simulated New York's fabled skyline, employing the beacon of architect William Van Allen's Chrysler Building or Shreve, Lamb & Harmon's about-to-be-completed Empire State Building as a symbol of energy and technological progress. What was curious in this type of gee-whiz description of floating cities was a recurrent theme of hygiene, suggesting that the autonomy of the suspended city would ensure freedom from all kinds of earthbound diseases, natural calamities, and social corruptions.[121] The discourses of lightness and height exhibited an affinity with those of purity and moral superiority.

Kathleen Church Plummer has shown how science fiction imageries nourished modernist designers' own brand of utopian thinking during the interwar period.[122] While speculative scientific fantasy remained its guiding force, Paul's airborne city brought to the fore a trademark fascination—beyond the funnies and pulp fiction—with aerialized architecture, a virtual reconstruction of the ascending human body itself. If gravity oriented architecture in certain telluric spatial relationships, then the City in the Air and Bel Geddes's Aerial Restaurant disclosed the possibilities of reconfiguring architecture, as well as its inhabitants, within a new type of visual field.

By offering an "aeroplane view"—as the news media dubbed it—of the fairground, Lake Michigan, and the Chicago skyline at large, Bel Geddes's Aerial Restaurant probed the theoretical limits of this recalibrated visual field. As the triple-decked restaurant made a complete revolution every thirty minutes, the patrons would be treated to a 360-degree vista, a novelty extolled by the architects Harvey Wiley Corbett and Raymond Hood during the Chicago World's Fair Architectural Commission's review of the project.[123] Bel Geddes himself highlighted the rotating aerial view in the *Chicago Daily News*: "I think this aerial restaurant is practicable, beautiful and worthwhile for Chicago to have, if for no other reason than to give its visitors a beautiful view of the city during the course of a meal."[124] The significance of such a view, beyond the lure of commerce and entertainment, however, must be pondered. By elevating restaurant patrons to an altitude of 278 feet, Bel Geddes sought to provide them with a grand aerial view of the fairground's master plan, which expressed the progressive ideals of the Chicago Fair. An expansive view would create an optical analogue of the fair's prophecies of progress by visualizing them in an ordered, axial, and functional ground plan.

Bel Geddes's flirtation with the exalted viewpoint from above found a systematic testing ground in his City of Tomorrow (1936–37) project, which laid the foundation for his more ambitious project for the 1939 New York World's Fair.[125] As a master showman well-known in the café society of New York, Bel Geddes had struck up personal friendships with many corporate leaders. Among them was Stanley Resor, president of New York–based J. Walter Thompson advertising agency, who provided him with various commissions for product design, including furniture for the Simmons Company, automobiles for the Graham-Paige Company, and a factory for the Toledo Scale Company. It was Resor who made a deal for Bel Geddes to develop a traffic-related advertising campaign for Shell Oil (Figure 3.19).[126] At the behest of Shell Oil Company, Bel Geddes began developing ideas for "the traffic conditions of the future," enlisting expert consultants for the project, including Miller McClintock, who promoted frictionless highways.[127] The contract with Shell Oil, signed in November 1936, required Bel Geddes to provide sixteen sketches showcasing potential solutions to urban traffic congestion in a typical American city circa 1960.

Figure 3.19. Shell advertisement from Life *magazine, 1937.*

Bel Geddes characteristically channeled the Shell advertisement campaign to look far beyond the corporate vision of unimpeded traffic movement, fleshing out his own ideas for a city based on mobility, efficiency, hygiene, and, finally, an aesthetic vision befitting the machine age. He proposed to create a scale model (eventually to be photographed for a Shell advertisement) depicting the pivotal role of interstate highway systems in the City of Tomorrow. He produced an intricate model, triangular in plan with 280 standard city blocks, with one of its six-foot sides representing approximately the width of Manhattan on a scale of one inch to one hundred feet. The triangular shape of the scale model allowed deep perspectives, which Bel Geddes further dramatized by using electric lighting to cast long shadows, producing a convincing impression of a realistic future city.

Drawing on, in particular, Harvey Wiley Corbett's 1923 proposal for the *Regional Plan of New York and Its Environs,* McClintock's corporatist ideas of frictionless automobile circulation, and, in general, the urban visions of Hugh Ferriss and Le Corbusier, Bel Geddes's project offered a grand vision of multilevel urban circulation through a skyscraper city. A September 1937 article in the *New Republic* described the project's appeal: "If the administration of New York City could start all over again from scratch and build a new metropolis according to the pattern of the Geddes model, it is estimated that five times as much vehicular traffic could be handled, and that traffic would move, on the average, about five times as fast as it does today."[128] In *Magic Motorways,* Bel Geddes juxtaposed the aerial views of Manhattan and the Shell Oil advertising model to promote his proposal's streamlined mobility through urban areas where "well-spaced towers rise amid light and air."[129] The Shell advertisement recalled, with striking visual resemblance, the exuberant cleanliness and geometrical grandeur of Le Corbusier's Ville Contemporaine and Ferriss's *Metropolis of Tomorrow* (1929), a book studied by Bel Geddes and his assistant Paul F. Berdanier, a Paris-trained visual artist who helped develop sketches for the Shell Oil project.[130]

What makes his proposal intriguing, however, was its filtration through another layer of representation: the photographic medium. From the beginning, Bel Geddes conceived the scale model for the City of Tomorrow as though it would be viewed through a suspended camera. The city's photogenic qualities, as seen from the air, became the project's overriding consideration. The model was photographed in an empty warehouse so that a camera could maneuver freely above it in a replication of Bel Geddes's own heroic vantage as he gazed down at the City of Tomorrow.[131] Peering down from an elevated platform or acrobatically perched at the top of a stepladder, Richard Garrison, a commercial photographer who knew Bel Geddes's penchant for showmanship and spectacle, created from the model a series of sublime urban views (Figure 3.20). Smoke bombs created the illusion of urban haze as well as clouds, which would testify to the camera's supposed airborne position. In the model,

Figure 3.20. The making of the City of Tomorrow model for Shell Oil (photographer Richard Garrison). Reprinted in Jeffrey L. Meikle, The City of Tomorrow: Model 1937 *(London: Pentagram Design, 1980), 24.*

Bel Geddes placed miniaturized versions of the Woolworth Building, Trinity Church, the National Archives, and Notre Dame de Paris among much taller skyscraper slabs. From July to November 1937, such well-circulated magazines as *Life* and the *Saturday Evening Post* carried Shell Oil's predictions on the city of the future—"as it might appear from about the level of a fifteenth floor"—accompanied by ebullient captions.[132] Garrison's hovering camera persuasively captured Bel Geddes's omnipotent gaze, seeking to solve the entire gamut of urban problems. The camera-reproduced gaze toward an ideal future was able to arouse the confidence of many powerful figures, including McClintock, who, in a slide lecture presented at the 1937 National Planning Conference in Detroit, praised Bel Geddes as a "master of functional shape and form" and extolled his project's ability to achieve "maximum facility for intercommunication."[133] Newspapers, too, bought into Bel Geddes's representation of an achievable utopia, calling him "the famous designer of the future." While on display at the auditorium of the J. Walter Thompson agency during the summer of 1937, the City of Tomorrow model was enthusiastically viewed by, among others, three highway engineers from the office of New York City Park Commissioner Robert Moses.[134]

During the course of the model's construction, the designer utilized a set of variously sized rectangular wooden blocks to represent skyscrapers—the tallest being fifteen inches—moving them around from above like a chess player intent on winning his game (Figure 3.21). This was a classic Bel Geddesian choreography that had a larger rhetorical underpinning, one in which he typically liked to have himself photographed and shown in control of determining the destiny of architecture and, by implication, the civilization it creates. From modest interior designs to large-scale projects, he represented himself in an audaciously imposing position to create a perfect world. His intense aerial gaze at a model

Figure 3.21. Norman Bel Geddes with his models. Harry Ransom Center, The University of Texas at Austin. Courtesy of the Edith Lutyens and Norman Bel Geddes Foundation.

became the symbol of that self-representation (Figure 3.22). A year later, in 1938, when working on the more ambitious Futurama project, Bel Geddes wrote to his wife about the quasi-divine experience of "walking around with pockets and hands full of skyscrapers" and laying them out for a "whole effect" from a bird's-eye view. In her reply, his wife said she wished that she had seen him "playing god . . . [and] plunking down skyscrapers where you want and spending millions as you choose."[135] This personal exchange had a larger resonance with the culture of modern planning. Behind the heightened drama of "playing god" lurked a persistent and enabling sense of self-aggrandizement that propelled the master builder's imagination. The experience of the City of Tomorrow foretold some of Futurama's key inquiries.

ENVISIONING SUPERMAN BUILDERS

If Bel Geddes's claim that the construction of the Futurama ensemble took eight months was true, then the process began sometime in late August 1938. The

Figure 3.22. Norman Bel Geddes working on a model. Harry Ransom Center, The University of Texas at Austin. Courtesy of the Edith Lutyens and Norman Bel Geddes Foundation.

final throes of a difficult decade revealed how the groundswell of anxiety and bitterness caused by the Great Depression provoked a range of social and cultural reactions: from ameliorative strategies (like the New Deal) and a renewed didactic focus on traditional mores as a buffer against adverse conditions (the theme of the 1936 film *San Francisco*) to future-minded fantasies (like the 1939 New York World's Fair).[136] Despite the distrust of the prevailing economic systems, there came anew a forceful desire to perpetuate and disseminate "those traditional values that emphasized personal responsibility for one's position in the world."[137] Amid the sense of disenfranchisement, the need to rearticulate what the nation traditionally venerated as the noblest socioracial standards in the shape of an ideal American character was strongly promoted, especially in popular culture. The aviating pop hero Superman was a poignant example of this sentiment. The superhero created by writer Jerry Siegel and artist Joe Shuster—two teenagers living in Cleveland, Ohio—debuted in the first *Action Comics* (June 1938), the title of which, "Superman, Champion of the Oppressed," alluded to Depression-era America's widespread disillusionment with establishment ideologies and, consequently, to the populist dream of a righteous builder of a just civilization.[138] Who would be better suited to rebuild an ailing America during panicky times than Superman, canonized as an embodiment of America's highest ideals?

Was Bel Geddes's idealization of Futurama's spectator—as if "playing god" from the aviator-builder's virtual cockpit to materialize a new America of 1960—logically analogous to the idealization of Superman? As somebody who inquired into the nature of visionary artists, geniuses, heroes, and Nietzsche's idea of the *Übermensch*—and who even presented himself as the "Man of Tomorrow" in an article published three weeks prior to the inauguration of the 1939 World's Fair—Bel Geddes was likely to be swayed by the Superman mythos when casting Futurama's spectator in the drama of the World of Tomorrow.[139] A fitting successor to Charles Lindbergh, as well as a crisp distillation of an American tradition couched in hero worship, individual endurance, and justice, Superman sparked the popular imagination at the same time that Futurama was being conceived.

The superman theme, however, was not new in American culture. Provocative projections of a superman, representing a superior human race, had already permeated American literature, media, and popular culture in the early twentieth century. As early as 1916, an advertisement for the Wright Flying School in *Flying* magazine sought to attract aviation students by dubbing the aviator "the Superman of Now" (Figure 3.23).[140] A eugenicist report in the *New York Times* in 1929, titled "Science Pictures a Superman of Tomorrow," suggested the role of human engineering in the collective progress of humanity: "Now science provokes serious thought by envisioning real probabilities of a superior race of human beings who in some respects may fulfill the dreams of fanciful

August, 1916 FLYING 273

The Aviator—The Superman of Now

The world has its eyes on the flying man. Flying is the greatest sport of red-blooded, virile manhood.

Make your vacation the greatest you ever had by joining the Wright Flying School. Live in the open—in the aviators' tent city. Convenient hotels for the fastidious.

A short course at the Wright camp will fit you to fly any type of machine. Expert instruction in flying, assembly, upkeep, motor-overhaul, etc. Dual controls. Pupil flies the first lesson. The school is located on Hempstead Plains—the greatest aerodrome in America.

Send for New Booklet

WRIGHT FLYING FIELD, Inc.
60 Broadway, New York

Figure 3.23. An advertisement for the Wright Flying School on Hempstead Plains, New York. Flying, *August 1916.*

writers. Will the man of tomorrow, endowed with the control of size, structure, characteristics and capacities that science is seeking, force evolution into new channels, establish a new mode of living and ordain a new, radical course for human development?"[141] While the superman theme had been present in American cultural and eugenic circles since the early twentieth century, Siegel and Shuster's formal introduction of the blue-suited, red-caped flying "Man of Tomorrow" in 1938 could not have offered a more appropriate citizen of the 1939 New York World's Fair's World of Tomorrow. Superman was, in fact, an integral part of the fair's popular iconography. Two issues of *World's Fair Comics* were printed, and Superman graced the covers of both (see Plate 11). The fair even marked the first public appearance of Superman (played by actor Ray Middleton) as part of its Superman Day, which "cracked all attendance records for any single children's event, drawing 36,000 of them at ten cents a head."[142]

The Superman story was essentially an urban plot, even though the superhero was raised by a kindly Kansas couple who endowed him with a homespun Jeffersonian rural goodness.[143] Superman, who is "able to leap tall buildings in a single bound," invariably resides in, or flies above, the metropolis, providing aerial angles to experience the urban morphology in its entire range of connectivity. Equipped with a magisterial gaze and X-ray vision, he negotiates increasingly complex urban spaces that expand both vertically and horizontally. His untethered and mobile existence in the sky above the metropolis is a narrative device to create a new vantage on the metropolis, while suggesting a new freedom of movement unencumbered by the urban grid at the ground level.[144] According to the urban historian Anthony Sutcliffe, "One may reasonably presume that the flying abilities of these heroes [e.g., Superman] were developed in the original comic-strips to facilitate a three-dimensional relationship with the skyscrapers of the New York location which (albeit thinly disguised as 'Metropolis' or 'Gotham City') pervaded these stories."[145] As a ubiquitous metropolitan performer, Superman became a towering monument to the modern metropolis itself. His aerial gaze on the metropolis seemed doubly therapeutic for the reader of the comic-book superhero's stories in the 1930s. On one hand, while Superman did not "produce an urban analysis that city planners can use, [these stories] nevertheless providc a compclling iconography of a rich urban imaginary, unfettered and uncanny."[146] On the other hand, his godlike perspective airbrushed away all the anxieties (except the criminals he pursued) that plagued the Depression-era metropolis, thus reducing it to a simplified pixilation of good and evil.

As the historian Lawrence Levine posits, it was against the foil of the Depression—when distrust of the establishment permeated all walks of life, and the old social verities no longer held sway—that the concept of an American hero shifted from the rugged frontiersman of the nineteenth century, such as the folk figures Davy Crockett and Mike Fink, to an urban operative such as Superman. Unlike the frontier folk heroes who could "cross rivers in a single stride, uproot trees with a single yank, conquer wild animals with their bare hands," 1930s heroes like Superman were often seen correcting urban problems and pursuing deviants with the aid of their lofty vantage point, as well as navigating the secret alleys of tenements and offices.[147] Levine further argues that a populist supersolver like Superman was almost born out of necessity, because during the Depression the conventional social tools at the disposal of ordinary mortals seemed just too inadequate to rectify the complex range of problems associated with the center of the economic system: the metropolis.

Thus, solving the colossal metropolitan problems called for superheroes with unimaginable cognitive evolution, physical prowess, and, as Umberto Eco proposed, an ahistorical position outside the conventional structures of temporality (i.e., Superman continued to serve humanity in episode after episode without ever growing old).[148] Therefore, Superman was frequently seen acting on metropolitan pathologies as if he was conscious of his peculiar ideological complicity in creating a controllable metropolis. Interestingly, in a 1939 issue of *Action Comics,* Superman even personifies the modernist planner (Figure 3.24). Convinced that congested, squalid, and disorderly tenements spawn juvenile delinquency, Superman destroys an entire urban slum to force the government to build ultramodern low-cost housing![149] In the May 1939 issue of *Action Comics,* he takes on hazardous urban traffic and confronts the mayor of Metropolis: "Why has our city one of the worst traffic situations in the country?"[150] Outraged by careless drivers who kill pedestrians, he takes control of a radio station and thunders: "The auto accident death rate of this community is one that should shame us all! It's constantly rising and due entirely to reckless driving and inefficiency!" Superman even punishes corrupt builders of skyscrapers who employ sinister means to outdo competing construction companies.[151] Superheroes were not just watching the world from on high. They also gravitated toward it and cured its ills.

The omniscient downward gaze on the futurist city, analogous to Superman's, offered a metonymic image of the master builder himself, intent on rectifying the physical as well as social disorder below. Such an analogy seems plausible because both the master builder and Superman steadfastly saw their own heroism as contingent on their mastering and reshaping the modern metropolis. As cultural theorist Scott Bukatman argues: "Superman seems to be an incarnation of [Le] Corbusier's panoramic authority based on perfect transparency, control, and knowledge. He is democratic, open, and idealistic, carving a space for the little guy. A walking, flying figure of utopian progress, Superman prefigures in his mode of perception and spatial negotiation the development of the city of tomorrow."[152] By choosing the metropolis as the focus of their sagacious gaze and as the battleground on which to rescue laissez-faire modernity from disastrous consequences, Superman and the master builder narrated remarkably similar stories of derring-do.

Figure 3.24. Jerry Siegel and Joe Shuster, "Superman in the Slums," Action Comics, *vol. 1, no. 8 (DC Comics, January 1939). Reprinted in Jerry Siegel and Joe Shuster,* Superman, the Action Comics Archives, *vol. 1 (New York: DC Comics, 1997), 41–54. From* Action Comics, *no. 8 copyright DC Comics. Reprinted with permission.*

In evoking these popular manifestations of fictional urban amelioration, we confront the issue of whether Futurama's spectator might also epitomize similar superheroic ambitions. Was Bel Geddes's conveyor belt in Futurama an assembly line for mass-producing supermen? Against the backdrop of the exhibit's heady conjecture of an exuberantly sanitized American utopia, the viewer in his or her mobile aerie conjured up a familiar image. In this image, Superman glides over his embattled metropolis intent on restoring order, or the master builder fixates his self-righteous eyes on his model of the future city (Figure 3.25). In the chiaroscuro interior of Futurama, the spectator summoned the image of a superhero who appeared convincing as both the builder and the guardian angel of that utopia. Just as superheroes occasionally came down to the earth to rectify its faults, at the end of their eighteen-minute ride Futurama visitors likewise descended toward the heart of the future city (Figure 3.26). In one way or another, "the exit into the world of tomorrow"—as Bel Geddes put it—was the heroic finale of Futurama's voyage extraordinaire. Whereas on arrival the spectators experienced only a life-size, vertically stratified traffic intersection (filled, not surprisingly, with the latest General Motors cars), the sense of "arrival" itself rallied their esprit de corps behind Bel Geddes's utopian ambition: to achieve the perfect World of Tomorrow as early as 1960. It was as if this realm awaited the hero's triumphant homecoming in order to experience its own magical birth.

In *What Is History?* (1961), E. H. Carr noted that "the cult of individualism is one of the most pervasive of modern historical myths."[153] And Walter Benjamin claimed, "The hero is the true subject of modernism."[154] As with all utopias, the

Figure 3.25. Harvey Wiley Corbett, Raymond Hood, and others inspecting a model of Rockefeller Center (photographer Walter Kilham Jr.). Reprinted in Rem Koolhaas, Delirious New York *(New York: Monacelli Press, 1994), 179.*

Figure 3.26. The life-size road intersection at the end of the conveyor-belt ride in Futurama. Harry Ransom Center, The University of Texas at Austin. Courtesy of the Edith Lutyens and Norman Bel Geddes Foundation.

myth of modern architecture was inseparable from the idea of its protagonist hero. As Andrew Saint and others have shown, the idea of a superman-like architect seeking to redeem a fallen world from the heights of authority he has claimed has been an enduring modernist myth.[155] The modernist apologists of the early twentieth century consistently championed the visionary architect-builder as a sort of secularized God—most famously immortalized in Ayn Rand's novel *The Fountainhead* (1943)—able to bring about a veritable paradise in this world.[156] Hinged on the remnants of nineteenth-century romantic individualism, this heroic image within twentieth-century design culture fashioned a master-builder persona. The underlying assumption was that there was an inherently causal relationship between planning and the condition of society and that the master builder could play a crucial role in making the world a better place. The master builder's all-seeing eyes could filter the messy world below into a utopian simplicity, affording him the illusion that he could impose a neat physical order on the world and create an ideal society. The same year that Futurama foretold the advent of the World of Tomorrow, Sigfried Giedion theorized modern architecture's heroic aspirations.[157] A new millennium was dawning, Giedion claimed, one that would manifest itself through the visual culmination of a functional, socially beneficial, and universal architecture. Lurking behind Giedion's adumbration was none other than the larger-than-life master builder, who would wage a protracted aesthetic battle against all sorts of disorder and effete traditionalism in architecture and city planning, ultimately delivering an ideal city attuned to modern science and technology. Le Corbusier's famous "hand into the picture frame" offers a poignant visual case in point (Figure 3.27). The symbolic extension of the architect's powerful hand over the paradisiacal mathematics of Ville Contemporaine signified not only the literal embodiment of the master builder's gaze but also a magical unveiling of an impending state of infinite progress, harmony, and happiness. Hovering over the gigantic model of Futurama, spectators occupied a position fraught with similar projections of the future. In a curious, and coincidental, resemblance, the superhero's flying posture above the roofscape of Metropolis was echoed in Bel Geddes's position over the model of Futurama (Figures 3.28 and 3.29).

In significant ways, Bel Geddes's creation was self-referential, a sort of auto-construction. In Futurama's spectator, he attempted a mass reconstruction of his own self: the Man of Tomorrow, the visionary builder of an ideal America. In other words, his trompe l'oeil choreography of an American Elysium also included an idealized spectator who was conceived to play a cavalier character similar to one Bel Geddes himself assumed in shaping it. Consider, for instance, this official Futurama publicity:

> With the imagination of a practical designer, the shape of the new world is spread out before you. Mother earth is the same, of

Figure 3.27. Le Corbusier's hand over the model of Ville Contemporaine. Copyright 2012 Artists Rights Society (ARS), New York / ADAGP, Paris / F.L.C.

> course, as to mountains and valleys, and the streams that go down to the seas. But here and there are evidences of how she has been harnessed and made to do man's work so as to increase his wealth, add to his comfort and give him more leisure. Great water projects, flood control stations, terrace lands to prevent erosion and intensified farming under glass. And connecting all this is the Norman Bel Geddes Motorway, a system of shining threads stretching across the continent.[158]

Futurama's conveyor ride, representing a twenty-four-hour odyssey across America, evoked the phantasmagoria of the time machine. Audio commentary, individually synchronized with each spectator's relative position over the model, further dramatized the unfolding of the grand aerial epic. The voice of the "disembodied angel," as *BusinessWeek* dubbed the sound system, was in many ways an aural analogue of the spectator's hyperfunctional eyes that surveyed objects, places, and regions.[159] Cleansed of filthy slums, marooned fringes, or any visual anomalies that could short-circuit the viewer's expectations of collective beauty and cohesion, an airbrushed vision of America came alive. The

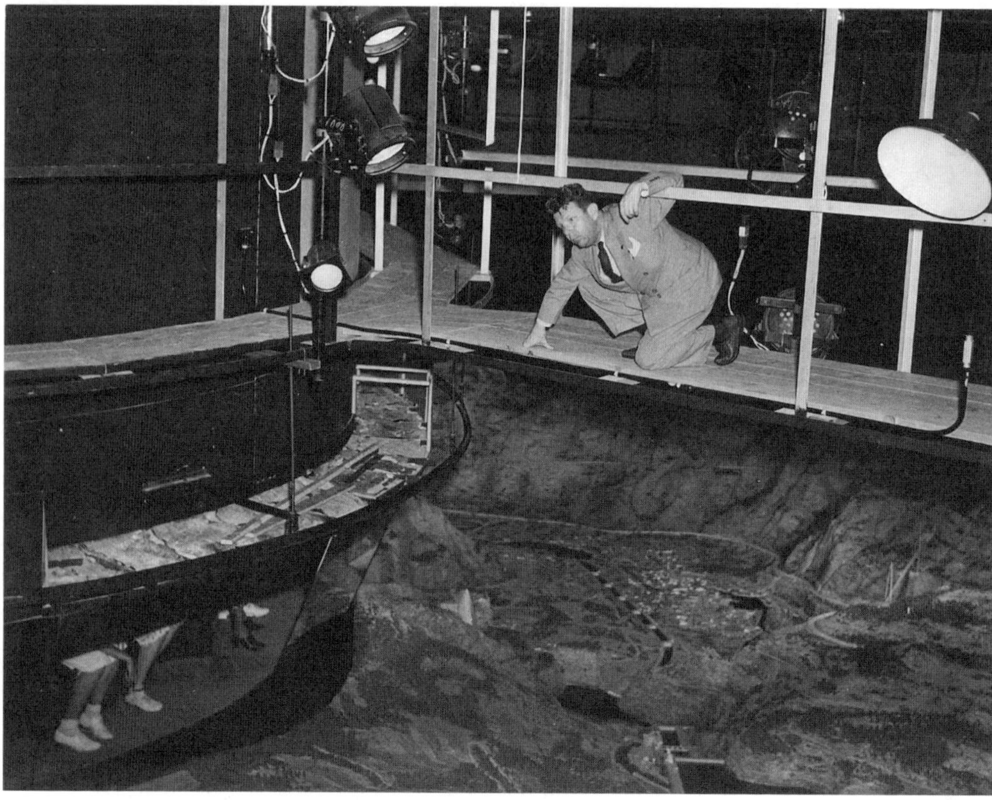

Figure 3.28. Norman Bel Geddes perching on a viewing platform above the Futurama model during a photo shoot. Harry Ransom Center, The University of Texas at Austin. Courtesy of the Edith Lutyens and Norman Bel Geddes Foundation.

visitors' eyes became a conceptual stage set on which the modernist drama of producing a veritable paradise was rehearsed:

> Man has forged ahead since 1940. New and better things have sprung from [Bel Geddes's] industry and genius. Since the beginning of civilization, transportation and communication have been keys to Man's progress,—his prosperity,—his happiness. Twenty years have passed since 1940. What wondrous changes and improvements have developed in our national highways! . . . Directly ahead is a modern experimental farm and dairy. The fruit trees bear abundantly under individual glass housings. Strange? Fantastic? Unbelievable? Remember, this is the world of 1960! . . . Just as improved highways have benefitted the farmer, so have they added to the comforts of living and economic welfare of those in industrial communities. In the foreground is a model airport . . . Railway trains run in and out on fast schedules, carrying products

Figure 3.29. Superman over Metropolis on the cover of the first Superman comic, Superman, *vol. 1, no. 1 (DC Comics, Summer 1939). Reprinted in Jerry Siegel and Joe Shuster,* Superman, the Action Comics Archives, *vol. 1 (New York: DC Comics, 1997), 41–54. SUPERMAN™ copyright DC Comics. Reprinted with permission.*

> of the community to consumers. A spectacular, thrilling composite of motor, air, and rail transportation in the world of tomorrow! . . . Who can say what new horizons lie before us if we but have the initiative and imagination to penetrate them—new economic horizons—new social horizons—new horizons in many fields, leading to new benefits for everyone, everywhere.[160]

Finally, the city of 1960 burst forth before the spectators' eyes: "Now we near the great metropolis of 1960. We will bank high over the city for a spectacular view of its many wonders. . . . The city of 1960 has abundant sunshine, fresh air, fine green parkways, recreational and civic centers—all the result of thoughtful planning and design" (Figure 3.30).[161] Bel Geddes described the experience of the modern city in the interwar era's typical planning parlance: "As the spectator circles high above the city, he is able to compare the congested, badly planned areas of the 1930's with the well-organized districts of the newer city."[162] Monumental skyscrapers, "sheathed in glass, the thin shafts of the tallest ones reaching more than a quarter of a mile in the air," dazzled below. Mimicking the geometricizing tenets of the Ville Contemporaine and setback rules applied in Ferrissian skyscrapers, these structures (equipped with landing pads for helicopters and autogyros) were arranged "on a unifying grid system" so as not to cast shadows on each other, thereby guaranteeing plentiful light, airflow, and a complete order of urban hygiene. One-third of the total

Figure 3.30. "The City of 1960," as depicted in Futurama. Reprinted in a General Motors publicity brochure.

city area, as the spectator witnessed, was devoted to urban breathing spaces or parks. "In contrast to the congested city," the spectator handily understood the benefits of "more green space and a more open distribution of buildings." Seen in its totality, Futurama's city of 1960, neatly zoned into residential, recreational, commercial, and industrial enclaves, represented "a splendid contrast between the old method of thoughtless planning and the careful and intelligent modern city designing."[163]

Futurama's narration took on a sanctimonious undertone. Having taken the Futurama ride, the American writer E. B. White summed up the feeling with a solemn satire in the *New Yorker* two weeks after Futurama's opening:

> A ride on the Futurama of General Motors induces approximately the same emotional response as a trip through the Cathedral of St. John the Divine. The countryside unfolds before you in five-million-dollar micro-loveliness. . . . The voice is a voice of utmost respect, of complete religious faith in the eternal benefaction of faster travel. The highways unroll in ribbons of perfection through the fertile and rejuvenated America of 1960—a vision of the day to come, the unobstructed left turn, the vanished grade crossing, the town which beckons but does not impede, the millennium of passionless motion. When night falls in the General Motors exhibit and you . . . hear the soft electric assurance of a better life—the life which rests on wheels alone—there is a strong, sweet poison which infects the blood. I didn't want to wake up. . . . It wasn't till I passed an apple orchard and saw the trees, each blooming under its own canopy of glass, that I perceived that even the General Motors dream, as dreams often do, left some questions unanswered about the future. The apple tree of tomorrow, abloom under its inviolate hood, makes you stop and wonder. How will the little boy climb it? Where will the little bird build its nest?[164]

A complete package of the good life, Futurama could have been a faultless, perpetually kinetic, and happy world, but White also shrewdly observed how total planning and overzealous reliance on technology could dehumanize life. In a way, White's criticism was directed at the regimental tenets of modern planning as well as at the infallible self-image of the master builder. By overdetermining the significance of efficiency and functionalism in the World of Tomorrow, modern planning principles missed out on the simple pleasures of life, like the little boy wanting to climb the apple tree.

Such ambivalences and minutiae could not figure prominently in the grandiloquent vision of Futurama. The dreamworld it sought to peddle to fairgoers could not be bogged down by nuances and everyday details. It had to be a mesmerizingly large and all-encompassing dream, without any blemish on its shiny garb. Perhaps the challenge of creating the exhibit was consciously made to climax with a particular set of spatio-social problems posed by the modern city. As deliberated by early twentieth-century urban planners—for instance, the delegates to the Congrès International de Architecture Moderne (CIAM) in 1933—these problems were thought to have stemmed from a chaotic mix of various urban functions (housing, work, recreation, and traffic), lack of traffic infrastructure and urban parks, and an unholy triumph of private interests over collective needs.[165] One could not comprehend this urban chaos, Bel Geddes and others seemed to suggest, when one was embroiled in the chaos; rather, one had to be elevated above it to be able to see the corrosive effects of chaos on urban life. Le Corbusier—the shepherd of CIAM's manifesto, the Charter of Athens—highlighted this phenomenon of urban critique when he memorably said, "The airplane indicts."[166] The flying machine, in Le Corbusier's estimation, became the all-seeing planner to indict the wrongful blending of various functional zones. In this oculocentric discourse, the airplane equipped the modern planner with the panoptic power to both inspect and remedy the problems of the modern city.

Futurama's climactic narration of the city of 1960 echoed the paradigmatic voice of the master builder, who, in the first decades of the twentieth century, looked indignantly at the disorderly industrial city as though it were a diagram to be inspected and corrected. (Le Corbusier's megaplanning in South America in 1929 was an example.) The imposition of a diagrammatic order would presumably cleanse what Bel Geddes thought were simply anomalies in the World of Tomorrow, the "outmoded business and undesirable slum areas." This was the type of righteousness with which Bel Geddes embraced the task of rebuilding America. But what was crucial in Futurama's aerial ride was that Bel Geddes wanted Futurama's spectators to see how *he* saw America and the challenge of transforming it into an earthly paradise. He "planted" his own master-builder eyes on the heads of Futurama's spectators.

Futurama's aerial narrative was a stunning visualization of the common modernist gospel of total planning, a master grid of terrestrial morphology reorganizing the entire country. However, its unique purpose was fully understood only by the spectators' theatrically orchestrated, aerialized, and mobile eyes. If Futurama was an epic parable of modernist planning, then visualizing it in the span of an eighteen-minute ride bordered on what James Gibson would call a divinely omnipotent act. As Gibson notes: "Seeing the world at a traveling point of observation, over a long enough time for a sufficiently extended set of paths, begins to be perceiving the world at all points of observation, as if one could be everywhere at once. To be everywhere at once with nothing hidden is to be all-seeing, like God."[167] To be godlike was also to inhabit an exclusive realm, detached from earthbound mortals. This detachment ultimately made it easy for the master builder to recommend wholesale demolition of existing cities and to propose grand spatial blueprints from which civilization would arise renewed, phoenixlike.

Interestingly, searching for a point of origin for his "new" America, Bel Geddes focused not on hyperurbanized coastal cities such as his hometown and host to the World's Fair, New York, but on America's geographic center, the relatively underdeveloped city of St. Louis, "one situated inland to be away from the peculiar conditions appertaining only to port towns."[168] Coastal cities, in Bel Geddes's formulation, had long been polluted by the shiploads of Old World colonialists and immigrants; civilizing the New World (figuratively, Futurama's World of Tomorrow) would now entail reaching the heart of the continent without having to endure the corrupting germs of the edge. The edge was too polluted. The center was still pure and "spiritually capable of looking ahead and being willing to scrap present-day systems and methods when these proved outmoded and economically unsound."[169] Futurama's replication of god-like mobility by aerial routes proposed an easy resolution of the search for a mythical American center.

There was a peculiar, and perhaps contradictory, frontier psychology in Bel Geddes's identification of St. Louis with a *pure* American hub—embedded in the grandeur of America's pioneer history, a conquerable frontier, capable of producing an ideal future. Superman, the Man of Steel and of Tomorrow, too, grew up in the center, in the pastoral comfort of Kansas, but brought some form of frontier justice to the crime-ridden world of the metropolis. With Bel Geddes's peripatetic career taking him from the small city of Adrian, Michigan, to big cities like Los Angeles, Chicago, and, finally, New York, his proposed return to the American heartland recalled the Turnerian longing of the frontier as key to America's historic development. Yet Bel Geddes's version was a new frontier, the future. Likewise, conquering this frontier called for a new breed of frontiersman, someone like Superman or the aviator.

THE COLLAGE BLITZ

Paul Garrett, General Motors Corporation's vice president for public relations, was instrumental in commissioning Norman Bel Geddes to create Futurama. Garrett hosted a high-profile party at the exclusive University Club of New York on October 16, 1939, to celebrate the success of the exhibit in the first season of the New York World's Fair. For the gala, Bel Geddes designed a Futurama publicity brochure that was issued in a strictly limited edition of one thousand. Copies were distributed among the invited guests, including a wide swath of upper-crust New York society.[170] On the booklet's first page, the fair's symbols—the Trylon and the Perisphere—provided a shapely backdrop for the name of each invited guest, who "is presented with this memento of the New York World's Fair's most talked about exhibit highways and horizons and the General Motors futurama." The brochure included a verbatim transcript of Futurama's audio commentary.

In addition, it presented a series of collages that Bel Geddes used not only to publicize his magnum opus but, more important, also to create a corporate camaraderie in the sense that Futurama could be conceived only through the collective eyes of America's power elite. The exhibit, Bel Geddes reasoned, was a logical and objective creation of great minds rather than the "eureka" project of a solitary genius. In a characteristic manipulation of an original Futurama photograph, Bel Geddes replaced the heads of Futurama's "common" spectators with those of New York and Detroit power wielders (Figure 3.31). The head of the female spectator on the right in the original photo was substituted with the gratified visage of Al Smith, New York's forty-second governor and 1929 Democratic presidential candidate. Charles F. Kettering, director of GM's Research Division, replaced the next spectator, and other plutocrats were collaged onto ordinary bodies to create a virtual gallery of powerful men, who seemingly approved of the futurist spectacle spread out below them. Bel Geddes placed himself, too, in the high pulpit of this exclusive fraternity (second from the left).

The technique of collage, as the surrealist Max Ernst defined it, brought "two distant realities on a plane foreign to them both," and within a collaged representation, unrelated narratives begin to complement each other across schisms of time and space, producing a wholly new, protean relationship.[171] Bel Geddes's collage engendered a dialogical relationship that worked both ways. First, there could be no more telling example of his association of the Futurama spectators with East Coast éminences grises, providing an example of how he orchestrated the symbolic transformation of ordinary visitors into political and corporate leaders who presumably had the executive power to design a favorable future. Second, his collage characterized political and corporate honchos as rightful shapers of the future: With their commanding gaze and high vantage point, these powerful men could contemplate their role in shaping an American utopia, just as Bel Geddes or modernism's protagonist builder could envision his own. His self-inclusion in the cadre of the ruling class demonstrated an ideological affinity that Bel Geddes felt toward corporate management, which he considered a convenient tool for creating the World of Tomorrow. At the same time, by placing the triumvirate of corporate, political, and media leaders in the master builder's virtual cockpit, he stroked their egos, hoping to land grander commissions.

The publicity booklet also contained a continuous band of photomontages at the tops of the pages, creating a visual theme parallel to Futurama's audio transcript. The band constituted a frontal illustration of the conveyor belt and its adjoining coupes of double seats. The composition of the page recalled the

Figure 3.31. A photograph from Futurama of "common" spectators viewing the model from the conveyor belt (above) *and a manipulated photograph in which the spectators have been replaced with powerful men from New York and Detroit.* Futurama, *General Motors publicity brochure (1940).*

exhibit's original viewing arrangement: spectators placed above a massive scale model for optimum viewing. All of the seats on the moving gallery were occupied by elegantly suited hand-drawn male figures collaged with the headshots of an American oligarchy (Figure 3.32). The headshots showcased the combined participation of corporate, political, and media regimes in Bel Geddes's projection of a resplendent America. Futurama could not be just his project; it was also the power elite's. His careful selection of faces represented, if self-servingly, the bigwigs of corporate, political, and media worlds, the three most useful sectors whose patronage a designer would covet for his career advancement. The list of faces was impressive and suggested methodical prior research. Among them were Grover Whalen, businessman and president of the 1939 New York World's Fair; Tom MacDonald, politician and chief of the Iowa State Highway Commission; Gerard Swope, president of General Electric Company; William S. Knudsen, president of General Motors; Henry Luce, publisher and editor in chief of *Time, Life,* and *Fortune* magazines; Tom Lamont, head of J. P. Morgan and Co. and a representative of the U.S. Department of Treasury; William S. Paley, chief executive officer of CBS; Will Hays, chairman of the Republican National Committee; Averell Harriman, Democratic Party politician; and many GM and other automobile company executives.[172] Here, too, Bel Geddes placed himself in the company of America's plutocrats, all seated on Futurama's commanding thrones. The elevated positioning of these men exhibited a peculiar affinity with how a master builder would envision a designed world from above. The hovering gallery can perhaps be seen as a theatrical reconstruction of the master builder's downward gaze at a tabula rasa that awaited his grand and moralizing intervention, except that, in the case of Futurama, the designer

Figure 3.32. The corporate industry leadership depicted above images of Futurama. Futurama, *General Motors publicity brochure (1940).*

Figure 3.33. Norman Bel Geddes, William S. Knudsen, and Richard H. Grant looking over the General Motors building for the 1939 New York World's Fair. Harry Ransom Center, The University of Texas at Austin. Courtesy of the Edith Lutyens and Norman Bel Geddes Foundation.

fully subscribed to a corporate vision of the future in terms of efficiency, functionality, and unimpeded mobility of capital.

Photographic manipulation remained central to the endurance of Futurama's heroic theme. If a key player was missing from a publicity photograph, Bel Geddes audaciously collaged him into a Futurama scene to make a case that the exhibit perpetually enjoyed universal support. If the exhibit was a jigsaw puzzle of collective patronage, a single missing part would defeat its mission to create a perfect world. Richard Garrison, a longtime Bel Geddes photographer, captured a crucial presentation of the Futurama proposal to General Motors executives sometime in early 1938, when the designer explained the project's life-size traffic intersection of the future to company president Knudsen and Richard H. Grant, vice president of sales (Figure 3.33).[173] In another photograph, shot during the same meeting, two key missing GM bigwigs—Alfred P. Sloan Jr., chairman, and Kettering—were pasted on to complete the pantheon of Futurama's corporate guardian angels (Figure 3.34). Sloan's picture was lifted from another unrelated photograph in which Kettering explained to him the operation of a machine (Figure 3.35).[174]

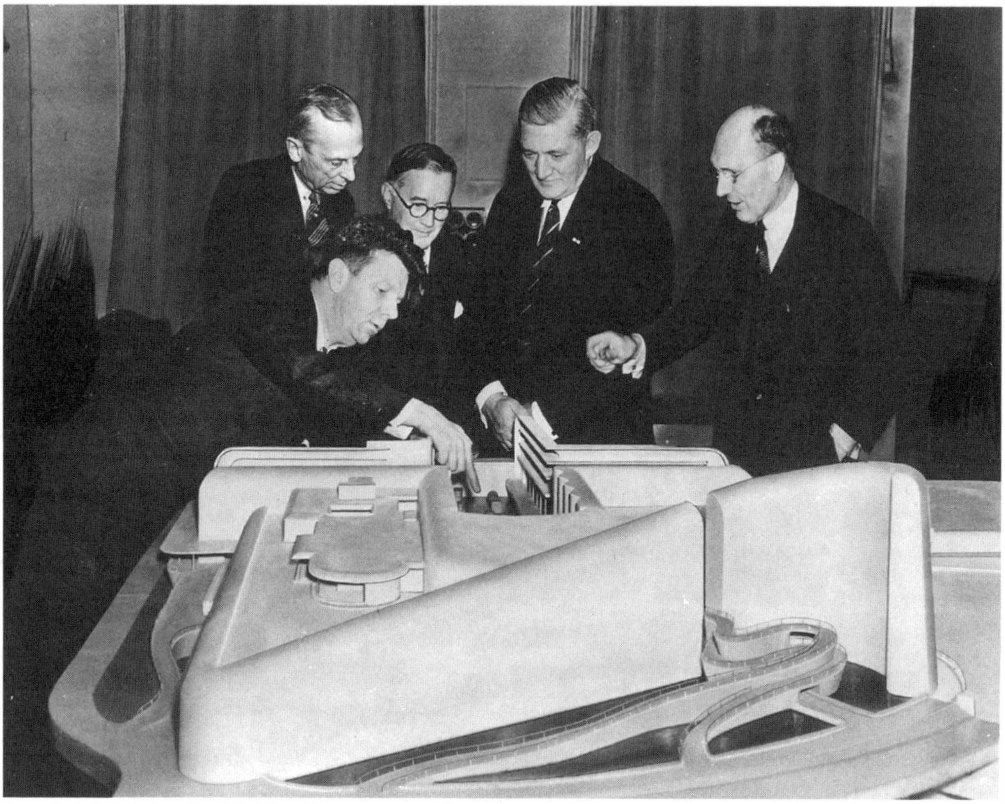

Figure 3.34. Norman Bel Geddes and the General Motors leadership in a collaged photograph. Harry Ransom Center, The University of Texas at Austin. Courtesy of the Edith Lutyens and Norman Bel Geddes Foundation.

In the collage, the executives' focused downward gaze, following Bel Geddes's (or mimicking an aviator's or Superman's), at the model of the General Motors Building epitomized what could be called corporate America's utopian practice—that is, as Susan Stewart has described it, the future could be miniaturized into a toy that they could put on a conference table and inspect from comfortable heights and from all angles.[175] It was a splendid toy that they could rotate, rearrange, and reshuffle at will. The space of the future was now transformed from the sublime abstract to a tactile object. In the photos, Bel Geddes and GM's bosses perhaps represented superenlargements of Futurama's spectators, now looking not at the miniature world of Futurama but at its container itself, the big toy that housed the toy World of Tomorrow. The enlargement of the human body and the miniaturization of the object allow the human to have total domination over the object. Stewart argues, "To toy with something is to manipulate it, to try it out within sets of contexts, none of which is determinative."[176] In a way, Bel Geddes's hand toying with the miniaturized GM Building

Figure 3.35. The original photograph of Charles F. Kettering discussing the operation of a machine with Alfred P. Sloan Jr. Reprinted in Michael W. R. Davis, General Motors: A Photographic History *(Charleston, S.C.: Arcadia Publishing, 1999), 62.*

became a measure of the project's provisional status and, ultimately, of the malleability of the particular brand of future Bel Geddes presented inside it.

This is how the character of Edgar, E. L. Doctorow's prepubescent visitor to the 1939 New York World's Fair, conceptualizes his impression of Futurama at the end of his visit to the popular exhibit: "Nothing compared with seeing it for myself: all the small moving parts, all the lights and shadows, the animation, as if I were looking at the largest most complicated toy ever made! . . . It was a toy that any child in the world would want to own. You could play with it forever."[177] Futurama was certainly an expensive toy, too, with an overshot budget of more than seven million dollars. But for the General Motors executives it was worth the money, because they understood the peculiar capacity of world's fairs to, as the anthropologist Burton Benedict argues, "take people out of their ordinary routines and thus remove them temporarily from their usual positions in the social structure."[178] This capability of building illusion on a mass scale fed the Depression-era public's appetite for fantasy as a reprieve from harsh economic times. As David Nye observes: "Edgar, like many of the fairgoers of the time, finds that the miniaturization of Futurama, the City of Light, and the Perisphere provides the psychological illusion of glimpsing another life from an Olympian height. . . . [Such projects] employ a representational strategy that permits each person to imaginatively enter a scene without impediment."[179]

Bel Geddes's strategic replacement of Futurama's common spectators with a power-wielding squad of men was part and parcel of a make-believe world in which ordinary citizens were led to think that they themselves were the shapers of their destiny, in the sense of what Frederik Polak characterized as active utopianism.[180] In this form of utopian imagination, the individual is seen as capable of taking control of his or her own development rather than submitting passively and wishfully to the laws of social dynamics. Within Futurama's illusory world, the spectator was momentarily given the opportunity to rehearse the lofty gaze of an active utopianist, to enter a future domain unfazed by everyday worries.

Futurama's success remained in its ability to seduce spectators to parrot Bel Geddes's dream of the World of Tomorrow and his role in actualizing it. In Futurama's final act, the life-size traffic intersection of the future and the corresponding enlargement of the exhibit's scale hardly diminished the self-aggrandizing experience the spectator had during the flight over a miniaturized world. Entering the life of the city entailed a change in perspective—more human than godlike—yet the conveyor belt's descent into a "realistic" future could only bolster the spectator's transcendent self-image, as if he, too, had participated in its realization. Doctorow's Edgar sums up this feeling:

> The amazing thing was that at the end you saw a particular model street intersection and the show was over, and with your i have seen the future button in your hand you came out into the sun and you were standing on precisely the corner you had just seen, the future was right where you were standing and what was small had become big, the scale had enlarged and you were no longer looking down at it, but standing in it, on this corner of the future, right here in the World's Fair.[181]

Futurama's spectacular finale, choreographed as the spectator's descent into the "corner of the future," was meant to ultimately transform the slogan "I have seen the future" to "I have built the future."

Futurama's collaging of the imaginary and the real—in the process inducing within the everyman an inchoate feeling of power—resonated with the ways 1930s America became identified, as Warren Susman has observed, with a transitional time.[182] During this time, the notion of culture and its relatedness to the average American changed from, say, a Matthew Arnoldian view preoccupied with highbrow art and individualist achievements to a popular acceptance that culture indeed constitutes a wider and multilateral operational grid, able to communicate broader behavioral patterns. This way, culture realigned the individual's relationship to, and value in, society. "It is not too extreme to propose," Susman asserts, "that it was during the thirties that the idea of culture was domesticated, with important consequences. It was during this period that we find, for the first time, frequent reference to 'an American Way of life.'"[183] The public as a driving force was "in abundant evidence in the rhetoric of the period, a fundamental image that appeared to speak deeply of the American consciousness."[184]

In one of most popular books of the 1930s, *The Epic of America* (1931), James Truslow Adams popularized the idea of the "American Dream," which implied an inclusive social system in which everybody could pursue his or her own dream, thereby renewing the potential of the individual.[185] The American Dream was a powerful cultural construct, ushering in the illusion that access to opportunities had been democratized. A cultural paradigm emerged in which the "promise of American life" was articulated as being part of a public domain and a communitarian feeling that considered the common people worth listening to. In fact, the idea of surveying the public to understand Americans' views became commonplace beginning in 1935, when George Gallup established the American Institute of Public Opinion. The decade's cultural cognizance of the people as a viable social force was encapsulated, for example, by the resilience of Ma Joad, John Steinbeck's matriarch in *The Grapes of Wrath* (1939): "We're the people that live. Can't nobody wipe us out. Can't nobody lick us. We'll go on forever. We're the people."[186] The rhetoric of the people reached a crescendo in the 1939 New York's World's Fair, when Grover A. Whalen, the fair's president, summed up the event with a pithy line in his welcoming remarks in the *Official Guide Book*: "This is your Fair, built for you and dedicated to you."[187] The message was unmistakable: The World of Tomorrow could not be built without you. You are the ideal citizen of utopia. You can plan and build the World of Tomorrow and carve out your own place in it. Against this enduring theme of the 1939 fair, Futurama's clairvoyant pronouncement at the end of the show, "I have seen the future," was a de facto celebration of each individual spectator.

This populist focus on the individual served disparate goals, depending on which patron of the people was in contention. For General Motors executives, putting Futurama's spectators in an aviator's cockpit to appreciate their idea of what worked best in the future (the wheels expanded the economy, which then benefited the everyman) revealed how the ideologies of consumer capitalism were advanced with captivating visuality and exhibition engineering at the 1939 fair. Elevating the people to heights where corporate leaders belonged and from which they outlined a streamlined cartography on the world was a shrewd corporate advertisement. Sharing their executive position with the people proved to be an act of public relations genius.

Furthermore, Futurama's conveyor belt—loaded with gratified people, composed in their streamlined roadway to the future—appeared to be an effective eugenic display. If the ideological construction of the interwar eugenic movement was buttressed by the pursuit of efficiency, hygiene, and the ideal physique (which was the logical outward manifestation of highly evolved cognition), then Futurama's simulated assembly line created the impression of mass-producing eugenic citizens, meeting corporate demands for efficient and able-bodied workers. Many American corporations and research organizations during the interwar period felt that they had a stake in the outcome of the eugenic movement because nothing, in their estimation, could serve corporate interests more effectively than a utopia filled with disciplined workers. It would not be wholly untenable to suggest that the covert sympathy of GM's chairman and president—Sloan and Knudsen, respectively—toward the German pursuit of an Aryan utopia during the 1930s would translate into a desire to exact some sort of an American eugenic utopia serving an economic oligarchy. Futurama and its future-gazing citizens promised, if momentarily, the fulfillment of that desire.

For Bel Geddes, Futurama's conveyor belt was a device extraordinaire to induce a superman feeling or a master-builder hubris among the masses who flocked to the exhibit. In their eighteen-minute excursion to the future, Futurama's voyeurs gawked at the world below through the moral filter of the costumed flying superhero they idolized or through the confident, panoptic perspective of the master builder. Bel Geddes's exhibit recalled the titillating polemics of the American Dream—that success was within everybody's reach—while seeking to restore the 1930s beleaguered "you" figure with what *Life* magazine, in its 1939 World's Fair coverage, called "the best energies in its citizens."[188] The streamlined and ascending body politic of the Futurama spectator was a powerful reminder of how the democratization of the aviator's holistic perspective would serve the rhetoric of the people as a powerful tool of social mobilization in New Deal America. Futurama's elevated viewing technique also recalled the modernist proclivity to embed the figure of the master builder in an evolutionary ideology and functionalist narrative, both intertwined with the discourse of the flying man. Bel Geddes sought to represent the Futurama spectator as Superman, but this was a superhero who, ironically, promoted a corporate vision of the good life, wherein a perennially kinetic state preassigned all citizens into roles that sustained a hyperfunctional civilization.

In an extreme portrayal of this civilization twenty years in the future, the cover illustration of the June 1, 1940, issue of the *New Yorker,* published during the second season of the New York World's Fair, depicted Futurama's aerial belt as being cannibalized by the highway system and its labyrinthine overpasses (see Plate 12). The geometrically pure cloverleaf road intersections that dotted Futurama's landscape had been transformed into mangled asphalt serpents with mechanical and fluorescent ants creeping along their overlapping bodies. The spectators were replaced not by New York and Detroit power wielders but by puzzled drivers—encased in their teardrop-shaped cars and without any avuncular voice to reassure them of a bright future just ahead. The *New Yorker* cover perhaps alluded to the pitfalls of overdetermining the power of planning and of the master builder. It was perhaps also a critique of prophecies too invested in the glories of technology and corporate expansionism masquerading as the good life.

EPILOGUE

The God's-Eye Vision

With his mind in the clouds, the master builder sought to create his utopian World of Tomorrow. Hugh Ferriss, Buckminster Fuller, and Norman Bel Geddes assumed, with a heavy dose of idealism, that the technologies of ascension had provided them with a powerful perspective that would, in turn, enable them to create a world of aesthetic perfection, technological superiority, and social advancement. They imagined themselves as the deserving master builders of their designed world. They were the heroes of their own narratives. The modern, technology-driven Worlds of Tomorrow that these visionaries imagined were as much about their protagonist creators as about their prophecies of the good life ahead. These master builders' exalted self-positioning in relation to the world they lived in shaped, to a great extent, their imagination of that world.

Design historians have usually examined modern utopias to reflect on the sociocultural and ideological structures that produced them, yet they have neglected the "heightened" presence of the utopias' self-aggrandized builders in the very ideation of the utopias. My central task in this book has been to fill that void. I have examined how the legend of the master builder as a solo creative genius found various conflicted expressions within a hybrid culture of technological utopianism, mystical religion, and heroic sentiments during the interwar years in America.

Ferriss, Fuller, and Bel Geddes were dreamers of the future. They aspired to inhabit a privileged sky, the province of gods, prophets, visionaries, future-gazers, and, of course, master builders. Their ideal tomorrow was premised on the conception that the constructability of a perfect social and spatial system called for the synoptic perspective of a master builder, who would also represent an advanced human race. This view meshed well with both a "crisis of civilization" sentiment and a future-minded, hero-worshipping American mood after World War I and particularly during the Great Depression. This trio of designers detected a common master-builder mythology cutting through Gurdjieff's mystical preaching in New York City, Lindbergh's sensational popularity, and the superhero themes of the 1930s. They embraced this mythology to fashion their own design personas.

But their aspirations were not without existential pitfalls. Ferriss, Fuller, and Bel Geddes were driven by exuberant visions and the certainty of their own agency in realizing these visions. Like superheroes, they felt destined to save the world. However, these three designers failed to consider any possible less-than-perfect outcomes of their individual utopias, as they missed seeing the messy conditions on the ground. They could not blemish the dainty surfaces of the Worlds of Tomorrow they created because their self-proclaimed mission was to shape a future without faults. In doing so, they at times airbrushed away or collaged out what they considered to be anomalies on the path to their utopian space. For them, the man on the street needed guidance in his journey toward an ideal world that only the master builder knew best how to reach and design.

Their argument eventually became embroiled in the superman discourses of the interwar period, many of which were directly or indirectly related to the prevalent evolutionary ideologies concerning the propagation of desirable human species and characteristics. An individual of mysterious provenance, Superman was a complex compilation of moral justice and social Darwinism, an airplane in a human body or, from the viewpoint of theosophists and the Gurdjieffite circle, the representation of highest spiritual consciousness. With the help of his all-encompassing vista he dispensed justice and ensured order in the world below. In their disparate ways, Ferriss, Fuller, and Bel Geddes internalized the Superman mythos, for they believed that the healing of an America allegedly mired in architectural banality, rampant consumerism, and spiritual degeneration called for the unimaginable power of a superhero. This imagined superhero was capable of creating the Metropolis of Tomorrow. He would be an architectural analogue of Dymaxion House, suspended from dominant architectural as well as social ideologies. Or he was a corporate CEO masquerading as a world's fair visitor and invoking the illusion of an empowered public.

Playing Superman was, however, ironic. To many proponents of inclusive modernism, it became a highbrow enterprise and seemed heavy-handed, antithetical to the ideals of a harmonious society that the World of Tomorrow promised. From Superman's aerial viewpoint, the chaotic world below disappeared behind broad outlines and silhouettes. This illusion of simplicity and vastness created a paradoxical effect on the aerial observer. Streets, people, houses dissolved into a lonely, utopian diagram, inspiring in the observer the master builder's hubristic fantasy of wholesale renewal that sometimes served, as in Bel Geddes's Futurama, the interests of corporate capitalism. Futurama's spectators experienced an inchoate sense of power when they witnessed from their mobile aerie a miniaturized America of 1960 that they could visually consume in its entirety. The pixilated urbanscape of nighttime New York City seen from the skyscraper observatory was Ferriss's ruse for his Metropolis of Tomorrow. Fuller imagined "nine chains to the moon" to reflect on the smallness of his own planet and the terrestrial plan he could impose on it for the maximum optimization of its resources. All these utopian reckonings had a robust allaying effect on the American people during the interwar period.

Yet, as prophecies of a designed world were delivered from above, these designers' views became tinged with exclusivity, grandiloquence, and detachment from life on the streets. This alienation gave them the impression that they were in a privileged position to reorder the world and eventually produce an ideal society. The miniaturization of geography made it easy for them to recommend, as Le Corbusier and other modernist planners did, indiscriminate demolition of existing cities and their regeneration through new spatial design.

In the righteous thoughts of Ferriss, Fuller, and Bel Geddes, the "above" denoted an exclusive domain to which they deserved to belong. From their sky perches, real or virtual, they saw the world as they wanted to see it so that they could shape it in their own image or with their own interests in mind. This phenomenon suggested a classic modernist metamorphosis. What the mind's eye had imagined in earlier historical eras now morphed not only into real spectacle—thanks to the technologies of ascension—but also, more important, into a testing ground for the master builder's operational eye. Fuller's first flight experience in 1922 and his aerial exploration of the coastline between New York and Maine later formed one of the conceptual bases for his aerial dropping of 4D towers and, alas, the inadvertent ecological violence that it wrought. Bel Geddes's Turnerian frontier nostalgia in creating *the* American metropolis in the heartland harked back to his practical deployment of the Fairchild aerial camera to produce Futurama's geographic realism.

The skyscraper view and the airplane eye that Ferriss, Fuller, and Bel Geddes deployed reverberated, unfortunately, in the misguided authoritarianism that would characterize "slum clearance" and urban renewal in postwar America. Armed with sophisticated technologies of aerial reconnaissance and power, the "surgeons" of postwar America unleashed a campaign of cosmetic cleansing, seeking to impose both visual and social order onto American cities. New York's self-declared master builder, Robert Moses, was at the forefront of this trend. It is ironic that only a year after Bel Geddes's magic moment for the "rebirth" of America in 1960, Jane Jacobs published *The Death and Life of Great American Cities,* in which she blamed the master builder's technocratic modality of "planning from above" for the atrophy of American cities. Within the master builder's big-picture perspective, she argued, the nuanced needs of mixed-use urban life and how people live in their neighborhoods and use their sidewalks seemed too insignificant for design consideration. Jacobs complained that planned cities, with their wide boulevards and spacious plazas, looked spectacular from above but felt dead in their lonely streets. Zoned cities that had promised rational solutions to congestion and urban chaos lacked the density necessary to create a dynamic everyday life. For Jacobs and other advocates of a "bottom-up" model of urbanism, a megascale urban vision might have been uplifting for its promoters, but ultimately it was the human scale that ensured urban livability.[1] The ideological battles between top-down and bottom-up models of urbanism continue today with a variety of political overtones.

In a way, Jacobs's fierce criticism of the master builder recalled Charles Baudelaire's albatross. With its wide wings, the bird flies majestically in the sky, but on the ground the albatross is "clumsy" and "piteous" because its long wingspan becomes its biggest liability. For Baudelaire, the albatross was an apt metaphor for the poet: "The Poet's like the monarch of the clouds / Who haunts like the tempest, scorns the bows and slings / Exiled on earth amid the shouting crowds / He cannot walk, for he has giant's wings."[2] Did the master builder experience the same fate as Baudelaire's poet?

It is curious that the last line of Le Corbusier's book *Aircraft* stated with a biblical righteousness that "the flock needs a shepherd" and the book's last photograph showed a parachute jumper who had just leapt from an airplane.[3] One is left wondering whether this was Le Corbusier's coded representation of the existentially torn master builder, mediating between the albatross and the poet: the big-vision domain of the sky and the ungainly minutiae on the ground. As we have seen, Ferriss, Fuller, and Bel Geddes were not immune to this existential dilemma.

Michel de Certeau's postmodernist generation criticized the master builder's logic of looking at the world and the "erotic" pleasure of consuming the entire spectacle. (Ironically, de Certeau's critique took shape with a Ferrissian gaze from the architect Minoru Yamasaki's now-gone Twin Towers in downtown Manhattan.) For many members of his generation, walking, not flying, offered an authentic experience of modern life. They claimed that the anonymous metropolitan man on the street—the so-called practitioner of everyday life—was a sort of "unheroic" hero. In their view, this hero tragically fell through the cracks of historical canons, for he existed merely to form a contiguous visual pattern to be observed from the elite pulpit of mortal gods. But this forgotten character now ought to be celebrated as the "real" hero of modernity, walking in "countless thousands on the streets."[4]

The postmodernist romance of the walking man offered an intellectual resistance to the complacency of the "voyeur-god," although at the risk of overglamorizing the walking man's purportedly embodied experience of streetscapes or his ground-level perspective as an engaged form of urban inquiry. For de Certeau and others, aerial viewing implied a heroic but ultimately disembodied system of epistemology, while walking the streets promised a rich urban phenomenology. In walking, the genius loci seemed to be within tantalizing proximity. Urban movements like the New Urbanism of the 1980s, in many ways, grew out of a critique of "planning from above" and a concern for the pedestrian's intimate experience of urban life.

These subsequent criticisms of the aerial eye should not, however, mislead us into thinking that the aesthetics of ascension was ultimately all about the naive idealism or elevation of the master builder and, therefore, unworthy of historiographical consideration. These criticisms do not fully expose the intricacies that define the modernity of this aesthetic consciousness. Even if this consciousness implied ideological naïveté on the part of the master builder, it was also a discursive product of how visionary culture worked during the interwar period. The aesthetics of ascension was not about seeing things from hitherto impossible heights, but rather about the *politics* of seeing things and a host of spiritualist, autobiographical, and corporatist arrangements of the world that could result from it. It was a consciousness that fused unlikely cultural elements, offering new insights into how visionaries of the period saw the world and themselves.

In hindsight, we may find Ferriss's blending of Orage's mystical pedagogy with his own pursuit of an ideal metropolis to be an improbable juxtaposition of ideas. To many people at the time, however, these ideas were not just complementary, they were essentially two expressions of *one* search for the "truth." Drawing on Lindbergh's temporary "deification" as an ideal American man, Dymaxion House's floating architecture created a tantalizing image of Fuller's own search for redemption. Bel Geddes's audacious collages, replacing the heads of Futurama's aerial spectators with those of the power elite, reveal how he imagined the popular exhibit's conveyor belt as the pulpit of corporate CEOs. Thus, to study the aesthetics of ascension is to reexamine the period's visual politics of representation.

The aesthetics of ascension was inspired by two iconic, superrational technological inventions of modern life: the airplane and the skyscraper. However, the great irony was that Ferriss, Fuller, and Bel Geddes expressed this aesthetic consciousness through a host of "irrationalities": illusions, myths, fantasies, mystical spiritualism, subjective excesses, and gee-whiz heroism. Their Worlds of Tomorrow can be viewed as refracting cultural mirrors that both dramatized and distorted interwar social conditions. On one hand, there was an innocent self-assurance about the grand rebirth of America that Ferriss's Metropolis, Fuller's Dymaxion House, and Bel Geddes's Futurama promised the American people during the 1920s and 1930s. On the other hand, these projects were crucial cultural artifacts that revealed an affiliation between *ascension* and a high modernist rationale of viewing the world. The self-aggrandizing, detached gaze of the master builder—masquerading as skyscraper observer, astronaut en route to the moon, or Futurama spectator—worked to dispel various social anxieties of interwar America; at the same time, it rendered most effectively the fantasy of an ideal World of Tomorrow. The heightened expectations of the aerial gaze, as reflected in the works of Ferriss, Fuller, and Bel Geddes, thus offered a populist manifestation of the master builder's modus operandi in the creation of an ideal future.

The aesthetics of ascension no longer held sway with the outbreak of World War II and the subsequent disillusionment with technology's promise to create a better future. Amid the ideological tensions of the Cold War, the innocence of utopia building slowly faded. The earlier idealism and visionary mentality associated with this aesthetic consciousness were replaced by practical aerial mapping geared toward the planning of functional cities, infrastructure development, and geological management. Aerial reconnaissance played an important role in the creation of two iconic modern cities of the postwar period: Chandigarh and Brasilia. Undertaking an aerial survey to pinpoint the site for a "showcase" city through which to construct a particular brand of national identity, as in the case of these two cities, somewhat recalled the master builder's ascension to the sky in his bid to introduce a new city and society. Yet, for Chandigarh and Brasilia, it was a different kind of master builder, less solo and more political and practical. Le Corbusier engaged in realpolitik to sway Prime Minister Nehru and the Indian bureaucracy to advance the Chandigarh project the way he wanted it. Lúcio Costa's "airplane-shaped" plan for Brasilia needed President Kubitschek's determined patronage.

In a 1947 article in the *Architectural Review* titled "The Architecture of Bureaucracy and the Architecture of Genius," Henry-Russell Hitchcock lamented the intellectual threat facing the genius master builder with the rise of a new regime of large-scale "bureaucratic architecture" that lacked "poetry."[5] The image of the master builder has undergone various shifts since the interwar period, but by no means is the master builder dead. In contemporary celebrity-obsessed culture, he has become a "starchitect." The advocates of "sustainable" architecture oppose a Howard Roarkian primacy of the architect in the design process, supporting instead collaboration among the public, design professionals, and policy wonks.

The wonderment of seeing cities and landscapes from skyscraper observatories and airplanes has not diminished. Today's urban tourists flock to the tallest buildings—the Empire State Building, the Petronas Tower, the Burj Khalifa—to get a 360-degree aerial view of the whole city, as did Ferriss, Dreiser, O'Keeffe, and Fitzgerald during the 1920s and 1930s in Manhattan. Tethered balloon rides next to historic landmarks offer tourists riveting views of built forms from the aviator's point of view. It is indeed impossible to grasp the land–water architectural composition of Angkor Wat in Siem Reap, Cambodia, without the help of a balloon ride. Helicopter rides over cities present modern tourists with a new type of commodified urban voyeurism.

The early twentieth-century high-intensity drama of the airplane and skyscraper as a key force propelling the World of Tomorrow seems like a saga of yore. Yet the pairing of these two ubiquitous symbols of modern life continues to have a place in our collective consciousness. The tragic collision of airplanes with the Twin Towers on September 11, 2001, remains among the darkest of expressions. The aerial vantage that was once reserved to the aviator has now been democratized and reaches stratospheric heights. The proliferation of high-tech devices enables anyone to access satellite views of the earth at any time, anywhere. Yet only the superrich can experience these views aboard a spacecraft, through the mushrooming space tourism industry. Thus, the elitism and exclusivity of the master builder's gaze reverberate in our own time.

Notes

INTRODUCTION

1. Gertrude Stein, *Picasso* (1938; repr., New York: Dover, 1984), 49–50.

2. Mary E. Burt, "Aeronautics Will Develop a Broader Vision," *Flying* (New York), October 1915; Marcella Holtkamp, "A Bit of Philosophy on Flying," *Air Travel News,* August 1929, 21–22; Clayton D. Russell, "Seeing Things from Above," *Craftsmen Aero News,* September 1937, 2, 7–8.

3. Thomas E. Tallmadge, *The Story of Architecture in America* (New York: W. W. Norton, 1927), 296. Tallmadge misquotes Percy Bysshe Shelley's poem "The Cloud" (1820).

4. Harry F. Guggenheim, *The Seven Skies* (New York: G. P. Putnam's Sons, 1930), 36.

5. Le Corbusier, *Aircraft* (New York: The Studio, 1935). DC-3 was an American fixed-wing and propeller-driven aircraft that brought about a new era of aerial transportation during the 1930s and 1940s. The image on the back cover is a smaller version of plate 98 inside. Manipulated identical photos of Eastern Airlines' Douglas DC-3 called *Florida Flyer* were most possibly pasted on an aerial photograph of New York City.

6. For a panoramic history of ascension in Christian thought, see Ernest T. Dewald, "The Iconography of the Ascension," *American Journal of Archaeology,* 2d ser., 19 (1915): 277–319. A thorough study of ascension, or *mi`rāj,* in Islam is found in James Winston Morris, "The Spiritual Ascension: Ibn 'Arabi and the Mi`rāj," *Journal of the American Oriental Society* 107, no. 4 (1987): 629–52.

7. W. Tudor Jones, *The Spiritual Ascent of Man* (New York: G. P. Putnam's Sons, 1917), 233.

8. Stephen Kern, *The Culture of Time and Space, 1880–1918* (Cambridge, Mass.: Harvard University Press, 1983), 242.

9. Ferriss was born in St. Louis, Missouri; Fuller in Milton, Massachusetts; and Bel Geddes in Adrian, Michigan.

10. Cecil Peoli, "How to Make a Model Aeroplane," *Scientific American,* October 14, 1911. "Winged Superchildren of Tomorrow" appeared as part of a Monarch advertisement in *Saturday Evening Post,* February 2, 1929, quoted in Joseph J. Corn, *The Winged Gospel: America's Romance with Aviation, 1900–1950* (New York: Oxford University Press, 1983), 113.

11. See, for instance, James H. Hare, "Aerial Photography—A New Art," *Flying,* May 1924; Crockett A. Harrison, "Making Maps from the Air," *MIT Technology Review,*

December 1930, 127–30, 149–50; Sherman M. Fairchild, "Aerial Photography: Its Development and Future," *Annals of the American Academy of Political and Social Science* 131 (May 1927): 49–55.

12. Le Corbusier, *Towards a New Architecture,* trans. Frederick Etchells (New York: Payson & Clarke, 1927), 109, 127.

13. Madison Grant, *The Passing of the Great Race* (1916; repr., New York: Charles Scribner's Sons, 1921); Edward M. East, *Mankind at the Crossroads* (New York: Charles Scribner's Sons, 1923); Raymond B. Fosdick, *The Old Savage in the New Civilization* (Garden City, N.Y.: Doubleday, Doran, 1929); Walter Lippmann, *A Preface to Morals* (New York: Macmillan, 1929); H. G. Wells, *The Open Conspiracy: Blue Prints for a World Revolution* (Garden City, N.Y.: Doubleday, Doran, 1928).

14. William E. Leuchtenburg, *The Perils of Prosperity, 1914–1932* (Chicago: University of Chicago Press, 1958), 142–45. For discussions of the bleak view of the period, see Lippmann, *A Preface to Morals*; Joseph Wood Krutch, *The Modern Temper* (New York: Harcourt, Brace, 1929).

15. William Harlan Hale, *Challenge to Defeat: Modern Man in Goethe's World and Spengler's Century* (New York: Harcourt, Brace, 1932), 193.

16. Hugh Ferriss, *The Metropolis of Tomorrow* (New York: Ives Washburn, 1929), 142; R. Buckminster Fuller, *Nine Chains to the Moon* (Philadelphia: J. B. Lippincott, 1938), 18–30; Norman Bel Geddes, "The Man of Tomorrow . . . Today," *Info* 1, no. 4 (April 7, 1939).

17. See, for example, Ralph E. Flanders, "The New Age and the New Man," in *Toward Civilization,* ed. Charles A. Beard (London: Longmans, Green, 1930), 21–23; Werner Graeff, "The New Engineer Is Coming," *G-Material zur elementaren Gestaltung,* July 1923, trans. Michael Bullock, reprinted in *Programs and Manifestoes on 20th-Century Architecture,* ed. Ulrich Conrads (Cambridge: MIT Press, 1971); Kasimir Malevich, *Suprematist Manifesto* (1924), excerpted in Conrads, *Programs and Manifestoes on 20th-Century Architecture,* 87–88. Later works exploring the theme include John R. Gold, *The Experience of Modernism: Modern Architects and the Future City, 1928–1953* (London: E & FN Spon, 1997), 35–36; Wolfgang Pehnt, "The 'New Man' and the Architecture of the Twenties," in *Social Utopias of the Twenties: Bauhaus, Kibbutz, and the Dream of the New Man,* ed. Jeannine Fielder (Wuppertal, Germany: Muller & Busmann Press, 1995).

18. The text of Burnham's address to the London Planning Conference of 1910, which was misquoted by Polk in 1912 as "Make no little plans" of 1907, is a clear error. For Burnham's work, see Peter Hall, *Cities of Tomorrow: An Intellectual History of Urban Planning and Design in the Twentieth Century* (Oxford: Blackwell, 1988), 174. Even though Burnham died in 1912, his master-builder persona continued to resonate with the visionary designers of interwar America.

19. Clarence S. Stein, "Toward New Towns for America," *Town Planning Review* 20, no. 3 (October 1949): 203.

20. Jean le Rond d'Alembert, *Preliminary Discourse to the Encyclopedia of Diderot,* trans. Richard N. Schwab and Walter E. Rex (Indianapolis: Bobbs-Merrill, 1963). For a detailed analysis of this phenomenon, see Martin Jay, *Downcast Eyes: The Denigration of Vision in Twentieth-Century French Thought* (Berkeley: University of California Press, 1993), 83–91.

21. For medieval depictions of God as the "Architect of the Universe," see Amos Funkenstein, *Theology and the Scientific Imagination: From the Middle Ages to the Seventeenth Century* (Princeton, N.J.: Princeton University Press, 1986); Katherine H. Tachau, "God's Compass and *Vana Curiositas*: Scientific Study in the Old French *Bible Moralisée*," *Art Bulletin* 80, no. 1 (March 1998): 7–33.

22. See Andrew Saint, "The Architect as Hero and Genius," in *The Image of the Architect* (New Haven, Conn.: Yale University Press, 1983); Julian Petley, "The Architect as *Ubermensch*," in *Picture This: Media Representations of Visual Art and Artists*, ed. Philip Hayward (London: J. Libbey, 1988); Pehnt, "The 'New Man' and the Architecture of the Twenties."

23. The American Philosophical Society in Philadelphia is the best source for information on the American eugenics movement in the early twentieth century. A few useful readings from the early twentieth century are James A. Field, "The Progress of Eugenics," *Quarterly Journal of Economics* 26, no. 1 (November 1911); Albert S. Beckham, "Applied Eugenics," *The Crisis* 28, no. 4 (1924): 177–78; Arnold White, "Eugenics and National Efficiency," *Eugenics Review* 1 (July 1909); Samuel J. Holmes, *The Eugenic Predicament* (New York: Harcourt, Brace, 1933). Mark Haller's *Eugenics: Hereditarian Attitudes in American Thought* (New Brunswick, N.J.: Rutgers University Press, 1963) provides a critical exploration of the American eugenics movement. For more recent work, see Edwin Black, *War against the Weak: Eugenics and America's Campaign to Create a Master Race* (New York: Thunder's Mouth Press, 2003); Christina Cogdell and Sudan Currell, eds., *Popular Eugenics: National Efficiency and American Mass Culture in the 1930s* (Athens: Ohio University Press, 2006); Marouf A. Hasian Jr., *The Rhetoric of Eugenics in Anglo-American Thought* (Athens: University of Georgia Press, 1996).

24. Jean Ferriss Leich, *Architectural Visions: The Drawings of Hugh Ferriss* (New York: Whitney Library of Design, 1980); Carol Willis, "Drawing towards Metropolis," in Hugh Ferriss, *The Metropolis of Tomorrow* (1929; repr., New York: Princeton Architectural Press, 1986). Recent scholarship includes Kevin R. McNamara, "Recentering the City: Hugh Ferriss and Urban Form," in *Urban Verbs: Arts and Discourses of American Cities* (Stanford, Calif.: Stanford University Press, 1996).

25. Giorgio Ciucci, Francesco Dal Co, Mario Manieri-Elia, and Manfredo Tafuri, *The American City: From the Civil War to the New Deal*, trans. Barbara Luiga La Penta (Cambridge: MIT Press, 1979); Rem Koolhaas, *Delirious New York: A Retroactive Manifesto for Manhattan* (1978; repr., New York: Monacelli Press, 1994).

26. Lewis Mumford, "The Sacred City," *New Republic*, January 27, 1926; Lewis Mumford, "The Sky Line: Mr. Rockefeller's Center," *New Yorker*, December 23, 1933; Ciucci et al., *The American City*, 447–51; Vincent Scully, *American Architecture and Urbanism* (1969; repr., San Antonio, Tex.: Trinity University Press, 2013).

27. Jay Baldwin, *Bucky Works: Buckminster Fuller's Ideas for Today* (New York: John Wiley, 1996); Alden Hatch, *Buckminster Fuller at Home in the Universe* (New York: Crown, 1974); Robert Marks and Buckminster Fuller, *The Dymaxion World of Buckminster Fuller* (1960; repr., Garden City, N.Y.: Anchor Books, 1973); John McHale, *R. Buckminster Fuller* (New York: George Braziller, 1962); Martin Pawley, *Buckminster Fuller* (London: Trefoil, 1990); Lloyd Steven Sieden, *Buckminster Fuller's Universe: An Appreciation* (New York: Plenum Press, 1989).

28. Reyner Banham, *Theory and Design in the First Machine Age* (1960; repr., Cambridge: MIT Press, 1980).

29. Stanford University acquired Fuller's archive in 1999, and a few years later it became accessible to scholars. Buckminster Fuller Institute, which moved from Sebastopol, California, to Brooklyn, New York, in 2004, has been instrumental in resuscitating scholarly interests in Fuller. Beginning in 2008, Buckminster Fuller Challenge, an annual international competition seeking design ideas based on Fuller's comprehensive approach to the world's natural resources, has drawn attention to Fuller's ecological worldview.

30. Yunn Chii Wong, "The Geodesic Works of Richard Buckminster Fuller, 1948–68" (Ph.D. diss., MIT, 1999); Joachim Krausse and Claude Lichtenstein, eds., *Your Private Sky: R. Buckminster Fuller, Discourse* (Baden, Switzerland: Lars Müller, 1999); Joachim Krausse and Claude Lichtenstein, eds., *Your Private Sky: R. Buckminster Fuller, the Art of Design Science* (Baden, Switzerland: Lars Müller, 1999); K. Michael Hays and Dana Miller, eds., *Buckminster Fuller: Starting with the Universe* (New York: Whitney Museum of American Art and Yale University Press, 2008). Other notable recent contributions are Federico Neder, *Fuller Houses: R. Buckminster Fuller's Dymaxion Dwellings and Other Domestic Adventures,* trans. Elsa Lam (Baden, Switzerland: Lars Müller, 2008); Hsiao-Yun Chu and Roberto G. Trujillo, eds., *New Views on R. Buckminster Fuller* (Stanford, Calif.: Stanford University Press, 2009).

31. Loretta Lorance, *Becoming Bucky Fuller* (Cambridge: MIT Press, 2009).

32. Donald J. Bush, *The Streamlined Decade* (New York: George Braziller, 1975); Jeffrey L. Meikle, *Twentieth Century Limited: Industrial Design in America, 1925–1939* (Philadelphia: Temple University Press, 1979); Roland Marchand, "The Designers Go to the Fair II: Norman Bel Geddes, the General Motors 'Futurama,' and the Visit to the Factory Transformed," *Design Issues* 8, no. 2 (Spring 1992): 23–40; Arthur J. Pulos, *The American Design Adventure, 1940–1975* (Cambridge: MIT Press, 1988); Arthur J. Pulos, *American Design Ethic* (Cambridge: MIT Press, 1983). The Italian journal *Rassegna* published a useful set of essays on Bel Geddes in 1994 (vol. 60).

33. Douglas Adams, "Norman Bel Geddes and Streamlined Spaces," *Journal of Architectural Education* 30, no. 1 (September 1976): 22–24; Vivien Arnold, "The Image of the Freeway," *Journal of Architectural Education* 30, no. 1 (September 1976): 28–30; Travis Brown Jr., "On an Aesthetic of Highway Speed," *Journal of Architectural Education* 30, no. 1 (September 1976): 25–27; Bruce Bliven Jr., "Metropolis: 1960 Style," *New Republic,* September 29, 1937; and Robert Coombs, "Norman Bel Geddes' Highways and Horizons," *Perspecta* 13/14 (1971): 11–27.

34. Christina Cogdell, *Eugenic Design: Streamlining America in the 1930s* (Philadelphia: University of Pennsylvania Press, 2004). I have written on the significance of human flight for Bel Geddes's Futurama: Adnan Morshed, "The Aesthetics of Ascension in Norman Bel Geddes's Futurama," *Journal of the Society of Architectural Historians* 63, no. 1 (March 2004): 74–99.

35. This understanding, as Darko Suvin argues, is a specific American contribution to the utopian literary tradition. Suvin writes: "The construction of a social system . . . is also the reconstruction of the hero . . . that a different kind of Man is indispensable if one wants a different World. It is an insight that came up precisely with the

American Revolution and its radical-democratic paradigm of dynamic changeability, of new Adamic figures (for example, Blake)." Darko Suvin, "Anticipating the Sunburst—Dream and Vision: The Exemplary Case of Bellamy and Morris," in *America as Utopia*, ed. Kenneth M. Roemer (New York: Burt Franklin, 1981), 60.

36. Even though it may appear to be a case of cultural essentializing in hindsight, Croly's thesis shines a spotlight on Progressive Era thoughts: "The Vision of a better future is not, perhaps, as unclouded for the present generation of Americans as it was for certain former generations; but in spite of a more friendly acquaintance with all sorts of obstacles and pitfalls, our country is still figured in the imagination of its citizens as the Land of Promise." Herbert D. Croly, *The Promise of American Life* (New York: Macmillan, 1909), 13. Also see Waldo Frank, *Our America* (New York: Boni & Liveright, 1919).

37. The suggestion that frontier values pervaded the description or experience of the new "frontier of the air" was a consistent theme in the popular magazines of aviation and the city during the 1920s. Will Rogers, celebrated American social commentator of the 1920s and 1930s, often described aviators as machine-age cowboys. See Stearns Clancy, "Aviation's Patron Saint," *Scientific American*, October 1929.

38. Frederick Jackson Turner, "The Significance of the Frontier in American History" (1893), in *The Frontier in American History* (1920; repr., Tucson: University of Arizona Press, 1986).

39. Charles A. Beard, "The Myth of Rugged American Individualism," *Harper's Monthly Magazine*, December 1931; Isaiah Bowman, *The New World: Problems in Political Geography* (Yonkers-on-Hudson, N.Y.: World Book, 1921); Percy Boynton, *The Rediscovery of the Frontier* (Chicago: University of Chicago Press, 1931); John Dewey, "The American Intellectual Frontier," *New Republic*, May 10, 1922; Robert Dripps, *New Pioneers for New Frontiers* (New York: Buffalo Bill American Association, 1924); Guy Emerson, *The New Frontier: A Study of the American Liberal Spirit, Its Frontier Origin, and Its Application to Modern Problems* (New York: Henry Holt, 1920); Waldo Frank, *The Re-discovery of America* (New York: Charles Scribner's Sons, 1929); Archer B. Hulbert, *Frontiers: The Genius of American Nationality* (Boston: Little, Brown, 1929); Oswald Villard, "The West: Tamed and Combed," *The Nation*, December 10, 1924; Benjamin F. Wright Jr., "American Democracy and the Frontier," *Yale Review* 20 (September 1930).

40. See Herbert Hoover, *American Individualism* (New York: Doubleday, 1922); Franklin D. Roosevelt, "Back to the Land," *Review of Reviews* 84 (October 1931); Franklin D. Roosevelt, *Looking Forward* (New York: John Day, 1933).

41. Quoted in Gilbert Seldes, "Transatlantic," *New Republic*, June 1, 1927. For more recent reflections on the topic, see James Oliver Robertson, *American Myth, American Reality* (New York: Hill & Wang, 1980), 200–201; Rem Koolhaas, "The Frontier in the Sky," in *Delirious New York*, 82–94.

42. Boynton, *Rediscovery of the Frontier*; Hulbert, *Frontiers*.

43. Robert Wohl, *A Passion for Wings: Aviation and the Western Imagination, 1908–1918* (New Haven, Conn.: Yale University Press, 1994), 61–66. In the Soviet Union, Communist Party officials embraced aviation as a crucial agent of collective change. See Scott W. Palmer, "Peasants into Pilots: Soviet Air-Mindedness as an Ideology of Dominance," *Technology and Culture* 41, no. 1 (January 2000): 1–26.

44. For a discussion of antiurbanism in American literary tradition, see Morton White and Lucia White, *The Intellectual versus the City: From Thomas Jefferson to Frank Lloyd Wright* (New York: Mentor Books, 1962).

45. Josiah Strong, *The Challenge of the City* (New York: Young People's Missionary Movement, 1907).

46. For an analysis of *View of Venice,* see John A. Pinto, "Origins and Development of the Ichnographic City Plan," *Journal of the Society of Architectural Historians* 35, no. 1 (March 1976): 35–50.

47. For an analysis of Nadar's balloon photography, see Shelly Rice, *Parisian Views* (Cambridge: MIT Press, 1997), 172–80.

48. Beaumont Newhall, *Airborne Camera: The World from the Air and Outer Space* (New York: Hastings House, 1969), 27.

49. Typical examples include "Airplane 'Eyes' Used for Map Making," *Scientific American,* April 1927, 256; C. H. Birdseye, "Aerial Mapping by the Geological Survey," *Aerial Age,* May 1923. Recent scholarship offers useful insights. See Allan Sekula, "The Instrumental Image: Steichen at War," *Artforum,* December 1975; John Welchman, "Here, There and Otherwise," *Artforum,* September 1988.

50. The chapters appear in this order based on the chronology of the subjects' signature projects: Ferriss's Metropolis of Tomorrow, Fuller's Dymaxion House, and Bel Geddes's Futurama.

51. E. H. Carr, *What Is History?* (New York: Macmillan, 1961), 24.

52. Northrop Frye, *The Educated Imagination* (Bloomington: Indiana University Press, 1964).

1. HUGH FERRISS AND THE "HARMONIOUS DEVELOPMENT OF MAN"

1. Georg Simmel examined this phenomenon in "The Metropolis and Mental Life" (1903), in *The Sociology of Georg Simmel,* trans. Kurt Wolff (New York: Free Press, 1950). On the literary front, Sinclair Lewis's novel *Arrowsmith* (written in the early 1920s and published in 1925) uses New York City as a foil for protagonist Martin Arrowsmith's idealistic pursuits and resulting conflicts. Henry Blake Fuller's *The Cliff-Dwellers* (1893) and James Oppenheim's *The Olympian: A Story of the City* (1912) deal with the dizzying environment of the modern metropolis.

2. Hugh Ferriss, *The Metropolis of Tomorrow* (New York: Ives Washburn, 1929). The book was reprinted by the Princeton Architectural Press in 1986 with an introductory essay by the architectural historian Carol Willis titled "Drawing towards Metropolis." Most of Ferriss's drawings are housed at the Hugh Ferriss Collection, Avery Library, Columbia University (hereafter HFC). Prior to the publication of *The Metropolis of Tomorrow,* some of these drawings had already been seen by the public at exhibitions or in articles and advertisements in magazines and newspapers.

3. On urbanization in America, see Carl Abbott, *Urban America in the Modern Age: 1920 to the Present* (Wheeling, Ill.: Harlan Davidson, 1987); Paul Kramer and Frederick L. Holborn, eds., *The City in American Life: From Colonial Times to the Present* (New York: Capricorn Books, 1970); Constance McLaughlin Green, *The Rise of Urban America* (New York: Harper Colophon Books, 1965); Peter Hall, *Cities of Tomorrow: An*

Intellectual History of Urban Planning and Design in the Twentieth Century (Oxford: Blackwell, 1988); Raymond A. Mohl, *The New City: Urban America in the Industrial Age, 1860–1920* (Arlington Heights, Ill.: Harlan Davidson, 1985); Anthony Sutcliffe, ed., *Metropolis 1890–1940* (Chicago: University of Chicago Press, 1984).

4. Sheldon Cheney, *The New World Architecture* (New York: AMS Press, 1930), 398–400. In a similar vein, Ada Louise Huxtable argued that Ferriss, "the extraordinary architectural illustrator . . . caught the period's substance and spirit and style." Ada Louise Huxtable, *Kicked a Building Lately?* (New York: Quadrangle Books, 1976), 291.

5. Cheney, *The New World Architecture,* 399. The drawing had appeared previously in *Pencil Points* 6, no. 5 (May 1925).

6. David Nye, "The Sublime and the Skyline," in *The American Skyscraper: Cultural Histories,* ed. Roberta Moudry (Cambridge: Cambridge University Press, 2005), 256.

7. *The Metropolis of Tomorrow* contains a total of sixty drawings. Among them, thirty-one are from a pedestrian point of view, and twenty-seven are bird's-eye views. The penultimate drawing is a plan of the core of Ferriss's proposed metropolis, and the final one is a mystical drawing suggesting the development of the city in the "image of man." Ferriss divides *The Metropolis of Tomorrow* into three sections. The first, "Cities of Today," includes "the significant structures which already exist" or are under construction, represented by such skyscrapers as architect Raymond M. Hood's Radiator Building and architect Arthur Loomis Harmon's Shelton Hotel. The second section, "Projected Trends," inventories the emerging trends in high-rise constructions and their purported role in materializing the city of tomorrow. The third section, "An Imaginary Metropolis," articulates Ferriss's vision of the future city, which is based on a civic center with a triangular arrangement consisting of a business zone, an art zone, and a science zone.

8. For discussion of the urban paintings of Ashcan artists, see Rebecca Zurier, Robert W. Snyder, and Virginia M. Mecklenburg, *Metropolitan Lives: The Ashcan Artists and Their New York* (New York: National Museum of American Art and W. W. Norton, 1995); Rebecca Zurier, *Picturing the City: Urban Vision and the Ashcan School* (Berkeley: University of California Press, 2006). Robert Henri and five of his followers—George Luks, William Glackens, John Sloan, Everett Shinn, and George Bellows—are among the best-known artists of the Ashcan School.

9. Nye, "The Sublime and the Skyline," 256.

10. Hugh Ferriss, "New York from a Studio Rooftop," *Christian Science Monitor,* December 3, 1923; Orrick Johns, "Architects Dream of a Pinnacle City," *New York Times,* December 28, 1924; "City Yearns to Sky in Architect's Vision," *New York Evening Post,* April 14, 1925.

11. Stuart Chase, *Men and Machines* (New York: Macmillan, 1929), 247.

12. Joshua C. Taylor, "The Image of Urban Optimism," in *America as Art* (Washington, D.C.: Smithsonian Institution Press, 1976).

13. The *Bicentennial Exhibition of American Art* at the National Collection of Fine Arts in Washington, D.C., included a section titled "The Image of Urban Optimism." "The Lure of the City" was one of four drawings that Ferriss contributed to the exhibition. The drawing, reproduced here as Figure 1.6, appears in Ferriss, *The Metropolis of Tomorrow,* 58. On the collision between horizontal and vertical frontiers, see William R.

Taylor, *In Pursuit of Gotham: Culture and Commerce in New York* (New York: Oxford University Press, 1992), 63–67.

14. Leo Marx, *The Machine in the Garden: Technology and the Pastoral Ideal in America* (New York: Oxford University Press, 1964).

15. Frederick Jackson Turner, "The Significance of the Frontier in American History" (1893), in *The Frontier in American History* (1920; repr., Tucson: University of Arizona Press, 1986).

16. Ferriss described this typical migratory pattern from the country to the city in the drawing's caption. See Ferriss, *The Metropolis of Tomorrow,* 59.

17. This pursuit of material well-being through perseverance is rooted in the work ethics of both the Puritans and the frontiersmen. See Richard Weiss, *The American Myth of Success: From Horatio Alger to Norman Vincent Peale* (New York: Basic Books, 1969).

18. See Peter J. Conn, *Literature in America: An Illustrated History* (New York: Cambridge University Press, 1989), 274. Also see Anne Gregory Terhune, *Thomas Hovenden: His Life and Art* (Philadelphia: University of Pennsylvania Press, 2006).

19. See Roland Marchand, *Advertising the American Dream: Making Way for Modernity, 1920–1940* (Berkeley: University of California Press, 1986).

20. Ferriss was born on July 12, 1889, in St. Louis, Missouri. He received his B.S. in architecture from Washington University in 1911 and an honorary master's degree in architecture from his alma mater in 1928. He received his professional licensure on February 13, 1930. Ferriss described himself as an Episcopalian and politically independent. Various biographical accounts portray Ferriss as poetic and versatile from early on in his life. One biographical note describes some of his early accomplishments: "By the time he received his bachelor's degree in architecture, he had been art editor of the college weekly; a member of four class societies; captain of the track team; president of the athletic association, his class, and the student body; and Consul of Tau Tau Chapter. Had there been a National Balfour Award in 1910, Hugh Ferriss would have been a front-line candidate for honors." "Hugh Ferriss, Nation's No. 1 Architectural Artist," HFC, Box 7, Biography.

21. Quoted in Rem Koolhaas, *Delirious New York: A Retroactive Manifesto for Manhattan* (1978; repr., New York: Monacelli Press, 1994), 110.

22. Copy of memo to N.Y. chapter of A.I.A., HFC, Box 7, Biography. Ferriss spent three years at Gilbert's office, mostly as a draftsman, drawing from Gilbert's blueprints visualizations of the buildings and their sections as they would appear upon their completion. In this way he played a part in the construction of Gilbert's masterpiece, the Woolworth Building. Later in his life, however, Ferriss recalled his time at Gilbert's firm as monotonous and unsatisfactory in terms of professional development. Also see Jean Ferriss Leich, *Architectural Visions: The Drawings of Hugh Ferriss* (New York: Whitney Library of Design, 1980), 20.

23. Ferriss's drawings appeared in *Vanity Fair, McCall's, Fortune, Saturday Evening Post,* and other magazines, as well as in such newspapers as the *New York Times, Chicago Tribune,* and *Christian Science Monitor.* His first one-man show took place at the Anderson Galleries in New York in 1925. From the mid-1920s, Ferriss was an in-demand speaker at Columbia University, Yale University, the University of Pennsylvania, and other institutions. See various clippings in HFC, Box 7, Biography.

24. See the exhibition brochure, *New York of To-Morrow*. Also see Leon Solon, "The Titan City Exhibition," *Architectural Record* 58 (January 1926).

25. "Titan City Exhibition," letter from Harvey Wiley Corbett to Ferriss, November 16, 1926, HFC, Box 10.

26. See "Letters/Comments," letter from Cass Gilbert, December 20, 1929, HFC, Box 7.

27. The drawing, reproduced here as Figure 1.9, appears in Ferriss, *The Metropolis of Tomorrow*, 18.

28. Albert Boime, *The Magisterial Gaze, Manifest Destiny and American Landscape Painting, c. 1830–1865* (Washington, D.C.: Smithsonian Institution Press, 1991), x.

29. According to Carol Willis, Ferris began living at the Architects Building at 101 Park Avenue in 1923. Slight confusion remains, as biographical memos show his home address as 35 East 9th Street. See HFC, Box 7, Biography. Also see Willis, "Drawing towards Metropolis," in Hugh Ferriss, *The Metropolis of Tomorrow* (1929; repr., New York: Princeton Architectural Press, 1986), 152.

30. Ferriss, "New York from a Studio Rooftop."

31. William R. Taylor suggests that there was a peculiar relationship between social or corporate career climbing and the symbolism of vertical structures. See Taylor, *In Pursuit of Gotham*, 67.

32. For discussion of the notion of success in American society, see Weiss, *The American Myth of Success*, 128–29. Weiss argues that the American myth of success, rooted in Protestant morality and agrarian society, required major reorientation as a result of the skepticism brought on by modern science and the rapid growth of urbanized culture.

33. Horatio Alger, *Ragged Dick and Mark, the Match Boy* (1867; repr., New York: Collier Books, 1962). Herman Melville, *Pierre, or the Ambiguities*, ed. Henry A. Murray (New York: Hendricks, 1949).

34. The first American to win the Nobel Prize in Literature, Sinclair Lewis created a protagonist who was "born to the prairies, never far from the sight of the cornfields, [and] was conveyed to blazing lands and portentous enterprises." Martin Arrowsmith's journey, fraught with personal tragedies and moral victories, from the American heartland to New York City, parallels Ferriss's own journey.

35. Roderick Nash, *The Nervous Generation: American Thought, 1917–1930* (Chicago: Rand McNally, 1970), 132–33, 153–63.

36. Edward W. Wolner examines the relationship between the skyscraper and the American mythology of success in "The City-within-a-City and Skyscraper Patronage in the 1920's," *Journal of Architectural Education* 42, no. 2 (Winter 1989): 10–23.

37. Thomas A. P. van Leeuwen, *The Skyward Trend of Thought: The Metaphysics of the American Skyscraper* (Cambridge: MIT Press, 1988), 64, 66.

38. Taylor, *In Pursuit of Gotham*, 65.

39. Irving K. Pond, *The Meaning of Architecture: An Essay in Constructive Criticism* (Boston: Marshall Jones, 1918), 112.

40. Claude Bragdon, "The Shelton Hotel, New York," *Architectural Record* 58 (July 1925).

41. "City Yearns to Sky in Architect's Vision."

42. Meir Wigoder, "The 'Solar Eye' of Vision: Emergence of the Skyscraper-Viewer in the Discourse on Heights in New York City, 1890–1920," *Journal of the Society of Architectural Historians* 61, no. 2 (June 2002): 152–69.

43. See Elisabeth Sussman and John G. Hanhardt, *City of Ambition: Artists and New York* (New York: Whitney Museum of American Art, 1996), 59; Merrill Schleier, *The Skyscraper in American Art, 1890–1931* (New York: Da Capo Press, 1983), 50–55; Wanda M. Corn, "The Artist's New York: 1900–1930" in *Budapest and New York: Studies in Metropolitan Transformation, 1870–1930,* ed. Thomas Bender and Carl E. Schorske, (New York: Russell Sage Foundation, 1994), 277; Mary N. Woods, *Beyond the Architect's Eye: Photographs and the American Built Environment* (Philadelphia: University of Pennsylvania Press, 2009), 22–28.

44. Thomas E. Tallmadge, "Today and Tomorrow," in *The Story of Architecture in America* (New York: W.W. Norton, 1927), 296.

45. For discussions of the American theosophical circle, see Jonathan Massey, *Crystal and Arabesque: Claude Bragdon, Ornament, and Modern Architecture* (Pittsburgh: University of Pittsburgh Press, 2009), 265–67; Roger Friedland and Harold Zellman, *The Fellowship: The Untold Story of Frank Lloyd Wright and the Taliesin Fellowship* (New York: HarperCollins, 2006); Kristina Wilson, "The Intimate Gallery and the *Equivalents*: Spirituality in the 1920s Work of Stieglitz," *Art Bulletin* 85, no. 4 (December 2003): 746–68. Ferriss was one of the inner-circle members.

46. In 1938, Ferriss wrote to Stieglitz to solicit O'Keeffe's painting for an exhibition that he was going to curate. Letter from Ferriss to Stieglitz, February 10, 1938, HFC. The Shelton Hotel had an iconic status as the first skyscraper hotel in New York City, completed in January 1924. Designed by Arthur Loomis Harmon in compliance with the new setback rules (mandating setbacks of one foot of air space for every four feet of height), the Shelton Hotel was later renamed the Halloran House. It is located on the east side of Lexington Avenue between 48th and 49th Streets. O'Keeffe and Stieglitz rented an apartment, Room 3033, at the hotel in 1925 and spent the next ten years there.

47. Anna C. Chave, "'Who Will Paint New York?' 'The World's New Art Center' and the Skyscraper Paintings of Georgia O'Keeffe," *American Art* 5, no. 1/2 (Winter/Summer 1991): 87–107. For an analysis of O'Keeffe's paintings of skyscrapers, especially the Radiator Building, designed by Raymond M. Hood and completed in 1924, from the perspective of gender politics, see Vivien Green Fryd, "Georgia O'Keeffe's 'Radiator Building': Gender, Sexuality, Modernism, and Urban Imagery," *Winterthur Portfolio* 35, no. 4 (Winter 2000): 269–89.

48. Chave, "'Who Will Paint New York?,'" 97.

49. Benita Eisler, *O'Keeffe and Stieglitz: An American Romance* (New York: Doubleday, 1991), 343. On O'Keeffe's spiritual quest in modern art, see Celia Weisman, "O'Keeffe's Art: Sacred Symbols and Spiritual Quest," *Woman's Art Journal* 3, no. 2 (1982–83): 10–14.

50. Eisler, *O'Keeffe and Stieglitz,* 343.

51. Letter from Stieglitz to Sherwood Anderson, December 9, 1925, Alfred Stieglitz Archive, Yale Collection of American Literature, Beinecke Rare Book and Manuscript Library, Yale University. Stieglitz and his photographer colleagues took a guarded view toward urban modernity. For a discussion of how the New York painters and

photographers viewed the rising city through the prism of an American fascination with sublime landscapes, see Corn, "The Artist's New York," 281–83. For an example of a Stieglitz photo shot from the skyscraper window, see his *Snapshot from My Window, New York,* which was published in *Camera Work* in October 1907.

52. Bragdon, "The Shelton Hotel, New York."

53. Ferriss, *The Metropolis of Tomorrow,* 18.

54. Willis, "Drawing towards Metropolis," 152.

55. Ibid.

56. Jane Jacobs, *The Death and Life of Great American Cities* (New York: Random House, 1961).

57. Reviewing the contributions of Ferriss to the *Titan City* exhibition, Mumford used these expressions in "The Sacred City," *New Republic,* January 27, 1926.

58. See Robert H. Wiebe, *The Search for Order, 1877–1920* (New York: Hill & Wang, 1967).

59. Morton White and Lucia White, *The Intellectual versus the City: From Thomas Jefferson to Frank Lloyd Wright* (New York: Mentor Books, 1962), 14.

60. Marshall Berman, *All That Is Solid Melts into Air: The Experience of Modernity* (New York: Simon & Schuster, 1982).

61. The protagonists of Alger and Melville, among other novelists, prefigured this ambiguity of the modern man in the nineteenth century. The dangerous uncertainties of city life paradoxically sharpened the entrepreneurial instincts of Alger's poor, orphaned protagonist Dick, who ultimately becomes a successful businessman in New York. As a boy growing up in the street, Dick knows how to negotiate the crooked urban paths efficiently and thus serves as a tour guide for a rural youth named Frank. Dick initiates Frank's acculturation into the norms, colors, and sounds of the city, decoding the city's mysteries for him. Melville is much less sanguine. Upon arriving in the city, Melville's Pierre is overwhelmed by its flurry of stimuli and their relentless signification. See Milton Rugoff, "Dealer in Daydreams: Horatio Alger, Jr.," in *America's Gilded Age: Intimate Portraits from an Era of Extravagance and Change, 1850–1890* (New York: Henry Holt, 1989); Marcus Klein, "Rags to Riches; or, Horatio Alger, Jr., and the Dangerous Classes of New York," in *Easterns, Westerns, and Private Eyes: American Matters, 1870–1900* (Madison: University of Wisconsin Press, 1994); Stephen Rachman, "Melville's Pierre and Nervous Exhaustion; or, 'The Vacant Whirlingness of the Bewilderingness,'" *Literature and Medicine* 16, no. 2 (1997): 226–49.

62. Ferriss, *The Metropolis of Tomorrow,* 15.

63. See Alan Trachtenberg, "Mysteries of the Great City," in *The Incorporation of America: Culture and Society in the Gilded Age* (New York: Hill & Wang, 1982).

64. John Bunyan, *The Pilgrim's Progress* (1678; repr., New York: Washington Square Press, 1957). Originally read by the poorer classes of England from the seventeenth century onward, the influence of this book on the American pilgrims and pioneers and, more generally, English-speaking America, was enormous.

65. Trachtenberg, "Mysteries of the Great City," 104.

66. Frank Luther Mott, *American Journalism: A History, 1860–1960* (New York: Macmillan, 1962), 436–49. In 1892, New York City had nine morning papers (among them the *Herald, Morning Journal, Sun, Times,* and *Tribune*) and seven evening papers

(including the *Daily News, Evening Post,* and *Evening Sun*). Also see Justin D. Edwards, "Henry James's 'Alien' New York: Gender and Race in *The American Scene,*" *American Studies International* 36, no. 1 (February 1998): 66–80; Trachtenberg, "Mysteries of the Great City," 122–24.

67. Jacob A. Riis, *How the Other Half Lives: Studies among the Tenements of New York* (1890; repr., New York: Penguin Books, 1997); Mott, *American Journalism,* 449; Edwards, "Henry James's 'Alien' New York"; Eric Homberger, *Mrs. Astor's New York: Money and Social Power in a Gilded Age* (New Haven, Conn.: Yale University Press, 2002), 27–34. Riis worked as a police reporter for the *Tribune.*

68. Lincoln Steffens, *The Shame of the Cities* (1904; repr., New York: Hill & Wang, 1957). Theodore Roosevelt called the urban reform crusade "muck-raking." Some of the key leaders in the movement were such magazines as *Collier's, Cosmopolitan,* and *McClure's.* See Mott, *American Journalism,* 57. Urban poverty and related degradation in living conditions were a key focus of muckraking journalism. See, for instance, Ernest Flagg, "The New York Tenement-House Evil and Its Cure," *Scribner's Magazine,* July 1894, reprinted in Robert A. Woods et al., *The Poor in Great Cities: Their Problems and What Is Being Done to Solve Them* (New York: Charles Scribner's Sons, 1895). Also see Lee Philpott, *The Slum and the Ghetto: Neighborhood Deterioration and Middle-Class Reform, Chicago, 1880–1930* (New York: Oxford University Press, 1978).

69. Josiah Strong, *The Challenge of the City* (New York: Young People's Missionary Movement, 1907); Albert Fein, "The American City: The Ideal and the Real," in *The Rise of an American Architecture,* ed. Edgar Kaufmann Jr. (New York: Praeger, 1970), 52.

70. Paul S. Boyer, *Urban Masses and Moral Order in America, 1820–1920* (Cambridge, Mass.: Harvard University Press, 1978), 285.

71. Josiah Strong, *Our Country: Its Possible Future and Present Crisis* (New York: Baker & Taylor for the American Home Missionary Society, 1885); Strong, *The Challenge of the City;* John Giffen Thompson, *Urbanization: Its Effects on Government and Society* (New York: E. P. Dutton, 1927).

72. Pitirim Sorokin and Carle C. Zimmerman, *Principles of Rural–Urban Sociology* (New York: Henry Holt, 1929).

73. Stephen Zoll, "Superville: New York—Aspects of Very High Bulk," *Massachusetts Review* 14, no. 3 (Summer 1973): 447–516, 533–38. Urban journalists sometimes identified skyscrapers with individualism and corporate profit-mongering. As Mona Domosh demonstrates, the growth of early skyscrapers in New York City in the late nineteenth century was intertwined with the development of newspapers and their owners' desire to use tall buildings as advertisements for the newspapers. Mona Domosh, "The Symbolism of the Skyscraper: Case Studies of New York's First Tall Buildings," *Journal of Urban History* 14, no. 3 (May 1988): 321–45. Articles on the skyscraper as part of an urban intrigue abounded in the popular magazines of the 1920s. Examples include H. A. Caparn, "The Riddle of the Tall Building: Has the Skyscraper a Place in American Architecture?," *The Craftsman* 4, no. 4 (July 1906); Wayne D. Heydecker, "Up in the Air: An Examination of Certain Arguments for Skyscrapers," *The American City,* February 1928.

74. Zoll, "Superville," 450.

75. Winston Weisman, "New York and the Problem of the First Skyscraper," *Journal of the Society of Architectural Historians* 12, no. 1 (March 1953): 20.

76. This was a common impetus in the early twentieth century. See, for instance, Charles Mulford Robinson's 1901 best seller, *The Improvement of Towns and Cities; or, The Practical Basis of Civic Aesthetics* (New York: G. P. Putnam's Sons, 1901), followed by his *Modern Civic Art; or, The City Made Beautiful* (New York: G. P. Putnam and Sons, 1903). Zoll writes of this reformatory zeal, "Part of the impetus was fed by indignation, nostalgia and idealism, the emotions of decency and the felt need for an improved environment toward which a perfectable citizenry could evolve." Zoll, "Superville," 451.

77. Reform of the existing city, a "modern Renaissance" as he gazes down on it from the heights, is a leitmotif of Ferriss's writings and lectures. See, for instance, Hugh Ferriss, "Foreword: Architecture of This Age," in *Machine-Age Exposition Catalogue* (exhibition held May 16–28, 1927, at 119 West 57th Street) (New York, 1927).

78. Koolhaas, *Delirious New York,* 117.

79. Westinghouse advertisement, "Civic Development Lags in the Twilight Zone," *The American City,* February 1931.

80. Christian Zapatka, "The Edison Effect: The History of Lighting in the American City," *Lotus,* no. 75 (1993): 63.

81. Westinghouse advertisement, "Planning Lighting . . . Means Looking Ahead," *The American City,* July 1930.

82. Other lighting companies with similar ads during the period included Elreco Poles, General Electric, and Union Metal.

83. David Nye, *Electrifying America: Social Meanings of a New Technology* (Cambridge: MIT Press, 1990); Wolfgang Schivelbusch, *Disenchanted Night: The Industrialization of Light in the Nineteenth Century,* trans. Angela Davis (Berkeley: University of California Press, 1988); Dietrich Neumann, *Architecture of the Night: The Illuminated Building* (Munich: Prestel, 2002); John A. Jackle, *City Lights: Illuminating the American Night* (Baltimore: Johns Hopkins University Press, 2001); Mark Caldwell, *New York Night: The Mystique and Its History* (New York: Scribner, 2005).

84. William Chapman Sharpe, *New York Nocturne: The City after Dark in Literature, Painting, and Photography* (Princeton, N.J.: Princeton University Press, 2008). Also see Mary Woods, "Photography of the Night: Skyscraper Nocturne and Skyscraper Noir in New York," in Neumann, *Architecture of the Night.*

85. John Dos Passos, *Manhattan Transfer* (1925; repr., Boston: Houghton Mifflin, 1943), 112.

86. Frank Lloyd Wright, *The Living City* (New York: Mentor Books, 1958), 59. This is a revised version of his book *The Disappearing City* (1932).

87. Dietrich Neumann, "Luminous Buildings—Architecture of the Night," *Lotus,* no. 75 (1993).

88. "Architects Ask for Aid in Illumination Field. Discuss Exterior Lights for Buildings at Session of the Edison Institute," *New York Times,* February 17, 1929, quoted in Neumann, *Architecture of the Night,* 58.

89. Neumann, *Architecture of the Night,* 58.

90. The booklet contains no information about the cover artist. Neumann confirms that Ferriss was not the illustrator of the cover, despite the strong resemblance in terms of visual projection. E-mail communication with Dietrich Neumann, July 21, 2010.

91. Wright, *The Living City,* 59.

92. Woods, "Photography of the Night," 73.

93. One observer in 1929 noted how "naturally the activities and habits of man gradually changed with the development of artificial illuminations and their extensive use." Charles J. Stahl, "Through the Dark Ages of Street Lighting," *The American City,* November 1929.

94. Washington University, St. Louis, Architecture, General Outline of Instruction, Catalogue Vol. II, No. 2, November 1906–1907, HFC, Box 7, Biography. One of Ferriss's teacher's, Wilbur Tyson Trueblood, contributed an article titled "The French Influence in Architectural Design in America" to the *Washington University Record* 6, no. 6 (April 1911). In this article, Trueblood outlined the impact of Beaux-Arts training on American architecture schools in the first decades of the twentieth century. James W. Fitzgibbon, another of Ferriss's teachers, wrote that one of the key Beaux-Arts rules is the synthesis of "dark-light, with focus at the intersection."

95. Douglas Haskell, "Architecture: The Bright Lights," *The Nation,* January 14, 1931, 55–56.

96. Ferriss, *The Metropolis of Tomorrow,* 62.

97. Josiah Strong, "The Modern City a Menace," in *The Challenge of the City,* 41–70.

98. George G. Foster, *New York by Gas-Light and Other Urban Sketches* (1856; repr., Berkeley: University of California Press, 1990), 69. A social commentator, poet, and urban connoisseur, Foster was a reporter for Horace Greeley's influential *New York Tribune.* Also see James D. McCabe Jr., *Lights and Shadows of New York Life; or, The Sights and Sensations of the Great City* (1872; repr., New York: Farrar, Straus and Giroux, 1970); Helen Campbell et al., *Darkness and Daylight; or, Lights and Shadows of New York Life* (Hartford, Conn.: A. D. Worthington, 1891).

99. Foster, *New York by Gas-Light,* 69.

100. Quoted in Homberger, *Mrs. Astor's New York,* 31.

101. Ibid.

102. McCabe, *Lights and Shadows of New York Life.*

103. Lewis A. Erenberg, *Steppin' Out: New York Nightlife and the Transformation of American Culture, 1890–1930* (Chicago: University of Chicago Press, 1981).

104. Caldwell, *New York Night,* 213.

105. Erenberg, *Steppin' Out,* xiii. For a general discussion of the development of personality in America, see Warren I. Susman, "'Personality' and the Making of Twentieth-Century Culture," in *New Directions in American Intellectual History,* ed. John Higham and Paul K. Conkin (Baltimore: Johns Hopkins University Press, 1979), 212–26.

106. Erenberg, *Steppin' Out,* xiv.

107. For useful histories of urban illumination, see Nye, *Electrifying America;* Jackle, *City Lights;* Jon C. Teaford, *The Unheralded Triumph: City Government in America, 1870–1900* (Baltimore: Johns Hopkins University Press, 1984), 229–31.

108. "The History of Street Lighting in New York City," *The American City,* March 1926; "Street Lighting in the Metropolis," *The American City,* January 1931.

109. Teaford, *The Unheralded Triumph,* 230. Also see Haskell, "Architecture"; John Allen Corcoran, "The City Light and Beautiful," *The American City,* July 1912, 46–47.

110. Nye, *Electrifying America,* 60.

111. Sharpe, *New York Nocturne,* 219.

112. Zapatka, "The Edison Effect," 70.

113. Haskell, "Architecture," 56.

114. Sharpe, *New York Nocturne,* 137. On the closure of the frontier, see Turner, "The Significance of the Frontier in American History."

115. Murray Melbin, *Night as Frontier: Colonizing the World after Dark* (New York: Free Press, 1987), 51.

116. Ibid., 29–52.

117. Advertisement for Graybar Street Lighting, "Our Cities No Longer Hide under Bushel," *The American City,* February 1930.

118. Gorham Munson, *The Awakening Twenties: A Memoir-History of a Literary Period* (Baton Rouge: Louisiana State University Press, 1985), 256.

119. For lists of attendees of the Gurdjieffite séances during the 1920s in New York, see Jean Toomer Papers, Box 68, and Muriel Draper Papers, Box 20, Folders 625 and 635, Beinecke Rare Book and Manuscript Library, Yale University. *Time* magazine also published a list of the Manhattan Gurdjieffites, including Ferriss; see "Harmonious Developer," *Time,* March 24, 1930.

120. For examples of Ferriss's poems, see "Random Thoughts," HFC, Box 7, Talks on Architecture. One of the reviewers of *The Metropolis of Tomorrow* wrote that Ferriss was "a poet among architects, an artist who can translate in terms of steel, the soaring aspirations of men." Quoted in Willis, "Drawing towards Metropolis," 148.

121. Hugh Ferriss, "Time for an Artistic Revival in Architectural Design," *AIA Journal* (February 23, 1955): 52. The paper was presented at the Columbia University Bicentennial Conference "The Role of the University in the Creative Arts," November 13, 1954.

122. Koolhaas, *Delirious New York,* 117.

123. Ferriss, "New York from a Studio Rooftop."

124. Cheney, *The New World Architecture,* 328.

125. Claude Bragdon, "From a Hotel Window," *Commonweal,* May 28, 1930, 99. Bragdon also wrote about skyscrapers; see his articles "Architecture in the United States, III, The Skyscraper," *Architectural Record* (July–December 1909); and "The Shelton Hotel, New York."

126. Claude Bragdon, "Mysticism and Architecture," *Inland Architect and News Record* 38, no. 1 (August 1901): 4–6; Claude Bragdon, "The New Mysticism," *The Reader* 4 (July 1904): 189–91.

127. Bragdon, "Mysticism and Architecture," 4. "Mysticism and Architecture" was first presented in Philadelphia in 1901 as a lecture at the Third Annual Convention of the Architectural League of America. The lecture was then developed into a book, *The Beautiful Necessity: Seven Essays on Theosophy and Architecture* (Wheaton, Ill.: Theosophical Publishing House, 1910). For discussion of Bragdon's mystical philosophy, see Massey, *Crystal and Arabesque.*

128. See H. P. Blavatsky, *The Key to Theosophy,* ed. Joy Mills (1889; repr., Wheaton, Ill.: Theosophical Publishing House, 1972); Robert S. Ellwood Jr., *Religious and Spiritual Groups in Modern America* (Englewood Cliffs, N.J.: Prentice Hall, 1973), 74–79;

Bruce F. Campbell, *Ancient Wisdom Revived: A History of the Theosophical Movement* (Berkeley: University of California Press, 1980), 2–8. Bragdon described his involvement with the Theosophical Society in his book *Episodes from an Unwritten History* (Rochester, N.Y.: Manas Press, 1910). Along with Nicholas Bessaraboff, he also translated from the Russian and wrote the introduction to Peter D. Ouspensky's influential theosophical treatise *Tertium Organum: The Third Canon of Thought, a Key to the Enigmas of the World* (New York: Alfred A. Knopf, 1922).

129. Campbell, *Ancient Wisdom Revived,* 62.

130. Theosophy influenced many leading figures, including Camille Flammarion (scientist), Rudolph Steiner (philosopher), Max Muller (scholar), Maurice Maeterlinck and Theodore Dreiser (writers), Alexander Scriabin (musician), and Wassily Kandinsky and Piet Mondrian (artists). See Peter Fingesten, "Spirituality, Mysticism and Non-objective Art," *Art Journal* 21, no. 1 (Autumn 1961): 2–6; Graham Livesey, "The Van der Leeuw House: Theosophical Connections with Early Modern Architecture," *Architronic* 7, no. 2 (1998); Claude Bragdon, *Theosophy and the Theosophical Society* (Rochester, N.Y.: Manas Press, 1909). Commenting on Bragdon's lifelong affiliation with theosophy, Lewis Mumford suggested in his *Roots of Contemporary American Architecture* (1952; repr., New York: Dover, 1972) that Bragdon's mystical writings based on a theosophical exploration of architecture ironically contributed to his being neglected in the modernist architectural canon, especially one that is defined by Sigfried Giedion's *Space, Time and Architecture: The Growth of a New Tradition* (1941; repr., Cambridge, Mass.: Harvard University Press, 1959).

131. Blavatsky, *The Key to Theosophy,* 159–65; Campbell, *Ancient Wisdom Revived,* 199. Rabindranath Tagore, the great Bengal poet, gave the honorific *mahatma* to Mohandas Gandhi, the political and spiritual leader of India's liberation movement against the British Raj.

132. Blavatsky, *The Key to Theosophy,* 159.

133. Quoted in Ellwood, *Religious and Spiritual Groups in Modern America,* 101.

134. Quoted in Campbell, *Ancient Wisdom Revived,* 198.

135. Ellwood, *Religious and Spiritual Groups in Modern America,* 90.

136. W. Tudor Jones, *The Spiritual Ascent of Man* (New York: G. P. Putnam's Sons, 1917), 233.

137. Mumford, "The Sacred City." Also see Herbert D. Croly, "The Skyscraper in the Service of Religion," *Architectural Record* 55 (February 1924); H. I. Brock, "Building to High Heaven," *The World's Work,* February 1929.

138. Croly, "The Skyscraper in the Service of Religion." Croly was editor of *Architectural Record* from 1900 to 1906.

139. Theodore Dreiser, "A Remarkable Art," *Great Round World* 19 (May 1902): 433, quoted in Wigoder, "The 'Solar Eye' of Vision," 163.

140. On Dreiser's theosophical orientation, see William Leach, *Land of Desires: Merchant, Power, and the Rise of a New American Culture* (New York: Vintage Books, 1993), 230. As a reporter for the *St. Louis Globe-Democrat,* Dreiser interviewed the American leader of theosophy, Annie Besant, whom he later quoted in his work. See Douglas C. Stenerson, "Some Impressions of the Buddha: Dreiser and Sir Edwin Arnold's *The Light of Asia,*" *Canadian Review of American Studies* 22, no. 3 (Winter 1991): 387–406.

141. Thomas A. P. van Leeuwen, *The Skyward Trend of Thought: The Metaphysics of the American Skyscraper* (Cambridge: MIT Press, 1988), 125–33.

142. Willis, "Drawing towards Metropolis," 157.

143. Some of the magazine and newspaper appearances of Ferriss's drawings are included in the HFC, Box 2.

144. Nina Purdy, "Prophet of Sky Lines: Hugh Ferriss Conceives Cities of the Future as Symbols of Strength and Beauty," *Personality,* November 1928, 33. A copy of this article is included in HFC, Box 7.

145. A comprehensive body of Gurdjieff's lectures compiled by two of his pupils—the New York socialite Muriel Draper and the author Jean Toomer—are in Muriel Draper Papers and Jean Toomer Papers, Beinecke Rare Book and Manuscript Library, Yale University. For key Gurdjieffite concepts, see Sophia Wellbeloved, *Gurdjieff: The Key Concepts* (London: Routledge, 2003).

146. Gorham Munson, "The Birthday of the Twenties," in *The Awakening Twenties,* 1–6. Ezra Pound is quoted in William E. Leuchtenburg, *The Perils of Prosperity, 1914–1932* (Chicago: University of Chicago Press, 1958), 142. For discussion of the culture of the 1920s, see Taylor, "The Image of Urban Optimism"; George E. Mowry, ed., *The Twenties: Fords, Flappers and Fanatics* (Englewood Cliffs, N.J.: Prentice Hall, 1963).

147. Waldo Frank, *The Re-discovery of America: An Introduction to a Philosophy of American Life* (New York: Charles Scribner's Sons, 1929), quoted in Munson, *The Awakening Twenties,* 176.

148. The literature on Gurdjieff's doctrines has generally been dominated by his pupils, hence its scholarly rigor and historical accuracy have remained questionable. Most historical accounts are anecdotal, often filtered through an air of loyalty. In this genre, Paul Beekman Taylor's *Gurdjieff's America: Mediating the Miraculous* (n.p.: Lighthouse Editions, 2004) is informative. More recently, James Webb's *The Harmonious Circle: The Lives and Work of G. I. Gurdjieff, P. D. Ouspensky, and Their Followers* (New York: G. P. Putnam's Sons, 1980) is an exception. Other useful sources include Ellwood, *Religious and Spiritual Groups in Modern America,* 159–64; Rom Landau, *God Is My Adventure: A Book on Modern Mystics, Masters and Teachers* (New York: Alfred A. Knopf, 1936); Andres Rawlinson, *The Book of Enlightened Masters: Western Teachers in Eastern Traditions* (Chicago: Open Court, 1997). For Orage's biography, see Wallace Martin, "Introduction," in *Orage as Critic,* ed. Wallace Martin (London: Routledge & Kegan Paul, 1974).

149. Ouspensky, *Tertium Organum.* Ouspensky reportedly met Gurdjieff in 1912 and became a convert to his ideas.

150. On Orage's acclaimed literary career, see Wallace Martin, *The New Age under Orage: Chapters in English Cultural History* (New York: Barnes & Noble, 1967).

151. Willis, "Drawing towards Metropolis," 183.

152. Webb, *The Harmonious Circle,* 11–15.

153. Gurdjieff discussed his early life and experience in his autobiography *Meetings with Remarkable Men* (London: Routledge & Kegan Paul, 1963) and his *View from the Real World* (London: Routledge & Kegan Paul, 1973). Other, secondary sources on Gurdjieff's life are Landau, *God Is My Adventure*; Rawlinson, *The Book of Enlightened Masters.*

154. A clipping of the February 10, 1924, article in *New York City America* is available in Muriel Draper Papers, Box 20, Folder 636, Beinecke Rare Book and Manuscript Library, Yale University. Other New York–area newspapers covered Gurdjieff's arrival as well, including the *New York City Tribune* (January 17, 1924), *New York City Times* (March 1, 1924), *New York City Post* (January 26, 1924), *New York City Evening World* (January 24, 1924), and *Brooklyn New York Eagle* (January 19, 1924).

155. This is part of a piece Wright contributed to the *Capitol Times* (Madison), September 12, 1934, following Gurdjieff's overnight visit to Taliesin. The typescript is in Jean Toomer Papers, Box 68, Folder 1548, Beinecke Rare Book and Manuscript Library, Yale University.

156. See the institute's manifesto, Jean Toomer Papers, Box 68, Folder 1537, Beinecke Rare Book and Manuscript Library, Yale University.

157. See, for instance, P. D. Ouspensky, *The Fourth Way: A Record of Talks and Answers to Questions Based on the Teaching of G. I. Gurdjieff* (New York: Alfred A. Knopf, 1970).

158. For a good analysis of the sacred dance, see Mel Gordon, "Gurdjieff's Movement Demonstrations: The Theatre of the Miraculous," in "Occult and Bizarre," special issue, *Drama Review* 22, no. 2 (June 1978), 32–44.

159. Rawlinson, *The Book of Enlightened Masters,* 290.

160. Jon Woodson, *To Make a New Race: Gurdjieff, Toomer and the Harlem Renaissance* (Jackson: University Press of Mississippi, 1999), 3.

161. On the eugenic politics of America, see Marouf A. Hasian Jr., *The Rhetoric of Eugenics in Anglo-American Thought* (Athens: University of Georgia Press, 1996); Donald K. Pickens, *Eugenics and the Progressives* (Nashville: Vanderbilt University Press, 1996); Christina Cogdell, *Eugenic Design: Streamlining America in the 1930s* (Philadelphia: University of Pennsylvania Press, 2004).

162. Raymond B. Fosdick, *The Old Savage in the New Civilization* (Garden City, N.Y.: Doubleday, Doran, 1929).

163. Claude Bragdon, "Introduction to the English Translation," in Ouspensky, *Tertium Organum,* 5–6.

164. Woodson, *To Make a New Race,* 14. Also see Mohammad-Hossein Tamdgidi, "Mysticism and Utopia: Towards the Sociology of Self-Knowledge and Human Architecture" (Ph.D. diss., Binghamton University, State University of New York, 2002), 13.

165. Alfred R. Orage, *Consciousness: Animal, Human and Superhuman* (London: Pembridge Design Studio, 1907).

166. Munson, "Orage in America," in *The Awakening Twenties,* 255.

167. Ibid., 255–58.

168. Gurdjieff's personality was complex, simultaneously galvanizing and alienating. Some disgruntled pupils complained about his mismanagement of funds and his conduct with women. See Webb, *The Harmonious Circle,* 328–33.

169. Munson, "Orage in America," 258.

170. Claude Bragdon, *More Lives than One* (New York: Alfred A. Knopf, 1938), 324, quoted in Paul Beekman Taylor, *Gurdjieff and Orage: Brothers in Elysium* (York Beach, Maine: Weiser Books, 2001), 61.

171. Taylor, *Gurdjieff and Orage,* 93–94. For the names of a cross section of attendees, see Munson, *The Awakening Twenties,* 260–62. They included the photographer Stieglitz and his wife O'Keeffe, Bragdon, Ferriss, psychologist C. Daly King, poets Melville Cane and Edna Kenton, and writers Jean Toomer and T. S. Matthews.

172. Bragdon, *More Lives than One,* 321, quoted in Louise Welch, *Orage with Gurdjieff in America* (Boston: Routledge & Kegan Paul, 1982), 1.

173. Ferriss, *The Metropolis of Tomorrow,* 142. It is likely that Ferriss borrowed the concept of the "harmonious development of man" from the name of Gurdjieff's institute.

174. Munson, *The Awakening Twenties,* 254.

175. On Muriel Draper's bohemian persona, see Betsy Fahlman, "The Great Draper Woman: Muriel Draper and the Art of the Salon," *Woman's Art Journal* 26, no. 2 (2005–6): 33–37. A contemporary of Ferriss, Draper was born in Haverhill, Massachusetts, in 1886 but lived in France, Italy, and England from 1905 to the mid-1920s. While in England, Draper hosted a salon that became a meeting place for avant-garde musicians, artists, and authors, including Arthur Rubenstein, John Singer Sargent, Pablo Casals, Henry James, and Osbert Sitwell. While in New York during the 1920s, she continued to support and encourage artists and authors as an eccentric but gracious hostess. As an interior designer, she published articles in *Harper's, Town and Country,* and *Vogue.* Draper maintained a list of the New York membership of the Gurdjieffite circle, coordinated meetings, collected membership fees, and hosted Orage's lectures on Monday nights. See Muriel Draper Papers, Box 20, Folder 635, at Beinecke Rare Book and Manuscript Library, Yale University. Other venues for Orage's discussion sessions included the salon of Carl Van Vechten, an avant-garde promoter and patron of the Harlem Renaissance and the author of the controversial novel *Nigger Heaven* (1926). For Munson's description, see *The Awakening Twenties,* 260–61.

176. Muriel Draper Papers, Box 20, Folder 625, Beinecke Rare Book and Manuscript Library, Yale University.

177. Welch, *Orage with Gurdjieff in America,* 53.

178. The heading of the four typed pages reads: "Notes on Talk by Orage—12/16/25 as taken down by Hugh Ferriss, architect." I am grateful to the Gurdjieff scholar J. Walter Driscoll for sharing these four pages with me. Ferriss's notes are part of the Carmel Note Books from the papers of Dora Hagemayer, which include notes—compiled by Mary Bulkey, Hugh Ferriss, Blanch B. Grant, Annette Herter, and Larry S. Morris—on nineteen dated lectures by Orage in New York and Carmel, California, between 1925 and 1931.

179. Ibid., 2.

180. David S. Thatcher, *Nietzsche in England, 1890–1914: The Growth of a Reputation* (Toronto: University of Toronto Press, 1970).

181. Orage, *Consciousness,* 70–86.

182. Ibid., 79.

183. Ibid., 80.

184. Jeanne de Salzmann, "The Look from Above," originally published as "The Awakening of Thought" in *Gurdjieff: Essays and Reflections on the Man and His Teaching,* ed. Jacob Needleman and George Baker (New York: Continuum, 1996).

185. Orage had come across Oswald Spengler's *Decline of the West* (New York: Alfred A. Knopf, 1927), which was originally published in German in two volumes, 1918 and 1922, and was translated into English in 1927.

186. Hugh Ferriss, "The Impact of Science and Materialism on Art Today," *AIA Journal* 22 (July 1954): 3. Although Ferriss made this statement in the 1950s, when he was deeply apprehensive regarding the destructive potential of the atomic bomb, he had fervently believed since the 1920s that the bifurcation of science and art had created an architecture devoid of human contents. See, for instance, his "Random Thoughts," HFC, Box 7.

187. Ferriss, *The Metropolis of Tomorrow,* 18.

188. Gurdjieff was mostly supported by donations from wealthy socialites and devotees. Munson, *The Awakening Twenties,* 279–81; Welch, *Orage with Gurdjieff in America,* 57.

189. Muriel Draper, *Music at Midnight* (New York: Harper & Brothers, 1929); Isa Glenn, *Transport* (New York: Alfred A. Knopf, 1929); T. S. Matthews, *An Autobiography* (New York: Simon & Schuster, 1960), 214. See Woodson, *To Make a New Race,* 14. Many of these publications were by African American authors who embraced, erroneously it seems, Gurdjieff's teachings as a paradigm of "postracial" conditions.

190. Welch, *Orage with Gurdjieff in America,* 60. As noted above, Muriel Draper's memoir *Music at Midnight,* a recollection of her childhood in New England, also resulted from this workshop. In addition, Orage's workshop fired up such established authors as Samuel Hoffenstein, Melville Crane, Jean Toomer, Mary Johnston, Isa Glenn, and John Riordan. In general, Orage's teaching inspired Gurdjieffite authors like C. Daly King, who wrote *Beyond Behaviorism: The Future of Psychology* (New York: Grant, 1927); *The Psychology of Consciousness* (New York: Harcourt, Brace, 1932); and *The Oragean Version* (New York: C. Daly King, 1951).

191. Taylor, *Gurdjieff and Orage,* 214.

192. The four unpublished drawings are cataloged in the Hugh Ferriss Papers, Section III: Drawing, Library, Center for Advanced Study in the Visual Arts, East Gallery, National Gallery of Art, Washington, D.C.

193. Ferriss, *The Metropolis of Tomorrow,* 142.

194. "Published Comments" on *The Metropolis of Tomorrow,* HFC, Box 7.

195. Ferriss, *The Metropolis of Tomorrow,* 16.

196. Ibid., 142.

197. For an excellent analysis of representations of God as the architect of the universe in medieval theology, see Katherine H. Tachau, "God's Compass and *Vana Curiositas*: Scientific Study in the Old French *Bible Morelisée,*" *Art Bulletin* 80, no. 1 (March 1998): 7–33. Also see Amos Funkenstein, *Theology and the Scientific Imagination: From the Middle Ages to the Seventeenth Century* (Princeton, N.J.: Princeton University Press, 1986).

198. Ferriss, set design for *New Year's Eve in New York,* Neighborhood Playhouse, *Theater Guild Magazine,* March 1930, reprinted in Leich, *Architectural Visions,* 32.

2. ASCENSION AS AUTOBIOGRAPHY

1. *Dymaxion* was a neologism derived from *dynamic, maximum,* and *tension* or *ion.* The term was coined by Waldo Warren, an employee in the advertising section at

Marshall Field Department Store in the Chicago Loop. Fuller presented the project's public debut at this department store in April 1929. At the Architectural League presentation the distinguished audience included, among others, the league's chairman, Harvey W. Corbett, and Raymond Hood.

2. Buckminster Fuller Archive (hereafter BFA), Series 2, Box 19, vol. 34, 1928. The Fuller archive is cataloged under the call number M1090 at the Special Collections of Green Library, Stanford University.

3. For descriptions of the house, see Knight Deacon, "Houses That Hang from a Pole!," *Modern Mechanix and Inventions,* September 1932; Douglas Haskell, "The House of the Future," *New Republic,* May 13, 1931, 344–45; "Architect Sees Hanging Homes, Moved at Will," *New York Herald Tribune,* November 6, 1929; "The Dymaxion House," *Architecture* (Chicago), June 1929; Theodore Morrison, "House of the Future," *House Beautiful,* September 1929; Lewis Mumford, "Mass Production and the Modern House," *Architectural Record* 67 (January 1930): 13–30. For more recent discussions, see Reyner Banham, *Theory and Design in the First Machine Age* (1960; repr., Cambridge: MIT Press, 1980), 325–26; Federico Neder, *Fuller Houses: R. Buckminster Fuller's Dymaxion Dwellings and Other Domestic Adventures,* trans. Elsa Lam (Baden, Switzerland: Lars Müller, 2008).

4. R. Buckminster Fuller, "Lightful Houses," in *Your Private Sky: R. Buckminster Fuller, Discourse,* ed. Joachim Krausse and Claude Lichtenstein (Baden, Switzerland: Lars Müller, 1999), 64–74.

5. See his letter to Vincent Astor (East 24th Street, New York City), BFA, Series 2, Dymaxion Chronofile, Box 29.

6. For the phenomenon of "an airplane in every garage," see Alexander Klemin, "An Airplane in Every Garage?," *Scribner's,* September 1935; "The Future American Country House," *Architectural Record* 64 (November 1928).

7. See Brian Horrigan, "The Home of Tomorrow, 1927–1945," in *Imagining Tomorrow: History, Technology, and the American Future,* ed. Joseph Corn (Cambridge: MIT Press, 1986).

8. The manifesto was reprinted in 1972. Buckminster Fuller, *4D Time Lock* (Albuquerque, N.M.: Lama Foundation, Biotechnic Press, 1972). The original copy of *4D Time Lock* is in BFA, Series 2, Dymaxion Chronofile, Box 20, vol. 35, 1928.

9. Le Corbusier, *Aircraft* (London: The Studio, 1935), 10.

10. R. Buckminster Fuller, *Nine Chains to the Moon* (Philadelphia: J. B. Lippincott, 1938), 165.

11. Allen Churchill, *The Year the World Went Mad* (New York: Crowell, 1961). See Scott Berg, *Lindbergh* (New York: Berkley Books, 1998); Kenneth S. Davis, *The Hero: Charles A. Lindbergh and the American Dream* (Garden City, N.Y.: Doubleday, 1959).

12. Frederick Lewis Allen, *Only Yesterday: An Informal History of the 1920's* (1931; repr., New York: John Wiley, 1997), 166. Also see John William Ward, "The Meaning of Lindbergh's Flight," *American Quarterly* 10 (Spring 1958), reprinted in *Red, White, and Blue: Men, Books, and Ideas in American Culture* (New York: Oxford University Press, 1969), 26–27; Alden Whitman, "Daring Lindbergh Attained the Unattainable with Historic Flight across Atlantic," *New York Times,* August 27, 1974.

13. Ward, "The Meaning of Lindbergh's Flight," 26–27.

14. See Charles L. Ponce de Leon, "The Man Nobody Knows: Charles A. Lindbergh and the Culture of Celebrity," in *The Airplane in American Culture,* ed. Dominick Pisano (Ann Arbor: University of Michigan Press, 2003); Daniel J. Boorstin, "From Hero to Celebrity," in *The Image; or, What Happened to the American Dream* (New York: Atheneum, 1962).

15. Markey Morris, "Young Man of Affairs," *New Yorker,* September 20, 1927; Markey Morris, "Young Man of Affairs—II," *New Yorker,* September 27, 1927.

16. Robert Wohl, *A Passion for Wings: Aviation and the Western Imagination, 1908–1918* (New Haven, Conn.: Yale University Press, 1994), 261.

17. Gladys Moon Jones, "Are You Fit to Drive an Airplane?," *Science News-Letter,* August 20, 1927, 113–14, 119.

18. See Gilbert Seldes, "Transatlantic," *New Republic,* June 1, 1927, 47; Ward, "The Meaning of Lindbergh's Flight," 30.

19. James Oliver Robertson, *American Myth, American Reality* (New York: Hill & Wang, 1980), 200–202.

20. Ibid., 201–2.

21. Reyner Banham, *Theory and Design in the First Machine Age* (1960; repr., Cambridge: MIT Press, 1980), 325–26. Also see Sean Keller, "Navigating Systems," *Artforum International,* November 2008, 285.

22. William H. Jordy has explored Le Corbusier's lifelong fascination with the airplane. See his "'I Am Alone': Le Corbusier, Bathrooms, and Airplanes," in *Symbolic Essence, and Other Writings on Modern Architecture and American Culture,* ed. Mardges Bacon (New Haven, Conn.: Yale University Press, 2005). Also see Adnan Morshed, "The Cultural Politics of Aerial Vision: Le Corbusier in Brazil (1929)," *Journal of Architectural Education* 55, no. 4 (May 2002): 201–10.

23. Sheldon Cheney, *The New World Architecture* (New York: AMS Press, 1930), 80.

24. Fuller, *Nine Chains to the Moon,* 310–11.

25. Fuller, *4D Time Lock,* 2.

26. Fuller, "Vertical Is to Live—Horizontal Is to Die," *American Scholar* 39, no. 1 (Winter 1969–70), reprinted in Krausse and Lichtenstein, *Your Private Sky: R. Buckminster Fuller, Discourse,* 280.

27. Like other modernists of the period, Fuller took deep interest in the fourth dimension. See Linda D. Henderson, *The Fourth Dimension and Non-Euclidean Geometry in Modern Art* (Princeton, N.J.: Princeton University Press, 1983), 235–36.

28. Ibid. Some examples of essays in popular magazines include Harold Jacoby, "Faith and the Fourth Dimension," *Popular Astronomy* (Northfield, Minn.), May 1908, 298–300; Samuel M. Barton, "The Fourth Dimension," *Popular Science Monthly,* October 1913, 381–93; H. Addington Bruce, "The Riddle of the Fourth Dimension," *Scientific American,* supplement, September 5, 1908.

29. Fuller wrote to Henderson (on October 2, 1979) that the subject piqued his interest when a professor named William James Siddis presented lectures on the fourth dimension at the Harvard mathematics department in 1914. Henderson, *The Fourth Dimension,* 235. Fuller later read the works of others on the topic. A letter from Fuller to Bragdon dated May 21, 1928, confirms that he read Bragdon's *Architecture and Democracy* (1918) and his translation of the Russian hyperspace philosopher P. D. Ouspensky's *Tertium*

Organum (1920). See Fuller, *4D Time Lock,* 46. Also see William J. Scheick, "The Fourth Dimension in Wells's Novels of the 1920's," *Criticism* 20, no. 2 (Spring 1978): 167–90.

30. Henderson, *The Fourth Dimension,* 235–36. Henderson also discusses the affiliation of the fourth dimension with aviation, 285–86.

31. Claude Bragdon, "New Concepts of Time and Space," *Dial,* February 1920, 187–91; Richard Eriksen, *Consciousness, Life and the Fourth Dimension: A Study in Natural Philosophy* (New York: Alfred A. Knopf, 1923); "Gravitation and the Fourth Dimension," *Literary Digest,* August 30, 1913.

32. Fuller, *Nine Chains to the Moon,* 96.

33. Ibid.

34. Richard Hofstadter, *Social Darwinism in American Thought* (1944; repr., Boston: Beacon Press, 2006). Also see Robert C Bannister, *Social Darwinism: Science and Myth in Anglo-American Social Thought* (Philadelphia: Temple University Press, 1979); Christina Cogdell, *Eugenic Design: Streamlining America in the 1930s* (Philadelphia: University of Pennsylvania Press, 2004); Donald J. Childs, *Modernism and Eugenics: Woolf, Eliot, Yeats, and the Culture of Degeneration* (Cambridge: Cambridge University Press, 2001); Lois A. Cuddy and Claire M. Roche, eds., *Evolution and Eugenics in American Literature and Culture, 1880–1940* (Lewisburg, Pa.: Bucknell University Press, 2003); Ellsworth Huntington and Leon F. Whitney, *The Builders of America* (New York: William Morrow, 1927).

35. See, for instance, Percy Williams Bridgman, *The Intelligent Individual and Society* (New York: Macmillan, 1938); William Harlan Hale, *Challenge to Defeat: Modern Man in Goethe's World and Spengler's Century* (New York: Harcourt, Brace, 1931); Dimitri Merejkowski, *The Romance of Leonardo da Vinci,* trans. Bernard Guilbert Guerney (New York: Modern Library, 1928); Alfred Sherwood Romer, *Man and the Vertebrate* (Chicago: University of Chicago Press, 1933); D'Arcy Wentworth Thompson, *On Growth and Form* (Cambridge: Cambridge University Press, 1942); Arnold Joseph Toynbee, *The Prospects of Western Civilization* (New York: Columbia University Press, 1950); Herbert George Wells, *The New World Order* (New York: Alfred A Knopf, 1940).

36. The book review was written by Clifton Fadiman, "Mr. Hale versus Spengler," *The Nation,* June 8, 1932. The review is in BFA, Dymaxion Chronofile, Box 26, vol. 42, 1932.

37. BFA, Series 2, Dymaxion Chronofile, Box 15, vol. 27, 1926. Fuller and his father-in-law, the New York architect James Monroe Hewlett, incorporated the company in 1923 in New York. The goal of their business enterprise was to introduce a new lightweight, fibrous building block. Although during Fuller's tenure as president of the company the Stockade Building System expanded into five franchises (in New York, New Jersey, Illinois, Massachusetts, and the District of Columbia), by the end of 1926 Fuller was in a tenuous position due to allegations of financial mismanagement. A letter supporting these allegations, dated January 25, 1927, is in BFA, Series 2, Dymaxion Chronofile, Box 17.

38. Letter dated July 1, 1939, in BFA, Series 2, Dymaxion Chronofile, Box 46.

39. The aviator as capable of "seeing the world from air" was a leitmotif in popular aviation, invention, and city-affairs magazines of the interwar period. See, for instance, Alan J. Cobhan, "Seeing the World from the Air," *National Geographic,* March 1928.

40. When Fuller was thirteen years old his father died. He was raised by his mother. Biographical materials are found in BFA, Series 25, Subseries 1, Box 1, Folders 1, 2; Box 5, Folder 1. Also see Calvin Tomkins's "profiles" of Fuller in "In the Outlaw Area," *New Yorker,* January 8, 1966; and in *Your Private Sky: R. Buckminster Fuller, the Art of Design Science,* ed. Joachim Krausse and Claude Lichtenstein (Baden, Switzerland: Lars Müller, 1999).

41. Elizabeth Kolbert, "Dymaxion Man: The Visions of Buckminster Fuller," *New Yorker,* June 9–16, 2008, 64.

42. Fuller studied Margaret Fuller-Ossoli's life extensively. BFA, Series 1, Family History, Box 5, Folder 1–2.

43. Joseph J. Corn, *The Winged Gospel: America's Romance with Aviation, 1900–1950* (New York: Oxford University Press, 1983), 114.

44. See, for example, *Aero Club of America Bulletin,* January 1912; Mary E. Burt, "Aeronautics Will Develop a Broader Vision," *Flying* (New York), October 1915; Cecil Peoli, "How to Make a Model Aeroplane," *Scientific American,* October 14, 1911.

45. "Buckminster Fuller Chronofile," in *The Buckminster Fuller Reader,* ed. James Meller (London: Jonathan Cape, 1970), 13.

46. Fuller, *4D Time Lock,* 43. Fuller speculated about how much a commercial airline company's stock market shares would gain in value if Lindbergh were the company's technical director.

47. See "4D Aviation," BFA, Series 25, Box 15, Folder 9. Also see Alden Hatch, *Buckminster Fuller at Home in the Universe* (New York: Crown, 1974), 73–75.

48. John A. Jackle, *The Tourist: Travel in Twentieth-Century North America* (Lincoln: University of Nebraska Press, 1985), 171. The first passenger airline in the United States was established in 1914, with flights between Tampa and St. Petersburg in Florida. Air travel and airmail service became tied during the 1920s, as the main profit came from the latter, while carrying passengers was supplementary.

49. See, for instance, "Fullers in Astor's Monoplane Fly to Bar Harbor," *New York Herald,* September 29, 1922; "Uses Astor Seaplane to Attend Weddings: Lieutenant Fuller Starts on Second Leg of Vacation in Air, in Which His Wife Will Join," *New York Times,* September 29, 1922.

50. "4D Aviation."

51. Hatch, *Buckminster Fuller at Home in the Universe,* 75.

52. The photographs are in BFA, Series 14, Box 27, Folder 24.

53. Le Corbusier, *Aircraft,* 11–12.

54. For a detailed description of Chronofile, see Hsiao-Yun Chu, "Paper Mausoleum, the Archive of R. Buckminster Fuller," in *New Views on R. Buckminster Fuller,* ed. Hsiao-Yun Chu and Roberto G. Trujillo (Stanford, Calif.: Stanford University Press, 2009), 6–22. Also see Krausse and Lichtenstein, *Your Private Sky: R. Buckminster Fuller, the Art of Design Science,* 13–14.

55. Leo Marx, "The Idea of 'Technology' and Postmodern Pessimism," in *Does Technology Drive History? The Dilemma of Technological Determinism,* ed. Merritt Roe Smith and Leo Marx (Cambridge: MIT Press, 1994).

56. Ibid., 251.

57. Mauro F. Guillen, *The Taylorized Beauty of the Mechanical: Scientific Management and the Rise of Modernist Architecture* (Princeton, N.J. Princeton University Press, 2006), 4–5.

58. Ibid., 4.

59. Fuller, *4D Time Lock,* 22.

60. Haskell, "The House of the Future," 344.

61. Fuller, *4D Time Lock,* 32.

62. Stephen Kern, *The Culture of Time and Space, 1880–1918* (Cambridge, Mass.: Harvard University Press, 1983), 242.

63. Fuller expressed these views in his original manuscript of *Nine Chains to the Moon,* 189, BFA, Series 8, Box 2, Folder 7.

64. Loretta Lorance examines Fuller's complicity with the construction of his own myth in *Becoming Bucky Fuller* (Cambridge: MIT Press, 2009).

65. Fuller, *4D Time Lock,* 111.

66. Lloyd Steven Sieden, *Buckminster Fuller's Universe: His Life and Work* (1989; repr., New York: Basic Books, 2000), 122–24. For discussion of the drawing, also see Robert Marks and R. Buckminster Fuller, *The Dymaxion World of Buckminster Fuller* (1960; repr., Garden City, N.Y.: Anchor Books, 1973), 77.

67. Le Corbusier, *Precisions: On the Present State of Architecture and City Planning,* trans. Edith Schreiber Aujame (Cambridge: MIT Press, 1991), 235–36 (published in French in 1930 as *Précisions sur un état présent de l'architecture et de l'urbanisme*). I explore Le Corbusier's aerial "discovery" of South America in Morshed, "The Cultural Politics of Aerial Vision."

68. Emily S. Rosenberg, *Spreading the American Dream: American Economic and Cultural Expansion, 1890–1945* (New York: Hill & Wang, 1982), 14–15.

69. Adnan Morshed, "A Tale of Two Symbols," *Thresholds: MIT Journal of Architecture* 23 (2001): 6–9.

70. BFA, Series 8, Box 3, Folder 3.

71. R. Buckminster Fuller, "Design for Survival—Plus" (talk delivered at the Illinois Institute of Technology, January 1949), in Meller, *The Buckminster Fuller Reader,* 260.

72. BFA, Series 8, Box 2, Folder 7. Fuller's personal library, called the "Live Book Squad," contained Dimitri Merejkowski's *The Romance of Leonardo da Vinci.*

73. Henry Ford and Samuel Crowther, "The Air," in *Today and Tomorrow* (1926; repr., London: Heinemann, 1988), 189–94.

74. Fuller, *Nine Chains to the Moon,* 196.

75. Ibid., 96.

76. Healthy at birth, Fuller's first daughter, Alexandra, died on November 14, 1922, of spinal meningitis and pneumonia at the tender age of four. See BFA, Series 2, Dymaxion Chronofile, vol. 23.

77. I thank the Fuller scholar and archivist Bonnie DeVarco for sharing with me a copy of "Cosmopolitan Home Corporation." Fuller conceived *Lightful* as a neologism sometime in early 1928 to use as a logo for his houses and other industrial products. Shortly thereafter, however, it was replaced by the supposedly more appealing trademark *4D,* which itself was changed in 1929 in favor of *Dymaxion.* The document is partially reproduced in Krausse and Lichtenstein, *Your Private Sky: R. Buckminster*

Fuller, Discourse, 64–75. Also see Yunn Chii Wong, "The Geodesic Works of Richard Buckminster Fuller, 1948–1968 (The Universe as a Home of Man)" (Ph.D. diss., MIT, 1999).

78. See Horrigan, "The Home of Tomorrow," 137; Haskell, "The House of the Future," 344–45; Gilbert Herbert, *The Dream of the Factory-Made House: Walter Gropius and Konrad Wachsmann* (Cambridge: MIT Press, 1984); Morrison, "House of the Future."

79. Horrigan, "The Home of Tomorrow," 137–38.

80. Gwendolyn Wright, *Building the Dream: A Social History of Housing in America* (New York: Pantheon Books, 1981); Horrigan, "The Home of Tomorrow," 138.

81. Wright, *Building the Dream,* 116–17.

82. The letter is included in Fuller, *4D Time Lock,* 78.

83. Krausse and Lichtenstein, *Your Private Sky: R. Buckminster Fuller, the Art of Design Science,* 28.

84. Davis, *The Hero,* 148–49.

85. The mimeograph starts with Fuller's letter written to Vincent Astor on August 28, 1928. The original version of *4D Time Lock* consists of two parts. The first part contains the chapters, and the second contains the cover letters to various recipients of the essay. The 1972 edition is divided into six parts. In the first part are nineteen chapters covering a wide range of topics, from philosophical ruminations on "the economic problem of this age" and "the new generation and the revolution of truth" to the nitty-gritty of the "weight in building as the new economic factor" and "building from the 'inside out' as opposed to 'building from the outside in.'" The second part comprises Fuller's "first announcement letters," which he sent out to a long list of recipients on May 21,1928. Fuller's drawings of the 4D House and the petition seeking a patent for it constitute the third section. The fourth section is a bit of a hodgepodge: it includes the letters Fuller received in response to his first 4D letters dispatched on May 21,1928, and his second-round responses; his letters, along with the 4D essay, to new recipients; and correspondence among other people discussing Fuller's "innovation." The fifth section includes his sketches explaining the philosophical ideas behind 4D housing schemes and transportation that would embody 4D principles. The final part is an "epic" letter Fuller wrote to a banker named George N. Buffington of Chicago, sent on August 31, 1928. The letter encapsulates the key ideas underlying 4D while seeking to convince the investment community to take Fuller's ideas seriously.

86. Fuller, "Cosmopolitan Home Corporation," 1.

87. Ibid., 4.

88. Fuller, "Lightful Houses," 64.

89. For a comprehensive explanation of this epistemological problem, see F. J. K. Soontiëns, "Evolution: Teleology or Chance?," *Journal for General Philosophy of Science* 22, no. 1 (1991): 133–41.

90. Charles Darwin, *Origin of Species,* 6th ed., quoted in Robert Scoon, "The Rise and Impact of Evolutionary Ideas," in *Evolutionary Thought in America,* ed. Stow Persons (New York: George Braziller, 1956), 24–25.

91. Georg W. F. Hegel *The Philosophy of History,* trans. J. Sibree (Buffalo, N.Y.: Prometheus Books, 1991).

92. Fuller, "Lightful Houses," 67.

93. Morrison, "House of the Future," 293. Also see Haskell, "The House of the Future."

94. Cogdell, *Eugenic Design,* 4. On streamlining as a sign of progress, see Donald J. Bush, *The Streamlined Decade* (New York: George Braziller, 1975); Claude Lichtenstein and Franz Engler, eds., *Streamlined: A Metaphor for Progress* (Princeton, N.J.: Princeton University Press, 1995); William Pretzel, "The Ambiguities of Streamlining: Symbolism, Ideology, and Cultural Mediator," in *Streamlining America: A Henry Ford Museum Exhibit,* ed. Fannia Weingartner (Dearborn, Mich.: Henry Ford Museum & Greenfield Village, 1986).

95. Fuller wrote about streamlining in his magazine *Shelter* in 1932.

96. Cheney, *The New World Architecture,* 80.

97. Fuller, "Lightful Houses," 67.

98. Le Corbusier, *Towards a New Architecture,* trans. Frederick Etchells (New York: Payson & Clarke, 1927), 109–10.

99. "Chronology," in Krausse and Lichtenstein, *Your Private Sky, R. Buckminster Fuller: The Art of Design Science,* 28.

100. Fuller, *4D Time Lock,* 46.

101. Ibid., 8.

102. Italo Calvino, *Six Memos for the Next Millennium* (New York: Vintage, 1993).

103. For discussion of various iterations of lightness in modern architecture, see Ulrich Conrads, ed., *Programs and Manifestoes on 20th-Century Architecture* (Cambridge: MIT Press, 1971); Wendy Mooran, "An Architecture of Lightness," *MoMA,* no. 20 (Autumn 1995); John Rajchman, "Lightness: A Concept in Architecture," *ANY,* no. 5 (March/April 1994); Rosemarie Haag Bletter, "The Interpretation of the Glass Dream—Expressionist Architecture and the History of the Crystal Metaphor," *Journal of the Society of Architectural Historians* 40, no. 1 (March 1981): 20–43.

104. R. Buckminster Fuller, "Universal Architecture Essay No. 1," *Shelter,* March 1932, 63.

105. Fuller, "Lightful Houses," 68.

106. "Architect Sees Hanging Homes, Moved at Will."

107. Fuller's second daughter, Allegra, was born on November 14, 1927.

108. R. Buckminster Fuller, "Dirigible," in Krausse and Lichtenstein, *Your Private Sky, R. Buckminster Fuller, the Art of Design Science,* 103.

109. Marks and Fuller, *The Dymaxion World of Buckminster Fuller,* 74–75; Sidney Rosen, *Wizard of the Dome: R. Buckminster Fuller, Designer of the Future* (Boston: Little, Brown, 1969), 57–59.

110. Beginning in the late nineteenth century, comic strips became increasingly popular in the American media as a pictorial form of social commentary on various ways the mass culture of consumption was created. Fuller kept a clip of the strip "Smitty—The Bargain Hunter" from the *Chicago Daily Tribune,* December 7, 1927. See BFA, Series 2, Dymaxion Chronofile, Box 16, vol. 29, 1927. On the history of American comic art, see Jerry Robinson, *The Comics: An Illustrated History of Comic Art* (New York: G. P. Putnam's Sons, 1974); Shirrel Rhoades, *A Complete History of American Comic Books* (New York: Peter Lang, 2008); Ian Gordon, *Comic Strips and Consumer*

Culture, 1890–1945 (Washington, D.C.: Smithsonian Institution Press, 1998); Bill Blackbeard and Marin Williams, *The Smithsonian Collection of Newspaper Comics* (Washington, D.C.: Smithsonian Institution Press, 1977); Russel Nye, *The Unembarrassed Muse: The Popular Arts in America* (New York: Dial Press, 1970).

111. James Silke, "Aviation Comic Strips," *Air Progress/Aviation Review,* August 1980, 18–23; Al Flick, "Flying in the Funnies," *Aviation Quarterly* 8 (1985): 4–47.

112. Silke, "Aviation Comic Strips," 20.

113. In Krausse and Lichtenstein, *Your Private Sky: R. Buckminster Fuller, the Art of Design Science,* 102.

114. Fuller idolized Leonardo as a Renaissance man whose appeal he thought was archetypal. He read Merejkowski's *Romance of Leonardo da Vinci*. After reading Fuller's 4D essay in May 1928, Fuller's friend Chicago architect Russell Walcott traced Fuller's ideas of the industrial house to Leonardo's notion of "prefabrication": "Let the houses be changed and arranged in order, and this will easily be done when they are first made in parts on the open places and then the framework can be fitted together on the site where they are to be permanent." From Leonardo's notebook, quoted in Edward McCurdy, *The Mind of Leonardo da Vinci* (New York: Dodd, Mead, 1928), 153. Walcott's letter is in Fuller, *4D Time Lock,* 48–49.

115. Sigmund Freud, *Leonardo da Vinci and a Memory of His Childhood,* trans. Alan Tyson (1910; repr., New York: W. W. Norton, 1964), 86–87.

116. For a survey of this mythology from the Renaissance to modern times, see Steve Poleskie, "Art and Flight: Historical Origins to Contemporary Works," *Leonardo* 18, no. 2 (1985): 69–88.

117. Harvey Goldberg, "Foreword," in *American Radicals: Some Problems and Personalities,* ed. Harvey Goldberg (New York: Monthly Review Press, 1957), ix.

118. Harvey Goldberg and William Appleman Williams, "Introduction: Thoughts about American Radicalism," in Goldberg, *American Radicals,* 7.

119. Thomas P. Hughes, *American Genesis: A Century of Invention and Technological Enthusiasm, 1870–1970* (New York: Viking Press, 1989), 24–25.

120. C. Wright Mills, *The Power Elite* (New York: Oxford University Press, 1956), 3–29. Mills defines the "power elite" thus: "By the power elite, we refer to those political, economic, and military circles which as an intricate set of overlapping cliques share decisions having at least national consequences" (18).

121. Fuller, *4D Time Lock,* 148.

122. Edward H. Carr, *The New Society* (London: Macmillan, 1951), 78.

123. The day after the *New York Times* advertisement on Fuller's exhibition ran, October 28, was Black Tuesday, when a record 16,410,000 stock market shares were sold and bought, precipitating the economic collapse and the Great Depression. See Thurman Arnold, "The Crash—What It Meant," in *The Aspirin Age, 1919–1941,* ed. Isabel Leighton (New York: Simon & Schuster, 1949), 214–31.

124. Rick Beard and Leslie Cohen Berlowitz, eds., *Greenwich Village: Culture and Counterculture* (New Brunswick, N.J.: Rutgers University Press, 1993), 49, 109, 359, 384. A native of Moldavia and immigrant in America, Romany Marie set up the first in a series of Village taverns in 1912.

125. Caroline Ware, *Greenwich Village, 1920–1930: A Commentary on American Civilization in the Post-war Years* (Boston: Houghton Mifflin, 1935), 5. Also see Anna Alice Chapin, *Greenwich Village* (New York: Dodd, Mead, 1917).

126. During this transient life in New York City, Fuller sometimes used Romany Marie's address as his contact address, revealing their relationship of mutual respect. See BFA, Series 2, Dymaxion Chronofile, Box 24, vol. 40, 1931. For discussion of the significance of Romany Marie in Fuller's life in New York, see Robert Schulman, *Romany Marie: The Queen of Greenwich Village* (Louisville, Ky.: Butler Books, 2006).

127. Sieden, *Buckminster Fuller's Universe,* 134.

128. Ibid., 141; Hatch, *Buckminster Fuller at Home in the Universe,* 120.

129. Fuller, *4D Time Lock,* 75. See Toomer's letter to Fuller's architect friend Russell Walcott, July 19, 1928, as well as correspondence between Fuller and Toomer, in BFA, Series 2, Dymaxion Chronofile, Box 19, vol. 34, 1928. From a letter Fuller wrote to Toomer on June 15, 1928, it appears clear that Fuller was well versed in contentious issues of the day and read important authors, including Gorham Munson, another Gurdjieffite, and the radical American thinker H. L. Mencken.

130. Rudolph P. Byrd, *Jean Toomer's Years with Gurdjieff: Portrait of an Artist, 1923–1936* (Athens: University of Georgia Press, 1990), 88.

131. Ibid.

132. Fuller, *4D Time Lock,* 75. Somewhat ambivalent about Fuller's mechanistic solution in the 4D House, Toomer wrote to Russell Walcott that "I told him [Ferriss] of Fuller and the new home, or better, the new house—for, after all, materials make a house; people make a home. He was very much interested in the idea. He'd like to meet Fuller."

133. BFA, M1019, Series 2, Dymaxion Chronofile, Box 24, vol. 40, 1931. A private edition limited to one thousand numbered copies and priced at three dollars, the book is described as "a philosophy of life, in three hundred definitions and aphorisms."

134. Henderson, *The Fourth Dimension,* 235–36. On Bragdon's biographical account and his mystical philosophy, see Jonathan Massey, *Crystal and Arabesque: Claude Bragdon, Ornament, and Modern Architecture* (Pittsburgh: University of Pittsburgh Press, 2009).

135. Fuller, *4D Time Lock,* 46.

136. Peter D. Ouspensky, *Tertium Organum: The Third Canon of Thought, a Key to the Enigmas of the World,* trans. Nicholas Bessaraboff and Claude Bragdon (New York: Alfred A. Knopf, 1922), 308.

137. Richard Maurice Bucke, *Cosmic Consciousness: A Study on the Evolution of the Human Mind* (1901; repr., Bedford, Mass.: Applewood Books, 2000). Bucke died as the result of an accident at the age of thirty-one.

138. Quoted in Ouspensky, *Tertium Organum,* 312–13.

139. Ibid., 308.

140. Corn, *The Winged Gospel,* 135.

141. See, for instance, Cobhan, "Seeing the World from the Air"; Marcella Holtkamp, "A Bit of Philosophy on Flying," *Air Travel News,* August 1929; Willis T. Lee, *The Face of the Earth as Seen from the Air: A Study in the Application of Airplane Photography to Geography* (New York: American Geographical Society, 1922); Clayton D.

Russell, "Seeing Things from Above," *Craftsmen Aero News,* September 1937; Carol Aronovici, "Space-Time Planning and Airmindedness," *The American City,* April 1930.

142. Fuller, "Design for Survival—Plus," 260.

143. Ouspensky, *Tertium Organum,* 324.

144. Ibid.

145. Fuller, *4D Time Lock,* 34–35.

146. Albert Einstein, "Religion and Science," *New York Times Magazine,* November 9, 1930. Fuller was captivated by Einstein's insight, a feeling he expressed in his writings on Einstein in the 1930s, especially in *Nine Chains to the Moon* (1938). Fuller's biographic section of Chronofile contains Einstein's picture.

147. Einstein, "Religion and Science."

148. The twenty-story Starrett-Lehigh Building was a joint venture of the Starrett Investing Corporation and the Lehigh Valley Railroad. BFA, Series 2, Dymaxion Chronofile, Box 24, vol. 40, 1931. The lease between Fuller and Starrett-Lehigh Building, Inc., at 601 West 26th Street, New York City, was signed on October 27, 1931. In the occupancy clause, the lease stated that the property was "to be used and occupied by the Tenant as an office and for no other purpose, for a term to commence December 1, 1931 and to end on November 30, 1934, . . . $450 for the first year, and $350 for each of the two (2) succeeding years, payable in equal monthly installments in advance on the first day of each and every month during said term." See Christopher Gray, "Streetscapes/Starrett-Lehigh Building; Time of Change for a Modern Industrial Landmark," *New York Times,* May 31, 1998, Real Estate; Barbara Nelson, "Historical Look: Starrett-Lehigh Building," *Real Estate New York,* May 24, 2006; David W. Dunlap, "For 1930's Behemoth, a New Upscale Life," *New York Times,* February 20, 2000.

149. The bulky building was an ingenious architectural response to the laments of 1920s New York City businesses about the costs they incurred from traffic delays within the city and the lack of a distribution system that could efficiently handle vast quantities of raw materials and merchandise.

150. See Terence Riley, *The International Style: Exhibition 15 and the Museum of Modern Art* (New York: Rizzoli/Columbia Books of Architecture, 1992), 170–75. The building was photographed by Berenice Abbott and designated a landmark in 1982. Lewis Mumford dubbed the building "a victory for engineering" and mentioned that "the contrast between the long, continuous red-brick bands and the green-framed windows, with sapphire reflections or depths, is as sound a use of color as one can see about the city." See Mumford's "Skyline" column in the *New Yorker,* 1931.

151. Sieden, *Buckminster Fuller's Universe,* 134. Also see "Chronology," 29.

152. BFA, Series 2, Dymaxion Chronofile, Box 24, vol. 40, 1931.

153. Carol Willis, "Pastoral Obsessions: The Garden in the Machine," *Skyline,* March 1983.

154. Ibid.

155. Fuller was eventually tricked out of the place by one of his party invitees, a colleague who took a liking to his rooftop quarter. Sieden, *Buckminster Fuller's Universe,* 138.

156. From archival sources it appears that Fuller met Evelyn Schwartz at one of his lectures on Dymaxion House and themes of housing as a fundamental need for

humanity. Schwartz must have been captivated by Fuller's endearing sense of purpose and commitment to his humanist cause. A large number of letters between Fuller and "Evy" written throughout 1931 are cataloged in BFA, Series 2, Dymaxion Chronofile, Box 24, vol. 40, 1931.

157. The Empire State Building formally opened on May 1, 1931. Alfred E. Smith, the president of the building, presidential candidate of the Democratic Party in 1928, and former governor of the state of New York, acted as the master of ceremonies for the opening festivities.

158. The letter, dated November 24, 1931, is in BFA, Series 2, Dymaxion Chronofile, Box 24, vol. 40, 1931.

159. See Jane Gallop, *Reading Lacan* (Ithaca, N.Y.: Cornell University Press, 1985).

160. On December 30/31, 1931, a distraught Fuller wrote a pleading letter to Schwartz, asking her to continue their "relationship." BFA, Series 2, Dymaxion Chronofile, Box 24, vol. 40, 1931.

161. Letter in BFA, Series 2, Dymaxion Chronofile, Box 24, vol. 40, 1931.

162. James Sterngold, "The Love Song of R. Buckminster Fuller," *New York Times,* June 15, 2008.

163. Roland Barthes, *Mythologies,* trans. Annette Lavers (New York: Hill & Wang, 1957), 111–17.

164. For a thorough historic account of the building of the Empire State Building, see John Tauranac, *The Empire State Building: The Making of a Landmark* (New York: Scribner, 1995). A more recent discussion is found in Mark Kingwell, *Nearest Thing to Heaven: The Empire State Building and American Dreams* (New Haven, Conn.: Yale University Press, 2006).

165. Quoted in Touranac, *The Empire State Building,* 228.

166. Beginning in the summer of 1935, Fuller claimed to have rewritten the manuscript eight times before turning it over to J. B. Lippincott Company in January 1938. The manuscript is in BFA, Series 2, Dymaxion Chronofile, Box 38, vol. 59, 1938.

167. Sinclair Lewis, review of *Nine Chains to the Moon,* by R. Buckminster Fuller, *Newsweek,* September 12, 1938.

168. Clifton Fadiman, review of *Nine Chains to the Moon,* by R. Buckminster Fuller, *New Yorker,* September 10, 1938.

169. Frank Lloyd Wright, review of *Nine Chains to the Moon,* by R. Buckminster Fuller, *Saturday Review of Literature,* September 17, 1938.

170. "Upset of 3-Wheeled Car Kills 1, Hurts 2; Noted British Aviator a Victim at Chicago," *New York Times,* October 28, 1933. The driver was an American named Francis T. Turner. Also see Phil Patton, "A 3-Wheel Dream That Died at Takeoff," *New York Times,* June 15, 2008.

171. Allegra Fuller Snyder with Victoria Vesna, "Education Automation on Spaceship Earth: Buckminster Fuller's Vision—More Relevant than Ever," *Leonardo* 31, no. 4 (1998): 289–92.

172. A chapter titled "E = MC^2 = Mrs. Murphy's Horse Power" builds on Einstein's fabled formula. Fuller, *Nine Chains to the Moon,* 70–82.

173. Ibid., xiii.

174. Ibid., 67.

175. Ibid., 68.

176. The science of rocketry was, by the 1930s, still primitive at best or mostly in the realm of fantasy. On the American romance with space exploration, see, for example, Howard E. McCurdy, *Space and the American Imagination* (Washington, D.C.: Smithsonian Institution Press, 1997); Willy Ley, *Rockets, Missiles, and Men in Space* (New York: Viking Press, 1968); Tom Crouch, "'To Fly to the Moon': Cosmic Voyaging in Fact and Fiction from Lucian to Sputnik," in *Science Fiction and Space Futures, Past and Present,* ed. Eugene Emme (San Diego, Calif.: American Astronautical Society, 1982); Frank H. Winter, *Prelude to the Space Age: The Rocket Societies, 1924–1940* (Washington, D.C.: Smithsonian Institution Press, 1983); Joel Makower, ed., *The Air and Space Catalog* (New York: Vintage Books, 1989).

177. As early as 1865 the French writer Jules Verne published *De la Terre a la lune (From the Earth to the Moon),* followed five years later by *Autour de la lune (Round the Moon)*. With similar scientific speculations on space travel, Russian author Konstantin Tsiolkovsky published *Investigation of Universal Space by Means of Reactive Devices* in 1911, and soon thereafter German physicist Hermann Oberth entered the field. Oberth offered scientific advice to film director Fritz Lang for his 1929 movie *By Rocket to the Moon*.

178. David Kyle, *A Pictorial History of Science Fiction* (London: Hamlyn, 1976), 38. Fuller understood aviation's potential to win wars from Wells's *The War in the Air* (1907). See Fuller, *Nine Chains to the Moon,* 293–94.

179. H. G. Wells, *The Shape of Things to Come* (New York: Macmillan, 1933), 348.

180. Wells was a student of Thomas Henry Huxley, an acolyte of Charles Darwin.

181. McCurdy, *Space and the American Imagination,* 16–20.

182. See Lew Holt, "Amazing Turbine Rocket to Explore Outer Space," *Modern Mechanix and Inventions,* December 1931. For a biographical account of Goddard, see Milton Lehman, *This High Man: The Life of Robert H. Goddard* (New York: Farrar, Straus, 1963). Also see McCurdy, *Space and the American Imagination,* 16–17. It is known from biographical sources that Goddard researched Verne's *From the Earth to the Moon* and Wells's story of Martian invasion in *War of the Worlds,* which was serialized in the *Boston Post*. As early as 1919, Goddard wrote a treatise on rocket propulsion titled "A Method of Reaching Extreme Heights," which was published by the Smithsonian Institution. Also see A. Scott Berg, *Lindbergh* (New York: G. P. Putnam's Sons, 1998), 223.

183. McCurdy, *Space and the American Imagination,* 17.

184. Berg, *Lindbergh,* 226. Also see Whitman, "Daring Lindbergh Attained the Unattainable."

185. McCurdy, *Space and the American Imagination,* 20.

186. For a similar explanation of Fuller's argument, see Betty Franks, "Futurists and the American Dream: A History of Contemporary Futurist Thought" (doctor of arts diss., Carnegie-Mellon University, 1985), 504.

187. William Kuhns, *The Post-industrial Prophets: Interpretations of Technology* (New York: Weybright and Talley, 1971), 232–33.

188. Ibid.

189. Fuller, *Nine Chains to the Moon,* 19.

190. Ibid., 41.

191. Ibid.

192. Ibid., 379.

193. Alexis Carrel, *Man, the Unknown* (1935; repr., New York: Harper & Brothers, 1939). Carrel's papers are in the Alexis Carrel Papers, Specials Collections Division, Joseph Mark Lauinger Memorial Library, Georgetown University.

194. Fuller quotes Carrel in *Nine Chains to the Moon,* 41.

195. See "Medicine: Carrel's Man," *Time,* September 16, 1935.

196. Fuller, *Nine Chains to the Moon,* 39.

197. Carrel, *Man, the Unknown,* xi. The book received much favorable attention in the print media during the 1930s, revealing how eugenic views and racial politics were very much part of the social fabric of the time. Critical assessment of Carrel's work began only recently. See, for instance, David Le Vay, *Alexis Carrel: The Perfectibility of Man* (Rockville, Md.: Kabel, 1996); Max Wallace, *The American Axis: Henry Ford, Charles Lindbergh, and the Rise of the Third Reich* (New York: St. Martin's Press, 2003); Andres Horacio Reggiani, *God's Eugenicist: Alexis Carrel and the Sociobiology of Decline* (New York: Berghahn Books, 2007); David Friedman, *The Immortalists: Charles Lindbergh, Dr. Alexis Carrel, and Their Daring Quest to Live Forever* (New York: Harper Perennial, 2008).

198. Fuller, *Nine Chains to the Moon,* 40.

199. Quoted in "Medicine: Carrel's Man," 42–43.

200. Carrel, *Man, the Unknown,* 303. See the chapter "Genius and Talent" in Fuller, *Nine Chains to the Moon.*

201. Fuller, *Nine Chains to the Moon,* 39.

202. Carrel, *Man, the Unknown,* xi.

203. Fuller, *Nine Chains to the Moon,* 165–66.

204. Ibid., 14–15. Fuller defines the "northwest spiral of civilization" thus: "A northwest spiraling motion independently maintained by each and every category of man's mechanical extensions to be noted as a seemingly unplanned advancement of each industrial art and language as it crops out to the west and north of the last development. Advanced arts (such as radio & aviation) finally trafficking across the pole, and thereafter developing as a radiant plan over the north pole."

205. Frederick Polak, *The Image of the Future,* 2 vols. (New York: Oceana, 1961), 36.

206. Carrel, *Man, the Unknown,* 278.

207. Fuller, *Nine Chains to the Moon,* xi.

208. In his study of Joseph Conrad's classic works *Heart of Darkness* and *Nostromo,* Edward Said argued that while Conrad remained a staunch humanist and opposed to the political philosophies of imperialism, he accepted the "inevitability" of imperialism, a tragic surrender to colonialist dogmas providing them, ultimately, tacit endorsement. "Dialog with Edward Said: The Pen and the Sword," interview conducted in January 1993 by David Barsamian of Alternative Radio, reprinted in *Design Book Review,* Summer/Fall 1993, 16.

209. Fuller, *Nine Chains the the Moon,* xi.

210. Quoted in Ira Bruce Nadel, *Biography: Fiction, Fact, and Form* (New York: St. Martin's Press, 1984), 7.

3. THE MASTER BUILDER AS SUPERMAN

1. World's Fair issue, *Time,* May 1, 1939. Also see Morton Eustis, "Big Show in Flushing Meadows," *Theatre Arts Monthly,* July 1939, 566–77.

2. Quoted in Larry Zim, Mel Lerner, and Herbert Rolfes, *The World of Tomorrow: The 1939 New York World's Fair* (New York: Main Street Press, 1988), 9. Roosevelt's opening remarks were televised to the New York metropolitan area, initiating broadcast television by RCA. See Robert Rosenblum et al., *Remembering the Future: The New York World's Fair from 1939 to 1964* (New York: Rizzoli, 1989), 22.

3. The New York elite were led by three influential members of the business community: Grover A. Whalen, president of 1939 New York World's Fair, Inc., and flamboyant "official greeter" of New York City; George McAneny, president of Title Guarantee and head of the Regional Planning Association; and Percy Straus, head of R. H. Macy's. The three joined forces in September 1935 in an incorporation meeting that also involved "the heads of twenty-three banking and trust companies, thirty corporations, fifteen Wall Street law firms, eight insurance companies and retail firms, and eight business associations." Joseph P. Cusker, "The World of Tomorrow: Science, Culture, and Community at the New York World's Fair," in Helen A. Harrison et al., *Dawn of a New Day: The New York World's Fair, 1939–40* (New York: Queens Museum/New York University Press, 1980), 3. There is a robust literature on the 1939 New York World's Fair, including Stanley Appelbaum, *The New York World's Fair 1939/1940* (New York: Dover, 1977); Barbara Cohen, Steven Heller, and Seymour Chwast, *Trylon and Perisphere: The 1939 New York World's Fair* (New York: Harry N. Abrams, 1989); E. L. Doctorow, *World's Fair* (New York: Fawcett Crest, 1985); Marco Duranti, "Utopia, Nostalgia, and World War at the 1939–40 New York World's Fair," *Journal of Contemporary History* 41, no. 4 (2006): 663–83; David Gelernter, *1939: The Lost World of the Fair* (New York: Free Press, 1995); Robert W. Rydell, *World of Fairs: The Century-of-Progress Expositions* (Chicago: University of Chicago Press, 1993); Ed Tyng, *Making a World's Fair* (New York: Vantage Press, 1958).

4. On the fair's planning, see *Official Guide Book, New York World's Fair: The World of Tomorrow, 1939* (New York: Expositions Publications, 1939).

5. See *Today at the Fair: Official Daily Program of the New York World's Fair 1939* (May 28, 1939). Also see General Motors press release, May 30, 1939, in Norman Bel Geddes Collection, Harry Ransom Humanities Research Center, University of Texas, Austin (hereafter NBG Collection, HRC), File 381.48.

6. See, for instance, "Success of Futurama Is Tribute to Bel Geddes," *New York Herald Tribune,* July, date n.a., 1939; "Smash Hit of the World's Fair," *Sunday News,* June 18, 1939; "*Life* Goes to the Futurama at World's Fair Where General Motors Shows the America of 1960," *Life,* June 5, 1939. Many architects wrote to Bel Geddes to congratulate him; some of these letters are in NBG Collection, HRC, File 381.43.

7. Doctorow, *World's Fair,* 323.

8. Bel Geddes was born in the final decade of the nineteenth century into an itinerant Michigan family with Scottish and German ancestry. His father died when he was just twelve, and his mother took teaching jobs in many midwestern schools, with the result that Bel Geddes had a peripatetic childhood, attending school irregularly in

Michigan, Ohio, Pennsylvania, Illinois, and Indiana. Reluctant to follow the dictates of conventional schooling (Bel Geddes was a self-taught designer), he later highlighted his expulsion from the ninth grade for drawing a caricature of the school principal as a sign of his rebel personality. Biographical materials on Bel Geddes are in NBG Collection, HRC, Files 954.16, 977.21a. His autobiography, covering his early life, was published posthumously as *Miracle in the Evening,* ed. William Kelley (New York: Doubleday, 1960). Other sources on his life include Geoffrey T. Hellman, "Profiles: Design for a Living—I: Norman Bel Geddes," *New Yorker,* February 8, 1941, 24–28; Geoffrey T. Hellman, "Profiles: Design for a Living—II: Norman Bel Geddes," *New Yorker,* February 15, 1941, 22–29; Geoffrey T. Hellman, "Profiles: Design for a Living—III: Norman Bel Geddes," *New Yorker,* February 22, 1941, 26–30, 33, 34; Jeffrey L. Meikle, "Norman Bel Geddes: A Portrait," *Rassegna* 60 (Winter 1994): 6–10; Arthur Pulos, "Design Horizons," *Rassegna* 60 (Winter 1994).

9. *Industrial Design,* circa July 1958, in NBG Collection, HRC, File 977.21a, "Photographs Personal Family."

10. Henry Dreyfuss, in *Industrial Design,* circa July 1958, in NBG Collection, HRC, File 954, Drawings, Caricatures, Cartoons, Oversized Box 16.

11. Norman Bel Geddes, "Streamlining," *Atlantic Monthly,* November 1934, 553, 556–58. Also see Donald J. Bush, *The Streamlined Decade* (New York: George Braziller, 1975); Jeffrey L. Meikle, *Twentieth Century Limited: Industrial Design in America, 1925–1939* (Philadelphia: Temple University Press, 1979).

12. For general discussions of Futurama, see Donald J. Bush, "Futurama: World's Fair as Utopia," *Alternative Futures* 2 (Fall 1979): 3–20; Barbara Hauss-Fitton, "Futurama, New York World's Fair 1939–1940," *Rassegna,* 60 (Spring 1994); Roland Marchand, "The Designers Go to the Fair II: Norman Bel Geddes, the General Motors 'Futurama,' and the Visit to the Factory Transformed," *Design Issues* 8, no. 2 (Spring 1992): 23–40.

13. *Official Guide Book,* 208. Also see NBG Collection, HRC, File 381.48. The model was constructed by George Wittbold Studios of Chicago.

14. Bel Geddes was an avid reader of these visionaries. His personal library (part of the NBG Collection, HRC) contains important books by Le Corbusier, Wright, and Wells. His customary orange pencil marks in them suggest that since the late 1920s, Geddes absorbed the ideas of these thinkers. Two instances suggest that Bel Geddes might have met Le Corbusier and Wells. During a trip to Paris in 1925 to design stage sets for the celebrated play *Jeanne d'Arc,* Bel Geddes probably crossed paths with the Franco-Swiss master or saw his Ville Contemporaine along with the Plan Voisin exhibited at the Pavillon de l'Esprit Nouveau, an important part of the Exposition Internationale des Arts Decoratifs et Industriels Modernes. In the essay "World of Tomorrow," *New York Times,* March 5, 1939, inaugural World's Fair section, Wells likened his futurist vision to that of Bel Geddes. No doubt Wells saw Futurama as the closest approximation of the "world of tomorrow" that he had earlier described in *The Shape of Things to Come* (New York: Macmillan, 1933), a book Bel Geddes researched extensively and kept in his library. The intellectual trajectories of Bel Geddes and Mumford intersected during the 1939 New York World's Fair, as both were involved in it. Futurama's didactic vision of the future was somewhat akin to *The City.*

15. Bel Geddes conducted research on McClintock's work on traffic solutions. See NBG Collection, HRC, File 381.3. McClintock collaborated with Bel Geddes on his Shell Oil advertisement project, the City of Tomorrow (1937). Also see Miller McClintock, "Street Traffic—Past and Future," *The American City,* September 1930. Bel Geddes must have also studied MacKaye's ideas, since the publication of MacKaye's seminal article "The Townless Highway," *New Republic,* March 12, 1930, coincided with Bel Geddes's interest in urbanization and highway traffic.

16. See, for instance, Douglas Adams, "Norman Bel Geddes and Streamlined Spaces," *Journal of Architectural Education* 30, no. 1 (1976): 22–24; Robert Coombs, "Norman Bel Geddes' Highways and Horizons," *Perspecta* 13/14 (1971): 11–27; Cliff Ellis, "Lewis Mumford and Norman Bel Geddes: The Highway, the City and the Future," *Planning Perspectives* 20, no. 1 (2005): 51–68; Paul Mason Fotsch, "The Building of a Superhighway Future at the New York World's Fair," *Cultural Critique* 48 (Spring 2001): 65–97.

17. Alfred P. Sloan Jr., "A Greeting to Our Guests," in *Futurama* (brochure) (Detroit: General Motors Corporation, 1940).

18. Norman Bel Geddes, *Magic Motorways* (New York: Random House, 1940), 211.

19. "St. Louis a Model City of the Future," 6, NBG Collection, HRC, File 381.48.

20. Bel Geddes, *Miracle in the Evening,* 23.

21. Charles A. Lindbergh, *"We": The Famous Flier's Own Story of His Life and His Transatlantic Flight, Together with His Views on the Future of Aviation* (New York: G. P. Putnam's Sons, July 1927), signed "Charles Lindbergh, October 1935," in NBG Collection, HRC.

22. Le Corbusier, *Aircraft* (London: The Studio, 1935), 13.

23. Ibid., 112.

24. For a cultural analysis of Superman, see Thomas Andrae, "From Menace to Messiah: The History and Historicity of Superman," in *American Media and Mass Culture: Left Perspectives,* ed. Donald Lazere (Berkeley: University of California Press, 1987).

25. Rydell, *World of Fairs,* 38–58.

26. The General Motors Building was designed by architect Albert Kahn. Jeffrey Meikle, however, suggested to me that Bel Geddes had actually designed it and that Kahn simply signed the construction documents as the "official" architect. Bel Geddes attempted to become registered as an architect multiple times during the 1930s, without any positive outcome. NBG Collection, HRC, File 314.

27. See "St. Louis a Model City of the Future," 2.

28. On Bel Geddes's experience in stage set design, see Norman Bel Geddes, *A Project for a Theatrical Presentation of the Divine Comedy of Dante Alighieri* (New York: Theatre Arts, 1924); Norman Bel Geddes, "Design for a New Kind of Theatre: A Complete Revolution—Artistic, Economic and Functional—Is Proposed for the Playhouse," *New York Times Magazine,* November 30, 1947; Frederick J. Hunter, "Norman Bel Geddes: The Renaissance Man of the American Theatre," *Texas Quarterly* (Winter 1962); William Condee, "Reform of the Performance," *Architectural Review* 185 (June 1989).

29. Adams, "Norman Bel Geddes and Streamlined Spaces," 22.

30. Condee, "Reform of the Performance," 75. The leaders of the New Movement were Adolphe Appia (1862–1928) and Edward Gordon Craig (1872–1966), who published

their works on new theater design around the beginning of the twentieth century and had a major influence on the theater scene in America.

31. Sheldon Cheney, *The New Movement in the Theatre* (New York: Benjamin Blom, 1914). Cheney's *Theatre Arts* became an important organ for the theater reform movement in America.

32. NBG Collection, HRC, File 954, Drawings, Caricatures, Cartoons, Oversized Box 16. Also see Bel Geddes, "Design for a New Kind of Theatre," 30. Cheney and Bel Geddes corresponded about theater design. Cheney's letter to Bel Geddes can be found in Sheldon Cheney, *Stage Decoration* (New York: John Day, 1928).

33. Jennifer Davis Roberts, *Norman Bel Geddes: An Exhibition of Theatrical and Industrial Designs* (Austin: Humanities Research Center, University of Texas, 1979), 12–13. Produced by Morris Gest and F. Ray Comstock, *The Miracle* played from January 15, 1924, through November 8, 1925.

34. For a description of the Futurama tour, see Eustis, "Big Show in Flushing Meadows."

35. For an entire transcript of the aural description of Futurama, see General Motors Corporation, *Futurama*.

36. "Technical Supplement," NBG Collection, HRC, File 381.31.

37. Descriptions of the conveyor belt and the sound system are in NBG Collection, HRC, File 381.19d, 30, 31, 32, 54.

38. Jean Pfaelzer, "The Impact of Political Theory on Narrative Structures," in *America as Utopia*, ed. Kenneth M. Roemer (New York: Burt Franklin, 1981), 117–32.

39. "Technical Supplement," 2.

40. NBG Collection, HRC, File 381.19e, Folder 51. Also see GM's publicity brochure *Futurama*.

41. NBG Collection, HRC, File 381.19e, 48.

42. A detailed transcript of Futurama's description is in NBG Collection, HRC, File 381.30.

43. NBG Collection, HRC, File 381.42.

44. Ibid. The order was submitted on August 4, 1938, and the photographs were delivered four days later.

45. "Notes Taken on Airplane Trip," November 2, 1938, NBG Collection, HRC, File 381.7. Whether Bel Geddes himself or his office staff took these notes on the airplane is unclear from archival sources.

46. "Caption Material," NBG Collection, HRC, File 381.48.

47. "General Caption," NBG Collection, HRC, File 381.48. Also see Marchand, "The Designers Go to the Fair II."

48. Frederick Jackson Turner, "The Significance of the Frontier in American History" (1893), in *The Frontier in American History* (1920; repr., Tucson: University of Arizona Press, 1986).

49. For a discussion of the early twentieth-century American city as a new frontier, see James Oliver Robertson, *American Myth, American Reality* (New York: Hill & Wang, 1980), 200–201.

50. Albert Boime, *The Magisterial Gaze: Manifest Destiny and American Landscape Painting, c. 1830–1865* (Washington, D.C.: Smithsonian Institution Press, 1991), 20.

51. General Motors Corporation, *Futurama*.

52. Warren I. Susman, "The People's Fair: Cultural Contradictions of a Consumer Society," in Harrison et al., *Dawn of a New Day*, 17–27.

53. Ibid., 22.

54. Quoted in ibid.

55. Marchand, "The Designers Go to the Fair II," 22–25. For an analysis of how advertisements/exhibits were created for mass appeal, see Roland Marchand, *Advertising the American Dream: Making Way for Modernity, 1920–1940* (Berkeley: University of California Press, 1986).

56. NBG Collection, HRC, File 381.3, 6.

57. NBG Collection, HRC, File 381.19a, 4.

58. For a useful description of Bel Geddes's presentation in GM's New York office, see Hellman, "Profiles: Design for a Living—III," 28–29.

59. A generation older than Bel Geddes, the GM executives had witnessed the progress of the fledgling aviation industry during their early careers. Kettering, who was Bel Geddes's ardent supporter among the GM leaders, was an avid aviation enthusiast who had learned to fly as early as 1912. In his office, he had mounted "an old, flimsy Wright airplane of early vintage." See "Business: All Change!," *Time*, January 9, 1933.

60. NBG Collection, HRC, File 381.19e, 48. For discussion of Knudsen's ideas on progress and civilization, see Norman Beasley, *Knudsen: A Biography* (New York: McGraw-Hill, 1947).

61. Walter Benjamin, "The Work of Art in the Age of Mechanical Reproduction" (1936), in *Illuminations: Essays and Reflections*, trans. Harry Zohn, ed. Hannah Arendt (New York: Schocken Books, 1968), 236.

62. Frank E. Manuel and Fritzie P. Manuel, *Utopian Thought in the Western World* (Cambridge, Mass.: Belknap Press, 1979), 4–5. For an excellent analysis of the role of the utopian narrator-hero, see Pfaelzer, "The Impact of Impact of Political Theory on Narrative Structures."

63. Manuel and Manuel, *Utopian Thought in the Western World*, 5.

64. Edward Bellamy, *Looking Backward, 2000–1887* (Boston: Ticknor, 1888). For an analysis of the character of Julian West as a narrative device, see Darko Suvin, "Anticipating the Sunburst," in Roemer, *America as Utopia*.

65. Interestingly, an editorial in the GM newsletter the *Bulletin* published in the wake of Futurama's immense popularity compared Futurama to Shangri-La. Lew Hahn, "The Future Is Beckoning," n.d., NBG Collection, HRC, File 653.52, "Photostats of Correspondence." On a more ominous side of the spectrum, Leni Riefenstahl's propaganda film *Triumph of the Will* (1934) demonstrates that the Nazi dream of an Aryan utopia began with the aerial gaze of an aviating Hitler.

66. The extravaganza of the fair is explored in Eustis, "Big Show in Flushing Meadows." On the symbolism of the airplane in the 1939 fair, see Gerald Wendt, *Science for the World of Tomorrow* (New York: W.W. Norton, 1939), 104–13.

67. *New York World's Fair Illustrated by Camera* (brochure) (New York: Manhattan Post Card Publishing, 1939), licensed by New York World's Fair. The centerpiece of the brochure is an airplane view of the fairground. The photographer is McLaughlin.

68. *Official Guide Book*, 8–9.

69. Transforming the dumping ground in northern Queens was a complex reclamation project for which Robert Moses secured state and federal funding. See Eugene A. Santomasso, "The Design of Reason: Architecture and Planning at the 1939/40 New York World's Fair," in Harrison et al., *Dawn of a New Day,* 30–31. Also see Robert W. Rydell, "Selling the World of Tomorrow: New York's 1939 World's Fair," *Journal of American History* 77, no. 3 (December 1990): 966–70.

70. Burton Benedict, *Anthropology of World's Fairs* (London: Scholar Press, 1983), 5.

71. The other two documentaries were General Motors' *New Horizons* and Westinghouse's *The Middleton Family at the Fair. The City*, sponsored by the Carnegie Corporation at a cost of fifty thousand dollars, was just over forty-three minutes long. Imbued with a socialist approach to planning, it was based on the environmentalist concerns of Pare Lorentz, director of *The Plow That Broke the Plains* (1936) and *The River* (1938). Aaron Copland was the film's music director. *The City* premiered at the fair with much fanfare and publicity on May 26, 1939.

72. See, for instance, Howard Gillette Jr., "Film as Artifact: *The City* (1939)," *American Studies* 18, no. 2 (1977): 71–85.

73. He was born in 1898 in Vienna, Austria, and emigrated to the United States in 1934. A pioneering graphic designer, he won many graphic design competitions organized by the Museum of Modern Art and the Art Directors Club New York and designed covers for *Fortune* and *Graphis* magazine.

74. A copy of the poster is in the Prints and Photographs Division of the Library of Congress, Washington, D.C.

75. On July 10, 1938, the thirty-three-year-old Hughes embarked on the round-the-world flight with a crew of four; he finished the journey in three days, nineteen hours, and eight minutes. The flight, which made Hughes a national hero, was a huge publicity stunt for the 1939 New York World's Fair. George J. Marrett, *Howard Hughes, Aviator* (Annapolis, Md.: Naval Institute Press, 2004), 38–39. The National Aeronautics Association named Hughes Aviator of the Year in 1939. He was also awarded the prestigious Colliers Trophy for his record-breaking flight around the world.

76. Grover A. Whalen, *Mr. New York: The Autobiography of Grover Whalen* (New York: G. P. Putnam's Sons, 1955), 191–93.

77. Michael Aronson, "Grover Whalen Welcome to New York City," *New York Daily News,* April 9, 1999. Other celebrity aviators and personalities Whalen greeted and feted were Wiley Post and Amelia Earhart, King Albert and Queen Elizabeth of Belgium, British prime minister David Lloyd-George, French premier Georges Clemenceau, U.S. president Woodrow Wilson, and the Prince of Wales. Whalen was appointed police commissioner in December 1928. Also see Whalen, *Mr. New York,* 112–28.

78. *Official Guide Book,* 117. The guide was sold for twenty-five cents.

79. Zim et al., *The World of Tomorrow,* 182.

80. The medal was sculpted by Rene Chambellan and minted by Medallic Art Company, USA, 1933.

81. For a description of the Theme Center, see "The Theme Center—Democracity," in *Official Guide Book,* 43–45. Also see Santomasso, "The Design of Reason," 33. The name Trylon derives from *tri,* for the three sides of the structure, and *pylon,* the Greek

word for a monumental gate. Perisphere is from *peri,* meaning beyond or all around. The Trylon and Perisphere were steel-framed structures clad in gypsum board.

82. Rydell, *World of Fairs,* 131.

83. Ibid.

84. *Official Guide Book,* 44.

85. Appelbaum, *The New York World's Fair 1939/1940,* 3. Also see Claudia Swan, ed., *1939: Music and the World's Fair* (New York: Eos Music, 1998).

86. *Official Guide Book,* 44–45.

87. Susman, "The People's Fair"; Marchand, "The Designers Go to the Fair II."

88. Gelernter, *1939: The Lost World of the Fair,* 26.

89. Quoted in Zim et al., *The World of Tomorrow,* 107.

90. Norman Bel Geddes, *Horizons* (Boston: Little, Brown, 1932).

91. Art historians, urbanists, architects, artists, and novelists have explored the links between the means of transportation and modern aesthetics. Among them are Sheldon Cheney, *Art and the Machine* (New York: Wittlesey House, 1936); I. F. Clarke, *The Pattern of Expectation, 1644–2001* (New York: Basic Books, 1979); Bruno Taut, *Modern Architecture* (New York: A. & C. Boni, 1929); Erich Mendelsohn, "The Problem of a New Arch," in *Erich Mendelsohn: Complete Works of the Architect* (1930; repr., New York: Princeton Architectural Press, 1992); Raymond Loewy, *Industrial Design* (New York: Overlook Press, 1979); Meikle, *Twentieth Century Limited*; Bush, *The Streamlined Decade.*

92. Bel Geddes, *Horizons,* 8.

93. Ibid., 4.

94. Ibid., 292.

95. Ibid., 19–20.

96. Ibid., 276–77.

97. Le Corbusier, *Aircraft,* 24.

98. Ibid., 76.

99. Loewy, *Industrial Design,* 74–76.

100. NBG Collection, HRC, File 381.27, "Air Terminals," Box 10a; Bel Geddes, *Magic Motorways,* 214.

101. Bel Geddes, *Horizons,* 280.

102. Ibid.

103. Ibid., 280–81.

104. Paul Virilio, *Open Sky,* trans. Julie Rose (London: Verso, 1997), 1–3.

105. Bel Geddes spelled out his concept of the "Man of Tomorrow" in an autobiographical essay, "The Man of Tomorrow . . . Today," *Info* 1, no. 4 (April 7, 1939). In Futurama press releases, Bel Geddes described himself as "visionary," "a contemporary man of tomorrow," a "man of imagination," and even "a specimen of health."

106. He presented a lecture, titled "Tomorrow's Airways," at the National Aviation Forum in Washington, D.C., in 1931. See NBG Collection, HRC, File 653.7.

107. See NBG Collection, HRC, File 328.1. The National Advisory Committee for Aeronautics responded to Bel Geddes's inquiry concerning Koller's aeronautical expertise on November 8, 1929.

108. For information on the design development of this project, see NBG Collection, HRC, File 328.17, 1–3.

109. The collection included, for instance, Grover Cleveland Loening, *Military Aeroplanes, Simplified, Enlarged: An Explanatory Consideration of Their Characteristics, Performances, Construction, Maintenance and Operation, Specially Arranged for the Use of Aviators and Students* (Boston: W. S. Best, 1918); Edward P. Warner, *Airplane Design: Aerodynamics* (New York: McGraw-Hill, 1927); Victor Page, *Modern Aircraft: Basic Principles, Operation, Application, Construction, Repair, Maintenance* (New York: Norman W. Henley, 1927); Juan de la Cierva, *Wings of Tomorrow: The Story of the Autogyros* (New York: Brewer, Warren & Putnam, 1931). In Warner's *Airplane Design,* Bel Geddes underlined this line on the opening page: "An Airplane is a vehicle supported by the dynamic reaction of the air against inclined surfaces, and the study of that reaction and of the laws which govern it is a branch of aerodynamics."

110. NBG Collection, HRC, File 328.1.

111. Virilio, *Open Sky,* 1–2. For examples of utopian architecture that defied gravity, see Ulrich Conrads and Hans G. Sperlich, *The Architecture of Fantasy: Utopian Building and Planning in Modern Times* (New York: Praeger, 1960); Christian W. Thomsen, *Visionary Architecture: From Babylon to Virtual Reality* (Munich: Prestel, 1994).

112. The idea of escape from architecture's earthly confinement has a history as old as that of architecture itself. Architecture imagined as a flying machine was part of post-Renaissance utopian thinking reflected in the fantastic island of Laputa in Jonathan Swift's *Gulliver's Travels* (1726), Jules Verne's nineteenth-century aerial vessels, and the flying cities of early twentieth-century expressionist architecture and science fiction. Bruno Taut introduced a synthesis of the aerial vessel and architecture in 1920 in his hypothetical cosmic-comical Aerial Pleasures, a fantastic sphere floating in space.

113. Calvino claimed that the conquest of gravity (and the attainment of lightness) became one of the great utopian dreams of the twentieth century. Italo Calvino, *Six Memos for the Next Millennium* (New York: Vintage Books, 1993), 7.

114. Quoted in Conrads and Sperlich, *The Architecture of Fantasy,* 18. El Lissitzky's *Wolkenbügel* (*Cloud Hanger,* 1924) had already expressed this dream of architectural flight.

115. For the design development of Aerial Restaurant, see NBG Collection, HRC, Files 169.1, 169.3. The project plan was submitted for approval by the Architectural Commission for the Chicago Century of Progress World's Fair. For the project description, also see Roberts, *Norman Bel Geddes,* 23.

116. The restaurant design was covered by *Creative Art,* August 1930; *Architecture,* July 1930; *Fortune,* n.d.; the *Chicago Daily News,* n.d., and *Popular Mechanics,* July 1930. The news clippings are in NBG Collection, HRC, File 169.5.

117. Martin Heidegger, "The Origin of the Work of Art," in *Poetry, Language, Thought,* trans. Albert Hofstadter (New York: Harper & Row, 1975), 17–87.

118. On Gernsback's role in the introduction of science fiction magazines in America in 1926, see Paul A. Carter, *The Creation of Tomorrow: Fifty Years of Magazine Science Fiction* (New York: Columbia University Press, 1977), 3–5; Sam Moskowitz, *Explorers of the Infinite, Shapers of Science Fiction* (Westport, Conn.: Hyperion Press, 1963), 225–42.

119. *Air Wonder Stories,* November 1929, Hugo Gernsback, editor, and Frank R. Paul, art director.

120. Ibid., 386.

121. A somewhat similar image of the aerial city conceived by Hugo Gernsback had already appeared in *Science and Invention,* February 1922. Again, the model was New York City, but it was also a city of environmental and social purification: "The city the size of New York will float several miles above the surface of the earth, where the air is cleaner and purer and free from disease carrying bacteria. Gravity-nullifying devices were pictured as the means of keeping these cities suspended."

122. Kathleen Church Plummer, "The Streamlined Moderne," *Art in America* 62, no. 1 (1974): 46–54.

123. Minutes of the review meeting (January 21, 1930) are in NBG Collection, HRC, File 169.1. Present at the meeting were architects Harvey Wiley Corbett, Raymond Hood, Paul Cret, Norman Bel Geddes, and one Mr. Walker.

124. Quoted in Tom Mead, "Revolving Café, 278 Feet in the Air, Is Plan for Fair: Unique Structure Designed for Century of Progress," *Chicago Daily News,* n.d., NBG Collection, HRC, File 169.

125. NBG Collection, HRC, File 356. Another important source of information on this project is Jeffrey Meikle, *The City of Tomorrow: Model 1937* (London: Pentagram Design, 1980). Also see Christopher Innes, *Designing Modern America: Broadway to Main Street* (New Haven, Conn.: Yale University Press, 2005), 173–78.

126. Meikle, *The City of Tomorrow,* 5–8.

127. Bel Geddes knew McClintock and collaborated with him. He read McClintock's article "Unfit for Modern Motor Traffic," *Fortune,* August 1936. Considered by many the "No. 1 man in U.S. traffic control," McClintock had earned his Ph.D. in traffic control from Harvard in 1924. During the development phase of the Shell Oil project, McClintock visited Bel Geddes's office a number of times. Bel Geddes also consulted McClintock while developing the ideas for Futurama. NBG Collection, HRC, File 381.3.

128. Bruce Bliven Jr., "Metropolis: 1960 Style," *New Republic,* September 29, 1937.

129. Bel Geddes, *Magic Motorways,* 217.

130. Bel Geddes's personal library included Ferriss's book. According to Meikle, it was Berdanier who "casually suggested photographing a scale model for the advertisements, a solution that eventually did produce some very good pictures." Meikle, *The City of Tomorrow,* 11.

131. Ibid., 22.

132. See, for instance, *Life,* July 5, 1937; *Saturday Evening Post,* July 31, 1937. Meikle, *The City of Tomorrow,* 23. Also see Bliven, "Metropolis," 211.

133. Quoted in Innes, *Designing Modern America,* 177.

134. Meikle, *The City of Tomorrow,* 29.

135. Both letters quoted in Meikle, *Twentieth Century Limited,* 208.

136. For general discussions of the Depression decade, see T. H. Watkins, *The Hungry Years: A Narrative History of the Great Depression in America* (New York: Henry Holt, 2001); Louise I. Gerdes, *The 1930s* (San Diego, Calif.: Greenhaven Press, 2000); Warren I. Susman, "The Thirties," in *The Development of an American Culture,* ed. Stanley Coben and Lorman Ratner (Englewood Cliffs, N.J.: Prentice-Hall, 1970); Amity Shelas, *The Forgotten Man: A New History of the Great Depression* (New York: Harper, 2008).

137. Lawrence W. Levine, "American Culture and the Great Depression," *Yale Review* 74 (1985): 208.

138. For a discussion of the origin of Superman, see John Kobler, "Up, Up and Away! The Rise of Superman Inc.," *Saturday Evening Post,* June 21, 1941.

139. Bel Geddes, "The Man of Tomorrow . . . Today." Bel Geddes's personal library contained at least three books by Nietzsche: *Beyond Good and Evil,* trans. Helen Zimmern (Edinburgh: T. N. Foulis, 1909); *Thus Spoke Zarathustra: A Book for All and None,* trans. Thomas Common (New York: Macmillan, 1911); and *Twilight of the Idols, or, How to Philosophise with the Hammer; The AntiChrist; Notes to Zarathustra and Eternal Recurrence,* trans. Anthony M. Ludovici (New York: Macmillan, 1911).

140. "The Aviator—The Superman of Now," *Flying,* August 1916.

141. Gordon Garvedian, "Science Pictures a Superman of Tomorrow: Biologists See Him a Man of Superior Intellect, Immune from Many Diseases, with a Greater Span of Life and Power to Choose the Sex of Children," *New York Times,* December 8, 1929.

142. Kobler, "Up, Up and Away!," 15.

143. On the rural origins of the Superman story, see Jeffrey K. Johnson, "The Countryside Triumphant: Jefferson's Ideal of Rural Superiority in Modern Superhero Mythology," *Journal of Popular Culture* 43, no. 4 (August 2010): 720–37.

144. For an analysis of this phenomenon, see Scott Bukatman, *Matters of Gravity: Special Effects and Superman in the 20th Century* (Durham, N.C.: Duke University Press, 2003), 196–202.

145. Anthony Sutcliffe, "The Metropolis in the Cinema," in *Metropolis, 1890–1940* (Chicago: University of Chicago Press, 1984), 169.

146. Bukatman, *Matters of Gravity,* 185.

147. Levine, "American Culture and the Great Depression," 220–21.

148. Umberto Eco, "The Myth of the Superman," *Diacritics* 2, no. 1 (Spring 1972): 14–22.

149. *Action Comics,* no. 8 (January 1939), reprinted in *Superman, the Action Comics Archives,* vol. 1 (New York: DC Comics, 1997), 41–54.

150. *Action Comics,* no. 12 (May 1939), reprinted in *Superman, the Action Comics Archives,* vol. 1, 97–110.

151. *Action Comics* (n.d.), reprinted in *Superman Archives,* vol. 1 (New York: DC Comics, 1989), 120–35.

152. Bukatman, *Matters of Gravity,* 198.

153. E. H. Carr, *What Is History?* (Harmondsworth, England: Penguin Books, 1964), 33.

154. Walter Benjamin, *Charles Baudelaire: A Lyric Poet in the Era of High Capitalism,* trans. Harry Zohn (London: NLB, 1973), 74

155. Andrew Saint, "The Architect as Hero and Genius," in *The Image of the Architect* (New Haven, Conn.: Yale University Press, 1983). See Julian Petley, "The Architect as *Ubermensch,*" in *Picture This: Media Representations of Visual Art and Artists,* ed. Philip Hayward (London: J. Libbey, 1988); Wolfgang Pehnt, "The 'New Man' and the Architecture of the Twenties," in *Social Utopias of the Twenties: Bauhaus, Kibbutz, and the Dream of the New Man,* ed. Jeannine Fielder (Wuppertal, Germany: Muller & Busmann Press, 1995).

156. Ayn Rand's *The Fountainhead* encapsulates the architect-as-hero sentiment, as the book salutes "the great profession of architecture and its heroes."

157. Sigfried Giedion, *Space, Time and Architecture: The Growth of a New Tradition* (1941; repr., Cambridge, Mass.: Harvard University Press, 1959). For an explanation of this phenomenon, see Colin Rowe, *The Architecture of Good Intentions: Towards a Possible Retrospect* (London: Academy Editions, 1994), 30–43.

158. NBG Collection, HRC, File 381.48, 2–3.

159. "Motoring at 100 M.P.H.," *BusinessWeek,* September 9, 1939, 27–28.

160. General Motors Corporation, *Futurama.*

161. Ibid.

162. "St. Louis a Model City of the Future," 6.

163. Ibid., 7.

164. E. B. White, "They Came with Joyous Song," *New Yorker,* May 13, 1939, 25–28.

165. Ulrich Conrads, ed., *Programs and Manifestoes on 20th-Century Architecture,* trans. Michael Bullock (Cambridge: MIT Press, 1971), 137–45. These deliberations on what was called "the Functional City" later became a landmark document of urban planning, the Charter of Athens (1941), under the guidance of Le Corbusier. For an extended history of CIAM's influence on twentieth-century urbanism, see Eric Mumford, *The CIAM Discourse on Urbanism, 1928–1960* (Cambridge: MIT Press, 2002).

166. Le Corbusier, *Aircraft,* 5. The subtitle of the book is *L'Avion accuse . . .* (the airplane indicts).

167. James J. Gibson, *The Ecological Approach to Visual Perception* (Boston: Houghton Mifflin, 1979), 197.

168. "St. Louis a Model City of the Future," 1.

169. Ibid.

170. Edward J. Orff World's Fair Collection, National Museum of American History, Washington, D.C.

171. Ernst is quoted in Roger Copeland, "Merce Cunningham and the Aesthetic of Collage," *Drama Review* 46, no. 1 (Spring 2002): 14–15.

172. Two notable exceptions were New York's mayor, Fiorello LaGuardia, and the city's parks commissioner, Robert Moses, with whom Bel Geddes reportedly disagreed with respect to highway planning. Moses contributed an article, "The City of Tomorrow," to the World's Fair section of the *New York Times,* March 5, 1939. In the article, Moses argued against the wholesale cleansing of existing cities, a stance considered untenable by many modernist planners, including Bel Geddes.

173. NBG Collection, HRC, File 381.7, Box 9a. Most photographs in the box are by Garrison. The meeting took place at the auto giant's New York offices. For a discussion of this meeting, see Hellman, "Profiles: Design for a Living—III," 28–29.

174. The photograph is in Michael W. R. Davis, *Images of Motoring: General Motors, a Photographic History* (Charleston, S.C.: Arcadia, 1999), 62.

175. Susan Stewart has explored the peculiar effect of miniaturization of landscapes and objects on the self-perception of an enlarged spectator (in relation to the object) in *On Longing: Narratives of the Miniature, the Gigantic, the Souvenir, the Collection* (Durham, N.C.: Duke University Press, 1983), 78.

176. Ibid.

177. Doctorow, *World's Fair,* 325.

178. Benedict, *Anthropology of World's Fairs,* 5–6.

179. David E. Nye, *Narratives and Spaces: Technology and the Construction of American Culture* (New York: Columbia University Press, 1997), 107.

180. Frederik Polak, *The Image of the Future,* vol. 1 (New York: Oceana, 1961), 287.

181. Doctorow, *World's Fair,* 325.

182. Susman, "The Thirties," 184.

183. Ibid.

184. Susman, "The People's Fair," 17.

185. James Truslow Adams, *The Epic of America* (Boston: Little, Brown, 1938), 415. The book was reprinted many times throughout the decade. In 1931 alone, it was reprinted eight times.

186. Quoted in Susman, "The People's Fair," 18.

187. *Official Guide Book,* 5.

188. "*Life* Goes to the Futurama," 81.

EPILOGUE

1. Jane Jacobs, *The Death and Life of Great American Cities* (New York: Random House, 1961).

2. Charles Baudelaire, *Selected Poems,* trans. Joanna Richardson (Harmondsworth, England: Penguin Books, 1975), 39.

3. Le Corbusier, *Aircraft* (London: The Studio, 1935).

4. Michel de Certeau, *The Practice of Everyday Life,* trans. Steven Rendall (Berkeley: University of California Press, 1984), 91–94.

5. Henry-Russell Hitchcock, "The Architecture of Bureaucracy and the Architecture of Genius," *Architectural Review* 101, no. 601 (January 1947): 3–6.

Index

ADNAN MORSHED is associate professor in the School of Architecture and Planning at the Catholic University of America.